Materials Horizons: From Nature to Nanomaterials

Series Editor

Vijay Kumar Thakur, School of Aerospace, Transport and Manufacturing, Cranfield University, Cranfield, UK

Materials are an indispensable part of human civilization since the inception of life on earth. With the passage of time, innumerable new materials have been explored as well as developed and the search for new innovative materials continues briskly. Keeping in mind the immense perspectives of various classes of materials, this series aims at providing a comprehensive collection of works across the breadth of materials research at cutting-edge interface of materials science with physics, chemistry, biology and engineering.

This series covers a galaxy of materials ranging from natural materials to nanomaterials. Some of the topics include but not limited to: biological materials, biomimetic materials, ceramics, composites, coatings, functional materials, glasses, inorganic materials, inorganic-organic hybrids, metals, membranes, magnetic materials, manufacturing of materials, nanomaterials, organic materials and pigments to name a few. The series provides most timely and comprehensive information on advanced synthesis, processing, characterization, manufacturing and applications in a broad range of interdisciplinary fields in science, engineering and technology.

This series accepts both authored and edited works, including textbooks, monographs, reference works, and professional books. The books in this series will provide a deep insight into the state-of-art of Materials Horizons and serve students, academic, government and industrial scientists involved in all aspects of materials research.

Review Process

The proposal for each volume is reviewed by the following:

1. Responsible (in-house) editor
2. One external subject expert
3. One of the editorial board members.

The chapters in each volume are individually reviewed single blind by expert reviewers and the volume editor.

Tanveer ul Haq · Yousef Haik

Electrochemical Water Splitting

Fundamentals, Challenges and Advances

Springer

Tanveer ul Haq
Department of Chemistry, College of Science
University of Sharjah
Sharjah, United Arab Emirates

Yousef Haik
Department of Mechanical and Nuclear Engineering, College of Engineering
University of Sharjah
Sharjah, United Arab Emirates

ISSN 2524-5384 ISSN 2524-5392 (electronic)
Materials Horizons: From Nature to Nanomaterials
ISBN 978-981-99-9859-3 ISBN 978-981-99-9860-9 (eBook)
https://doi.org/10.1007/978-981-99-9860-9

This Springer imprint is published by the registered company Springer Nature Singapore Pte Ltd.
The registered company address is: 152 Beach Road, #21-01/04 Gateway East, Singapore 189721, Singapore

Contents

About the Authors

Dr. Tanveer ul Haq received his Ph.D. in sustainable energy from Texas A&M University, USA. He completed his M.S. in Chemistry from the Lahore University of Management and Science (LUMS), Lahore, Pakistan. He currently serves as Assistant Professor in the Department of Chemistry at the University of Sharjah, UAE. His research primarily focuses on the electronic, surface, and structural engineering of electrocatalysts for direct seawater electrolysis, decoupled and tandem water electrolysis, and direct methanol fuel cells (DMFCs). He is keen on utilizing advanced electronic and spectroscopic characterization tools to explore surface and structural defects in catalysts, study phase transformations during electrocatalysis, and identify active centers for electrochemical reactions.

Yousef Haik is Internationally Recognized Scholar in nanotechnology. His research focus is in the synthesis and characterization of nanomaterials for a myriad of applications including hydrogen production, biomedical sensors, diagnostics, and imaging. His scholarly output includes over 300 peer-reviewed scientific articles, over 80 issued patents and patent applications, and a number of textbooks. His academic appointments include Professor of Mechanical Engineering, Professor of Biomedical Engineering, Professor of Nanoscience and Nanoengineering, and Professor of Medicinal Chemistry. His academic affiliation includes Harvard University, Texas A&M University, University of North Carolina, Florida State University, and the University of Sharjah. Professor Haik has over 25 years of academic experience in senior academic administrative positions ranging from Program Director, Department Chair, Dean, Associate Provost, Associate Vice President, Provost, and Vice Chancellor. Prof. Haik has B.Sc., M.Sc., and Ph.D. in Mechanical Engineering and a Juris Doctor degree.

Abbreviations

2D	Two-Dimensional
3D	Three-Dimensional
Ag/AgCl	Silver-Silver Chloride Electrode
CCE	Controlled Current Electrolysis
CE	Counter Electrode
CNT	Carbon Nanotube
CPE	Controlled Potential Electrolysis
CV	Cyclic Voltammogram
DFT	Density Functional Theory
DOE	Department of Energy
DOS	Density of States
EDS	Energy-Dispersive spectroscopy
EDX	Energy-Dispersive X-ray spectroscopy
EELS	Electron Energy Loss Spectroscopy
EIS	Electrochemical Impedance Spectroscopy
EPR	Electron Paramagnetic Resonance
EWS	Electrochemical Water Splitting
FC	Fuel Cell
FTIR	Fourier Transform Infrared Spectroscopy
FTO	Fluorine-Doped Tin Oxide
HEMFC	High-Temperature Proton Exchange Membrane Fuel Cell
HER	Hydrogen Evolution Reaction
Hg/HgCl	Mercury-Mercury Chloride Electrode
Hg/HgO	Mercury-Mercury Oxide Electrode
HOMO	Highest Occupied Molecular Orbital
HT-SOFC	High-Temperature Solid Oxide Fuel Cell
ICP-MS	Inductively Coupled Plasma Mass Spectrometry
LDH	Layered Double Hydroxide
LUMO	Lowest Unoccupied Molecular Orbital
MEA	Membrane Electrode Assembly
MOF	Metal-Organic Framework

NAC	Nanostructured Catalyst
NCs	Nanocluster
OER	Oxygen Evolution Reaction
ORR	Oxygen Reduction Reaction
PEC	Photoelectrochemical
PEM	Proton Exchange Membrane
RE	Reference Electrode
RHE	Reference Hydrogen Electrode
SAC	Single-Atom Catalyst
SEM	Scanning Electron Microscope
SRB	Sulfate-Reducing Bacteria
STM	Scanning tunneling microscopy
TEM	Transmission Electron Microscopy
TOF	Turnover Frequency
TON	Turnover Number
UPS	Ultraviolet photoelectron spectroscopy
UV-Vis	Ultraviolet-Visible Spectroscopy
WE	Working Electrode
XPS	X-ray Photoelectron Spectroscopy
XRD	X-ray Diffraction

Chapter 1
Electrocatalysis Fundamentals for OER and HER

1.1 Hydrogen Energy: A Sustainable Future

Traditional fossil fuels, like natural gas, coal, and oil, have devastated the environment, and their continued use threatens the planet's future. Climate change is one of the traditional fossil fuels' most significant adverse impacts. Burning fossil fuels releases large amounts of carbon dioxide and other greenhouse gases into the environment, which trap the entire heat and contribute to global warming. As the earth's temperature rises, we see more frequent and severe weather events, rising sea levels, and destroying ecosystems and habitats. The impacts of climate change are far-reaching, affecting not only the natural world but also human health, food security, and economic stability. Another negative impact of traditional fossil fuels is air pollution. Burning fossil fuels releases various toxic pollutants, including sulfur dioxide, nitrogen oxides, particulate matter, and volatile organic compounds. These pollutants have been linked to respiratory and cardiovascular diseases, cancer, and other health problems. In addition, burning fossil fuels produces smog, which can exacerbate asthma and other respiratory conditions. Air pollution also harms ecosystems and contributes to acid rain, damaging forests, lakes, and other natural resources. The negative impact of traditional fossil fuels on air quality is particularly severe in developing countries, where people often rely on polluting fuels for cooking and heating, leading to indoor air pollution that poses a significant health risk. Carbon-free energy: Unlike traditional fossil fuels, hydrogen is a carbon-free energy source. When hydrogen is burned, it only produces water vapor as a by-product, rendering it an environmentally friendly substitute for fossil fuels. This makes hydrogen a critical component of efforts to reduce greenhouse gas emissions and combat climate change. By replacing fossil fuels with hydrogen, we can significantly reduce our carbon footprint and move toward a more sustainable energy system [1].

Versatility: One of the most significant advantages of hydrogen energy is its versatility. Hydrogen can be used in various applications, from powering vehicles to generating electricity to heating homes and buildings. This versatility makes it a valuable

T. u. Haq and Y. Haik, *Electrochemical Water Splitting*, Materials Horizons: From Nature to Nanomaterials, https://doi.org/10.1007/978-981-99-9860-9_1

alternative to traditional fossil fuels and a key component of efforts to transition to a more sustainable energy future. One of the most promising applications of hydrogen is in transportation. Hydrogen fuel cell vehicles are already on the market, and many automakers have invested in developing this technology. Fuel cell vehicles have the potential to be much more efficient than traditional gasoline-powered cars and trucks, and they emit only water vapor as a by-product. This makes them an excellent option for reducing greenhouse gas emissions and improving air quality in urban areas. In addition, hydrogen can also be used to power trains, ships, and other forms of transportation, further reducing emissions and promoting sustainability. Another vital application of hydrogen is in electricity generation. Hydrogen can generate electricity in a fuel cell, producing only water and heat as by-products. This makes it a clean and efficient alternative to traditional fossil fuels like natural gas and coal. Hydrogen can also be used to store energy from sustainable resources like wind and solar power, helping to ensure the supply of stable energy. Additionally, hydrogen can be used to heat buildings directly or by generating electricity to power heat pumps. This could help reduce the reliance on natural gas for heating, a significant source of greenhouse gas emissions. Overall, the versatility of hydrogen energy makes it an essential component of efforts to transition to a more sustainable energy future. Its ability to be used in various applications, from transportation to electricity generation to heating, makes it a valuable alternative to traditional fossil fuels. Investing in hydrogen and promoting its use in multiple sectors can reduce our carbon footprint, improve air quality, and ensure a more stable and prosperous energy system for future generations [2].

Energy efficiency: Another significant advantage of hydrogen energy is its energy efficiency. When burned in a fuel cell, hydrogen has the capability to convert as much as 60% of its energy into electricity, compared to just 20–25% for gasoline-powered engines. This makes hydrogen an excellent option for powering electric vehicles, which are becoming increasingly popular as a way to reduce emissions from transportation. In addition, hydrogen fuel cell technology is incredibly efficient regarding weight and space. A fuel cell can provide the same amount of power as a traditional internal combustion engine, but it takes up much less space and weighs significantly less. This makes fuel cell vehicles more efficient overall, requiring less energy to move the same distance. In addition to transportation, hydrogen energy can also be used for electricity generation. Hydrogen fuel cells can be used to generate electricity, either on-site or for grid distribution. Fuel cells can generate power in various settings, from small-scale residential applications to large-scale industrial or commercial settings. In addition, because hydrogen may be generated from number of resources, including renewable sources like wind and solar power, it has the potential to provide a stable and sustainable energy supply [3].

Economic benefits: Hydrogen energy has several financial benefits, making it an attractive alternative to traditional fossil fuels. One of hydrogen's most significant economic benefits is its potential to create jobs and drive economic growth. The hydrogen industry is still in its early stages but has already created thousands of jobs in manufacturing, research and development, and other areas. As the industry continues to grow, it is likely to create even more jobs, particularly in regions that the decline of the fossil fuel industry has hard hit. In addition, because hydrogen can be produced from a range of sources, including renewable sources like wind and solar power, it has the potential to create new economic opportunities in rural areas where renewable energy resources are abundant.

Another crucial economic benefit of hydrogen energy is its potential to reduce energy costs over the long term. Although the initial investment in hydrogen infrastructure can be significant, once that infrastructure is in place, the cost of producing and using hydrogen can be much lower than that of traditional fossil fuels. This is particularly true in areas where natural gas prices are high, as hydrogen can be produced from natural gas at a lower cost than traditional gasoline or diesel. In addition, because hydrogen fuel cell vehicles are much more efficient than conventional gasoline-powered vehicles, they require less fuel overall, which can lead to significant cost savings over time [4].

1.2 Hydrogen Production Technologies

Hydrogen is a versatile energy carrier produced from various sources, including fossil fuels, renewable energy, and waste materials. Each source has advantages and disadvantages, and the choice of source will depend on factors such as availability, cost, and environmental impact. One of the most common sources of hydrogen is natural gas. Natural gas is a fossil fuel composed primarily of methane, often used as a feedstock for the generation of hydrogen. The most common method of hydrogen production from natural gas is steam methane reforming (SMR), which involves reacting natural gas with steam for the production of carbon dioxide and hydrogen. SMR is a mature technology and is currently the most cost-effective method of hydrogen production, accounting for over 95% of global hydrogen production. However, it is also a significant source of greenhouse gas emissions, as the production process generates carbon dioxide as a by-product [5].

Another source of hydrogen is coal gasification. Coal gasification involves reacting coal with oxygen and steam for hydrogen production, carbon monoxide, and carbon dioxide. The hydrogen can then be separated from the other gases and purified. Coal gasification is a more complex process than natural gas reforming. As a result, it is less efficient but has the advantage of using a more comprehensive range of feedstocks, including low-quality coal and coal waste. However, like SMR, coal gasification is also a significant source of greenhouse gas emissions [6].

Biomass gasification is another method of hydrogen production that involves reacting organic materials, such as wood chips or agricultural waste, with steam to produce a mixture of carbon dioxide, hydrogen, and carbon monoxide. The next step involves separating and refining the hydrogen for practical application. Biomass gasification has the advantage of using renewable materials as a feedstock and producing less greenhouse gas emissions than fossil fuel-based methods. However, it is less efficient than natural gas reforming and can be more expensive [7].

Hydrogen can also be produced from waste materials through a process called gasification. Gasification involves reacting waste materials, such as municipal solid or agricultural waste, with steam to produce a mixture of carbon dioxide, hydrogen, and carbon monoxide. Then hydrogen can be separated and purified for use. Gasification has the advantage of using waste materials as a feedstock, reducing the need for landfill space, and reducing greenhouse gas emissions. However, it is also less efficient than natural gas reforming and can be more expensive [8].

1.3 Electrochemical Water Splitting

One of the most beneficial approaches to hydrogen production is through water electrolysis, a process that splits water into hydrogen and oxygen by utilizing an electrical current [9]. A primary benefit of this approach is its compatibility with renewable energy sources like wind and solar power, rendering it an eco-friendly and sustainable means of producing hydrogen. As the demand for renewable energy sources grows, electrolysis is becoming increasingly attractive as a method of producing hydrogen. Another advantage of water electrolysis is that it can produce hydrogen on-site, eliminating the need for transportation and storage of hydrogen. This is particularly beneficial for industries that require hydrogen for their processes, such as the refining and chemical industries. These industries can reduce their carbon footprint and increase their energy independence by producing hydrogen on-site. In addition, water electrolysis can be used to produce hydrogen at different scales, from small-scale systems for residential use to large-scale designs for industrial use, making it a versatile method of hydrogen production. Overall, the advantages of water electrolysis make it an essential method of producing hydrogen for a clean energy future.

Electrochemical water splitting involves breaking down water into its constituent elements, hydrogen and oxygen, using an electric current. The process requires two half-cell reactions, one at the anode and other occurring at the cathode. The two half-reactions required for water electrolysis at the cathode and anode are the oxygen evolution reaction (OER) and the hydrogen evolution reaction (HER) respectively. The OER presumably involves oxidation of two water molecules under acidic circumstances, releasing four electrons, four protons, and one oxygen molecule. In contrast, the hydrogen evolution reaction (HER) involves the reduction of four protons by employing four electrons to produce two molecules of hydrogen [10].

HER (Reaction at the cathode)

$$4H^+ + 4e^- \rightarrow 2H_2(g); \quad E^o_{reduction}\ 0.0\,V$$

OER (Reaction at the anode)

$$2H_2O(l) \rightarrow O_2(g) + 4H^+ + 4e^-; \quad E^o_{oxidation} - 1.23\,V$$

where the standard reduction ($E^{o}_{reduction}$) and oxidation ($E^{o}_{Oxidation}$) potentials of HER and OER, concerning the fixed-pH standard hydrogen electrode (SHE), are provided at pH = 0.0 and at 298.15 K temperature, respectively. However, the HER reduces four molecules of water in alkaline media for generation of two hydrogen molecules and four hydroxide ions. In contrast, OER involves four hydroxide ions oxidation to generate four electrons, two water molecules, and one oxygen molecule. In contrast, the OER comprises the oxidation of four hydroxide ions to produce four electrons, two water molecules, and one oxygen molecule [11].

HER (Reaction at the cathode)

$$4H_2O(l) + 4e^- \rightarrow 2H_2(g) + 4OH^-; \quad E^{o}_{reduction} \; 0.83\,V$$

OER (Reaction at the anode)

$$4OH^- \rightarrow O_2(g) + 4e^- + 2H_2O(l); \quad E^{o}_{reduction} - 0.40\,V$$

Due to the differing standard states for balanced reactions involving OH^- (pH = 14) and H^+ (pH = 0), it is important to acknowledge the varying pH values between these two reaction sets. Despite this distinction, the second group of reactions (with balanced OH^-) will maintain E^{o}_{red} at 0 V and E^{o}_{Ox} at 1.23 V relative to the SHE at pH = 0; however, at pH = 14, E^{o}_{Ox} is 0.40 V and E^{o}_{red} around 0.83 V, correspondingly, compared to the SHE [12]. This illustrates that the redox potentials of both half-cell reactions are contingent on pH but not on how the reactions are balanced (whether with OH^- or H^+) (a pH-independent scal) when measured against the SHE. Nernst equation expresses the relation between potential and pH as

$$\begin{aligned} E_{RHE} &= E_{SHE} - \frac{2.303RT}{F}\text{pH} \\ &= -0.059\text{pH} \end{aligned}$$

where

F Faraday constant
T Absolute temperature
R Universal gas constant.

The RHE employs the identical electrolyte solution like working electrode, ensuring their matching pH levels. Consequently, the hydrogen evolution reaction involving protons around pH = 14 is expected to be 0.0 V reduction potential compared to the RHE. This equilibrium arises because the pH sensitivity inherent in the RHE potential offsets the pH sensitivity of the HER potential. Specifically, at pH = 14, the ERHE value is established at 0.83 V. Using RHE as a reference at pH 14 implies that proton presence will not influence the oxidation potential of the OER. As a result, at pH 14, both the RHE potentials for OER and HER, balanced with H^+, are envisioned to be 1.23 and 0.0 V, correspondingly. This leads to calculating

$E_{water\text{-}splitting}{}^{\circ}$ as $E_{water\text{-}splitting}{}^{\circ} = E_{red,cathode}{}^{\circ} - E_{ox,anode}{}^{\circ} = -1.23$ V. The negative sign underscores the external potential required to drive the process.

The standard potentials associated with the HER as well as OER will remain constant, precisely at 0 and 1.23 V compared to RHE, regardless of alterations in pH levels. This constancy persists because the reduction potentials for both HER and OER remain unaffected by utilizing H^+ or OH^- ions for equilibrium in the reactions. Whether we adopt the standard hydrogen electrode (SHE) or the RHE as the point of reference, this principle holds. The reactions $4H^+ + 4e^- \rightarrow 2H_2(g)$ and $4H_2O(l) + 4e^- \rightarrow 2H_2(g)^+$ exhibit thermodynamic equivalence. Nevertheless, their kinetics differ due to the interplay between H^+ and OH^- ions, which interchange roles as reactants and products. This interchange introduces pH-dependent concentration influences in the corresponding rate equations. The energy imperative for cleaving a water molecule is 1.23 eV per transferred electron. This value remains valid under conditions of equilibrium and at a temperature of 298.15 K.

Thermodynamically, 1.23 V of potential is required across an electrochemical cell to split the water molecule. In practice, though, a potential higher than 1.23 V is needed. This extra potential is termed "overpotential," defined as the difference between the potential that must be applied for water splitting to occur and its equilibrium voltage (1.23 V). The overall overpotential value in electrochemical water splitting is the sum of overpotential associated with both anodic and cathodic half-cell reactions. Numerous factors contribute to the overpotential, including the need for activation barriers and thermodynamic free energy for the fundamental processes that make up the OER and HER, potential mass transport restrictions in the solution phase, and resistance to current flow within photoabsorber (s)/electrodes as well as at junctions and interfaces (caused by the junction potential). In addition, the fluid dynamics of the evolved O_2 and H_2 bubbles, which are influenced by processes like electrolyte penetration, intermediate adsorption, bubble/gas nucleation, growth, coalescence, and desorption, also have an impact on the rate of water splitting by affecting the electrolyte's ohmic resistance, effective catalytically active area, and speed of mass transport [13].

1.4 Mechanism of Hydrogen Evolution Reaction

HER is a kinetically favorable two-step cathodic reaction. Under acidic conditions, HER proceeds with the Volmer step. Initially, hydrogen ions convert to a hydrogen atom (H^*) intermediate by absorbing an electron and adsorbed on the catalyst surface, which is later separated from it by the chemical combination of Tafel or electrochemical Heyrovsky processes (Fig. 1.1) [14]. As a result, there are two ways that HER proceeds in an acidic and an alkaline environment, and they are as follows [15]:

Volmer–Heyrovsky mechanism
Volmer–Tafel mechanism.

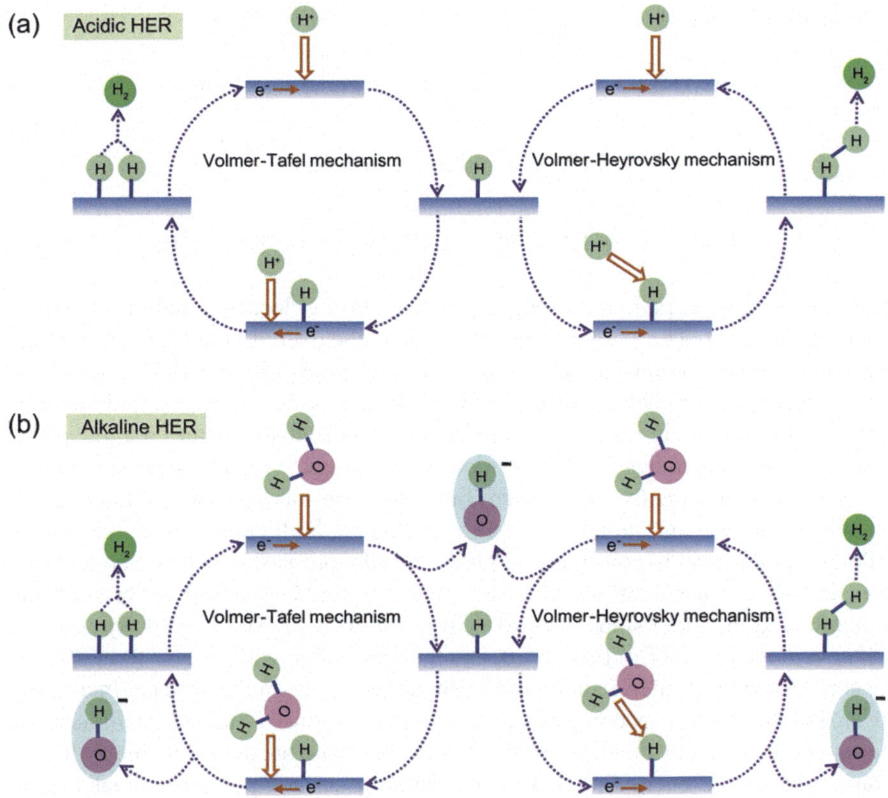

Fig. 1.1 HER pathways in the **a** acidic and **b** alkaline environment. Reproduced with permission from Ref. [14]. © 2021, Elsevier

The HER route in an alkaline environment is comparable to that in an acid environment, except that water molecules create intermediate atomic hydrogen in an alkaline environment. In contrast to an acidic environment, the electrochemical reduction of H_2O into absorbed OH and H^* occurs in an alkaline environment during the Volmer process. In these circumstances, breaking the H–O–H bond in the solution is more challenging before H^* adsorption at the catalyst surface. Because of its H^+ level, the acidic environment is more suitable for the HER. The HER reaction mechanism in different electrolyte is described below [16].

In neutral and basic electrolyte

$$H_2O + * + e^- \rightarrow H^* + OH^-$$

$$2H^* \rightarrow H_2 \text{ (Tafel Reaction)}$$

$$H_3O^+ + e^- + H^* \rightarrow H_2 + H_2O \text{ (Heyrovsky Reaction)}$$

In acidic electrolyte

$$H_3O^+ + * + e^- \rightarrow H^* + H_2O \text{ (Volmer Reaction)}$$

$$2H^* \rightarrow H_2 \text{ (Tafel Reaction)}$$

$$H_3O^+ + e^- + H^* \rightarrow H_2 + H_2O \text{ (Heyrovsky Reaction)}$$

All these steps and paths are strongly dependent on the electronic and surface structure of the catalyst. The Volmer step will regulate the overall kinetics of HER if the hydrogen adsorption on the catalyst surface is very poor, with the Tafel slope being approximately 120 mV/dec. An electrode with more cavities, edges, and surfaces will be more efficient in adsorbing more hydrogen and enhancing the charge transfer process if the Volmer stage is the rate-determining step. If the hydrogen adsorption is strong enough on active sites, then Tafel step controls the overall kinetics. As a result, if the concentration of H^* on the surface is high enough, the two nearby hydrogen atoms tend to combine and produce hydrogen gas, which is then released from the surface. In this instance, the Tafel reaction would regulate the overall kinetic response, and the Tafel slope is equivalent to 30 mV/dec. The Heyrovsky reaction will govern the overall reaction speed, and the Tafel slope will then be 40 mV/dec if the concentration of intermediate hydrogen atoms is low on the surface. Increasing the area of the reaction, accomplished by increasing surface roughness, will increase the water-splitting rate if the Heyrovsky or Tafel reactions are the RDS. The hydrogen atoms' Gibbs free energy (ΔG_H) on the catalyst surface plays a crucial role in determining the electrocatalytic activity. The volcano curve, which represents the exchange current density curve in terms of ΔG_H, is used to estimate the inherent electrocatalytic activity. A promising HER electrocatalyst must form a suitable bond with the adsorbed hydrogen, in which the electron–proton transfer process is easily carried out. On the other hand, this bond should be weak enough to cause the bond breaking to quickly release H_2 because the adsorption and desorption of the H atoms (H_{ads}) on the electrode surface are a competitive process. As a result, an ideal catalyst is in the coordinates around the volcano curve's vertex, where ΔG_H is close to zero [17].

1.5 Mechanism of Oxygen Evolution Reaction

Water oxidation is a multistep multielectron process with a high thermodynamic and activation barrier that impedes the overall reaction kinetics of electrochemical water splitting. An adsorbed water ($^*OH_2$), hydroxo ($^*OH$), oxo ($^*O$), hydroperoxo ($^*OOH$), peroxo ($^*O_2$), superoxo ($^*O_2$), or dioxygen ($^*O_2$) species, as well as a vacated active adsorption site (*), are the different intermediates involved in O–O coupling with oxygen valance state of 2, 2, 2, 1, 1, 1/2, and 0, respectively. The various proposed mechanisms on TM surfaces for O–O coupling are given below [18].

$$M + OH^- \rightarrow M - OH + e^-$$

$$M - OH + OH^- \rightarrow M = O + e^- + H_2O$$

$$M = O + OH^- \rightarrow M - OOH + e^-$$

$$M - OOH + OH^- \rightarrow M + H_2O + e^- + O_2$$

$$M = O + M = O \rightarrow M - O - O - M$$

$$M - O - O - M \rightarrow 2M + O_2$$

In a single-site mechanism, one active metal site mediates the production of the O–O bond through the nucleophilic attack of OH^- ions from the electrolyte on the M=O intermediate, which is thought to be the more energy-demanding step in OER, while two active metal sites are involved in O–O interaction and promote the oxidation reaction in the dual-site mechanism. Researchers demonstrated that bifunctional routes could reduce the energy barrier related to the OOH intermediate formation. In a bifunctional mechanism, a single metal site mediates the production of an O–O bond by attacking hydroxyl ions, while a neighboring acceptor site actively participates in the coordinated reaction by taking protons. The metal ion's oxidation state at the active site may change at each stage; the metal may undergo oxidation/reduction or remain unchanged. Additionally, the overall oxidation state of the $^*O_2$ moiety may be 0, 1, 2, or 3, depending on the active site [19]. One or more of the mechanisms may be simultaneously favored by the catalyst in several paths for O–O coupling. It is essential to highlight that the fundamental steps within the mechanism of the oxygen evolution reaction (OER) can encompass processes involving electron transfer (considered electroactive steps) or exclude such transfer (referred to as non-electroactive steps, involving proton–electron pairs). While steps not involving electron transfer inherently depend on thermal energy, rendering them potentially less favorable at room temperature, even when subjected to substantial external potentials, the distinguishing feature of electron transfer steps lies in the possibility of diminishing their free energy demand (or activation energy) through the application of an external electric potential [20].

1.6 OER Mechanism with Consideration of Spin

OER intermediates and evolution processes are encountered during the conversions of OH to O_2 (in an alkaline solution) and H_2O to O_2 (in an acid solution). It can be thought of as a process that releases an O_2 product with six valence electrons, two of which are occupied * orbitals with the same spin direction, by depleting the oxygen valence electrons in the OH/H_2O reactants with eight valence electrons per oxygen atom that satisfy the octet rule. Here, due to charge transfer to active sites, intermediates such as *OH (1), *O (2), *OOH (3), and *OO (4) associated through chemisorption with active sites in matching electrolyte–electrocatalyst interfaces are fundamentally electron deficient. As a result, it is possible to roughly calculate the overpotential by comparing the energy gap between the O 2p valence bands and the vacant 3d conduction bands. When empty energy levels in conduction bands of active sites are decreased below the O 2p band centers of adsorbates by a positive potential applied on electrodes/electrocatalysts during electrocatalytic water splitting, the actual depletion of electrons from oxygen adsorbates occurs.

Theoretically, the in-situ monitoring of transient valence states of active sites and associated particular oxygen species can prove electron transfer during each step. In the following sections, the three fundamental OER routes of different electrocatalysts shown in Fig. 1.2 will be used to clarify the spin function. In general, various electrocatalysts (or different local crystal plane surfaces within a catalyst) favor different reaction pathways based on the minimization of activation energy. For instance, the Eley–Rideal (ER)-type mechanism and the Langmuir–Hinshelwood (LH)-type mechanism are two distinct reaction pathways based on the adsorbate evolution mechanism (AEM) (Fig. 1.2) [21]. Orbital overlaps, chemisorptive bond formation, and interfacial charge transfer all contribute to the adsorption of reactants or intermediates on electrocatalyst surfaces. The Sabatier principle has historically suggested that "just appropriate" adsorptive interactions between catalysts and adsorbates are expected. Therefore, improving active site adsorption capabilities can potentially lower the surface reaction energy barrier. Due to spatial overlaps and energy similarities across electronic states, adsorption capabilities for 3d transition metal-based oxides with octahedron coordination correlate to e.g., occupancy. Furthermore, Shao-Horn et al. demonstrated that OER activity had volcano-shaped dependence on the quantity of e.g. electrons for transition metal perovskite oxides. This has led us to assume that spin must play a universal role during OER despite the complexity of local surface properties like reconstructions, space charges, polarity, or segregation. This is because the descriptor e.g. electron number, is a characteristic of the spin configuration of an electrocatalyst [22].

Figure 1.3a illustrates one potential instance of outer electron behaviors considering spin in the ER-type AEM pathway (using single cation active sites). For example, one active place will initially choose a spin-down electron from the six paired OH (0) electrons in the electrolyte present in three low-energy orbitals to generate *OH (1). Due to shared chemisorptive linkages, it continues to be shared by the adsorbate and the active surface site. It will then be removed from the interface through bulk transport to external circuits, freeing active sites to convert *OH (1) to *O (2). The following electron during the transformation of *OOH (3) to *OO (4) should be spin-down for the evolution of O_2 if the electrons extracted during the transformation of *OH (1) to *O (2) are spin-down and spin-up for the transformation of *O (2) to *OOH (3). In contrast, if spin-down electrons were extracted during the transformations of *OH (1) to *O (2) and *O (2) to *OOH (3), the next two electrons

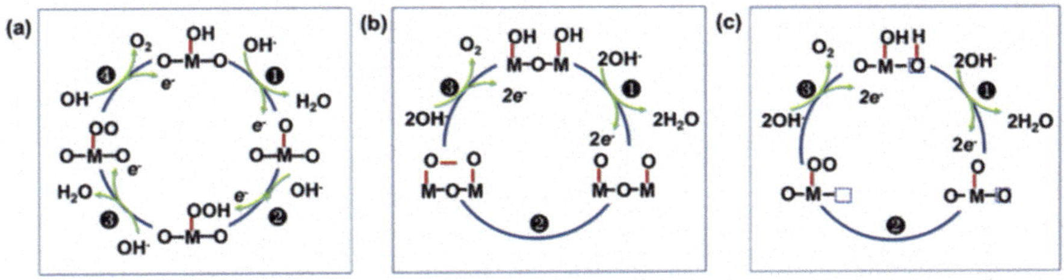

Fig. 1.2 Various OER electrocatalysts in alkaline media have proposed well-established reaction routes. These include: **a** the Eley–Rideal (ER) type adsorbate evolution mechanism (AEM) which involves a single metal cation active site and the evolution from OH-(0) reactant to *OH (1), *O (2), *OOH (3), *OO (4) intermediates, and O_2 (5) product, **b** the Langmuir–Hinshelwood (LH) type adsorbate evolution mechanism (AEM) which involves two adjacent metal cation active sites and the evolution from OH-(0) reactant to *OH (1), *O (2), OO (4) intermediates, and O_2 (5) product, bypassing *OOH (3), and **c** the lattice oxygen mechanism (LOM) which involves one metal cation active site and one oxygen in the lattice and the evolution from OH-(0) reactant to *OH (1), *O (2), OO (4) intermediates, and O_2 (5) product, bypassing *OOH (3). The number following each oxygen species is indicated for better illustration. Reproduced with permission from Ref. [21]. © Springer Nature

should have opposite spin axes. This indicates that three out of every four electrons removed from the reactant and intermediates have the same spin direction in all circumstances. Because the spin choice during electron transfer is determined by the current availability of active site orbitals, additional scenarios are also possible. For example, some electrons will experience spin flipping during transfer if active sites cannot extract three spin-up and one spin-down electron, as required by the theory behind oxygen formation. Due to the rate-limiting step of OOH* (3) being avoided by the cooperation of two adjacent cation active sites, the transformations of *O (2) to *OO* (4) in LH-type AEM and LOM pathways (involving two active sites) are replaced by the transformation of *O (2) to *OO* (4) with a relatively lower energy barrier. The significant overlap of metal 3d with oxygen 2p causes the Fermi level to be lowered to the oxygen 2p band, which is the fundamental physics of the LOM [23]. A catalyst that has adopted the LOM and has significantly smaller lattice parameters and shorter surface oxygen separations, the kinetics of the LOM is generally faster than that of the ER-type AEM due to non-concerted proton–electron transfer processes [24]. Even so, insufficiently close O p-band centers to the Fermi level is a need for favorable LOM kinetics, which can potentially lead to unstable electrocatalysts [24]. In conclusion, the LOM route differs from LH-type AEM in that lattice oxygen is activated as an active site.

Additionally, spin explanations based on the LOM and LH-type AEM pathways agree with what is known about the ER-type AEM pathway (which only has one active site) (Fig. 1.3b). This research concludes that 3/4 of the electrons transported during OER are in the same spin direction and may be universally applied to any suggested process. Theoretically, if the associated electrocatalyst cannot offer accommodating spin configurations, spin flipping may need additional energy, invariably delaying

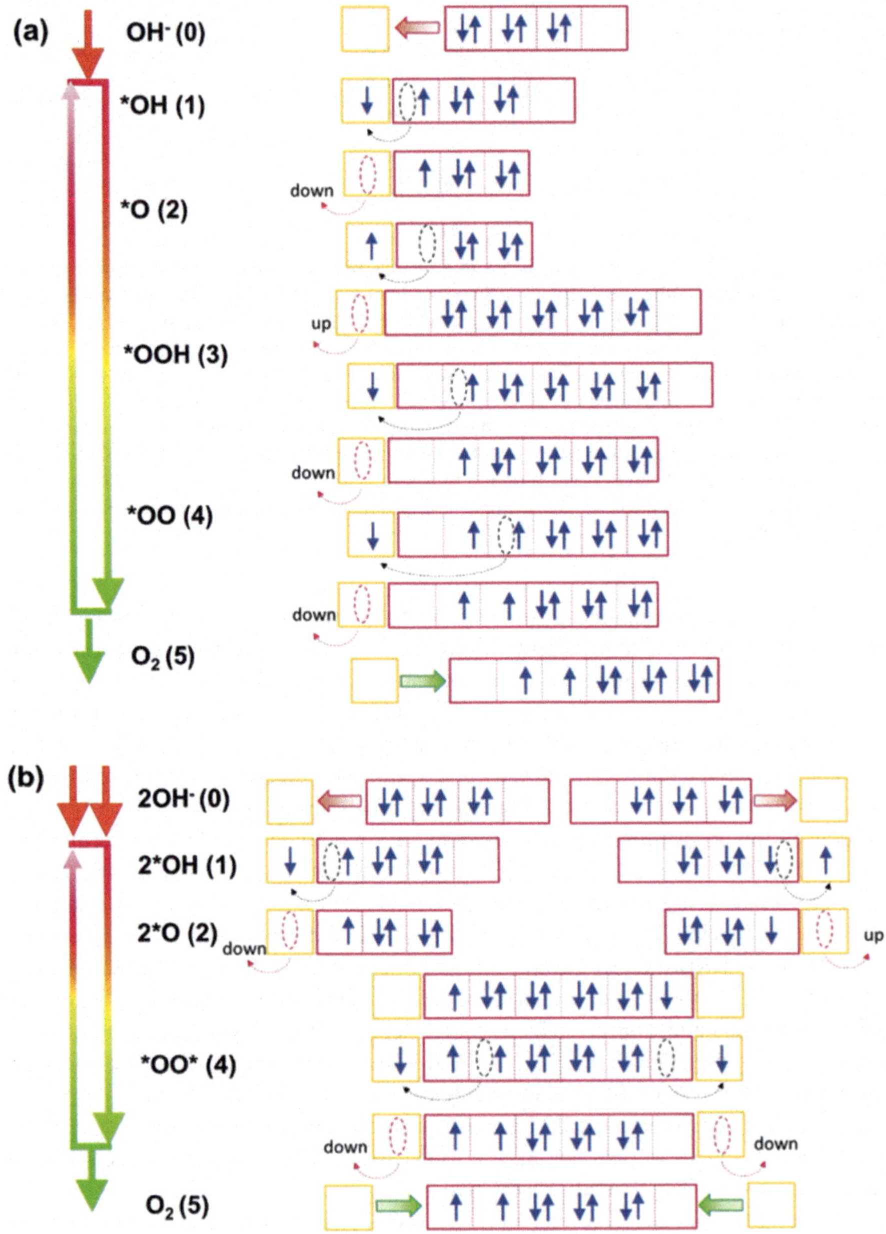
(a)
OH- (0)
*OH (1)
*O (2)
*OOH (3)
*OO (4)
O2 (5)
down
up
down
down
(b)
2OH- (0)
2*OH (1)
2*O (2)
OO (4)
O2 (5)
down
up
down
down

◄**Fig. 1.3** During the process of OER, there are different possible behaviors of outer electrons. These include: **a** ER-type AEM which involves a single active site mechanism and **b** LH-type AEM and LOM which involve two active sites mechanisms. The red squares in the figure represent hybrid orbitals of oxygen species, which are simplified without energy level contrasts. Black dashed circles and lines with arrows depict electron transfer or hopping between adsorbates and active sites through chemisorptive bonding. Red dashed circles and lines represent electron depletion as delivered away from the surface to bulk. In this case, the spin direction of the four electrons is down-up-down-down, but other combinations are also possible if three out of four electrons are in the same spin direction. Reproduced with permission from Ref. [21]. © Springer Nature

kinetics. Therefore, depending on the relevant rate-limiting steps of the various electrocatalytic systems, the contribution of spin to total OER performance may vary. However, spin has a unique function in OER that must be considered while defining actual processes and developing fresh electrocatalysts. Through a 40% increase in OER efficiency for some ferromagnetic electrocatalysts when an external magnetic field is applied, which can favor the parallel alignment of spin during O–O bond formation, the importance of spin has been experimentally justified [25].

1.7 d-Band Theory for HER

The assessment of hydrogen (H) adsorption strength and the change in Gibbs free energy (ΔG_H) for catalysts based on transition metals is intrinsically linked to the electronic configurations on their surfaces, particularly the energy levels of d-orbitals of the metal atoms. These d-orbital levels are closely associated with the Fermi energy, and theoretical calculations using density functional theory (DFT) provide insights into chemisorption energy, dissociation energy, and the activation barrier concerning the central position of the metal's d-band. The concept of the d-band center theory (d), pioneered by Norskov and other researchers, revolves around the binding energy of adsorbed hydrogen (H), as depicted in Fig. 1.4a. During the HER process, breaking bonds leads to energy level adjustments. Moreover, the interaction between the H orbital and the metal d-orbitals results in a deeply situated filled bonding state (low energy) and an empty or partially filled antibonding form (high energy). In this scenario, the metal-hydrogen (MH) bond's strength relies on the antibonding state's occupancy (*), where reduced occupancy corresponds to heightened bonding strength. Consequently, the adsorption of H and the ΔG_H for a given catalyst based on transition metals can be qualitatively predicted and substantiated by comparing the calculated d-band states of the metal's surface with its Fermi energy level. Additionally, research conducted by Santos and collaborators underscores that the positioning of the d-band center influences the activation energy and the barrier for bond breaking in the HER, as illustrated in Fig. 1.4b [26]. The d-band location fluctuates through the proton absorption, bond-breaking saddle point, and following product production due to the verified occupancy of the bonding and antibonding orbitals [27].

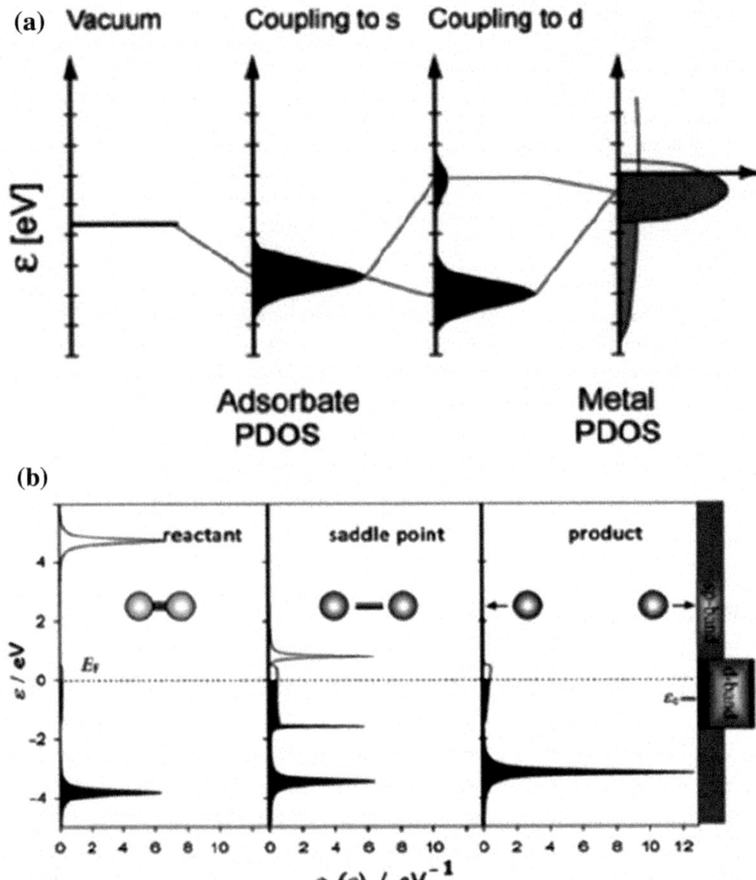

Fig. 1.4 **a** Energy level hybridization and realignment process for the formation of a chemical bond between an adsorbate and a transition metal surface is illustrated schematically. **b** Density of states (DOS) is shown at equilibrium for the H_2 molecule, at the saddle point, and after bond breaking. Reproduced with permission from Ref. [26]. © ACS

1.8 d-Band Theory for OER

When adsorbates interact with the surface of transition metal (TM) catalysts, their valence states undergo an expansion and downward shift. This interaction is closely intertwined with the d-orbital states of the TM, leading to the formation of both filled bonding states and partially occupied antibonding states. These antibonding states are situated above the d states, and their population plays a pivotal role in determining the strength of the chemical bond. This relationship is elucidated by the positioning of the band center concerning the Fermi level (EF), as illustrated in Fig. 1.5a. When the d states possess higher energy relative to the EF, the energy of the antibonding states becomes elevated, subsequently enhancing the strength of the chemical bond. Achieving an ideal balance between the adsorption and desorption of OER intermediates is essential within the OER domain. This necessitates precise tuning of the electrocatalyst's energy level (Ed), a factor that crucially impacts the promotion of OER activity. Similar to the alloying effect observed in transition metals, the strategic incorporation of dopants into TM-based compounds emerges as a highly effective technique for influencing the behavior of the d states. Notably, in compounds featuring a robust Ru–O bond, such as RuO_2, the consistently high free energy of the rate-determining step (RDS) is always observed [28].

On the contrary, introducing a small amount of copper (Cu) as a dopant causes a shift in the center of the d-band away from the Fermi energy (EF). This shift reduces the number of antibonding states and weakens the strength of the bond between ruthenium (Ru) and oxygen (O), as depicted in Fig. 1.5b. In contrast, when examining the properties of nickel phosphide (Ni_2P), its energy level labeled Ed is relatively lower. This lower energy level contributes to a higher likelihood of attracting and retaining intermediate species during the OER due to the lower energy

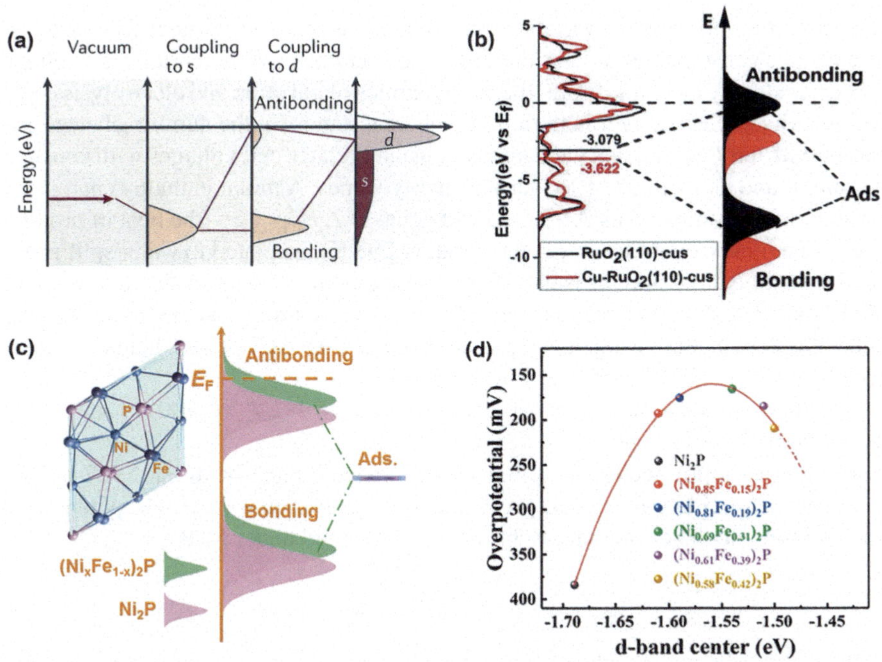

Fig. 1.5 Bond formation at **a** transition metal surface is shown schematically. The more packed the antibonding states are, and the weaker the adsorption bond is, the lower the d states are in energy in relation to the EF; **b** RuO_2, Cu–RuO_2, and the related schematic illustration of the creation of bonds between the reaction surface and adsorbate are depicted in density of states (DOS) plots; **c** Ni_2P and Fe-doped Ni_2P d-band centers; **d** associations between the estimated d-band centers and Fe-doped Ni_2P experimental overpotential. Reproduced with permission from Ref. [28]. © Springer Nature

needed for adsorption. Notably, research conducted by Chen and coworkers has demonstrated that the introduction of iron (Fe) doping adjusts the Ed energy levels to come closer to the Fermi energy, as illustrated in Fig. 1.5c [29]. This shift elevates the energy levels of antibonding states, thereby intensifying the interaction between adsorbates and the catalyst's surface. Consequently, this enhancement improves the catalyst's capacity to adsorb and manage the intermediates involved in the catalytic process of water oxidation. A noteworthy observation arises from the alteration of iron (Fe) concentrations, revealing a distinctive performance curve resembling a volcano concerning the d-band center, as depicted in Fig. 1.5d. This intriguing pattern indicates optimal catalyst performance is achieved when the Ed energy level strikes a delicate equilibrium. This equilibrium enables the catalyst to proficiently capture and release the intermediate species ($^*OH$, $^*O$, and $^*OOH$), which plays a pivotal role in the stepwise conversion of water into oxygen.

1.9 Thermodynamics of Electrochemical Water Splitting

The set of thermodynamic variables for a chemical reaction without free charges, such as an electric potential E or the difference between two potentials, the voltage V, is expanded by the equilibrium thermodynamics of galvanic and electrolysis cells. The quantity of electricity transferred nF, where n stands for the number of electrons transported and F represents the Faraday constant, relates cell voltages with constant pressure P and temperature T to thermal energy values. A molar enthalpy changes of reaction, thus designated as $\Delta H^{o}{}_{ws,298}$ and equal to $\Delta H^{o}_{f,H_2O,298}$ (the heat of production of one mole of liquid water at temperature 298 K), is related to water splitting at standard temperature and pressure (STP). These quantities have an absolute value of 286 kJ mol^{-1} or 2.96 eV. The amount $\Delta H^{o}{}_{ws,298}$ is also known as the higher heating value of one mole of hydrogen, $H^{o}{}_{HHV,298}$. The reaction $\Delta G^{o}{}_{ws,298's}$ change in Gibbs free energy is + 237 kJ mol^{-1} or 2.46 eV [12].

An approximate conversion of energy units is 1 eV = 100 kJ mol^{-1} = 25 kcal mol^{-1}.

The energy input needed to split a water molecule must be at least 237 kJ/mol (2.46 eV). The Gibbs free energy is related to the minimal electrode potential at equilibrium, which is necessary to break the O–H bond under STP as

$$\Delta G = 2FV^{0}_{rev,298}$$

By considering the absolute values of ΔG and F, the above equation becomes

$$2.46\,\text{eV} = 2\left(96{,}485\,\text{C mol}^{-1}\right)V^{0}_{rev,298}$$

Or

$$2.46\left(1.602 * 10^{-19}\,\text{C}\right) * \left(6.022 * 10^{23}\right)\text{V} = 2\left(96{,}485\,\text{C mol}^{-1}\right)V^{0}_{rev,298}$$

Or

$$V^0_{\mathrm{rev},298} = 1.229\,\mathrm{V} = 1.23\,\mathrm{V}$$

This value (1.23 V) is stated as thermodynamic or reversible electrolysis cell voltage. From the free energy equation, the enthalpy or greater heating value for water splitting can be computed.

$$\Delta G^0_{\mathrm{rev},298} = \Delta H^0_{\mathrm{HHV},298} + T\Delta S^0_{\mathrm{rev},298}$$

The entropy (heat required from the environment) is represented by the term $T\Delta S^0$, and at STP, it has a value of 50 kJ mol^{-1}. Thus, the equation can be used to determine the total amount of energy needed for water splitting at STP.

$$237\,\mathrm{kJ\,mol^{-1}} = \Delta H^0_{\mathrm{rev},298} + 50\,\mathrm{kJ\,mol^{-1}}$$

Since $V^{\mathrm{o}}{}_{\mathrm{rev},298}$ and $V^{\mathrm{o}}{}_{\mathrm{HHV},298}$ have different voltages, water electrolysis below 1.48 V and above 1.23 V is an endothermic process at STP. So, if we do not count ohmic and other loss processes, an electrolysis cell with a voltage below 1.48 V would cool down during the electrolysis process or would need a heat source to work in a steady state.

The above formula was based on the idea that hydrogen and oxygen gas could be made by electrolysis in a dehydrated state. Considering that there is a water vapor partial pressure P_{w} at a total pressure P of hydrogen and oxygen, the total energy needed to split one mole of water into humidified gases can be calculated. According to the thermoneutral voltage V_{TN}, this energy equals the potential difference between the two gases at the same temperature.

$$\begin{aligned}\Delta H_{\mathrm{TN},T,P} &= \Delta H_{\mathrm{HHV},T,(P-P_{\mathrm{w}})+1.5\frac{P_{\mathrm{w}}}{P-P_{\mathrm{w}}}\left(H_{\mathrm{H_2O(l)},T,P_{\mathrm{w}}}-H^0_{\mathrm{H_2O(l)},298}\right)}\\ &= 2FV_{\mathrm{TN},T,P}\end{aligned}$$

The first part of the expression is the stoichiometric water evaporation from STP to operating conditions. The second part is the HHV of the water electrolysis process at gas pressure $(P - P_{\mathrm{w}})$ and temperature T. Due to different energy and efficiency losses, the cell voltages in practical electrolysis cells are usually above both the higher heating value V_{HHV} and thermoneutral voltage V_{TN}. Too much heat energy usually causes the electrolysis cell to heat up. An isothermal electrolysis cell that works at a cell voltage of $V^{\mathrm{o}}{}_{\mathrm{op},T}$ is efficient if the amount of chemical energy it produces is equal to the amount of electrical energy it uses, which is $2FV^{\mathrm{o}}{}_{\mathrm{op},T}$. There are two ways to look at the amount of energy output. First, the maximum reversible work (Gibbs free energy) ΔG from reconverting hydrogen and oxygen, like in a fuel cell, may be of interest, and so a faradaic (voltage) efficiency of electrolysis, $\eta^{\mathrm{o}}{}_{298,\mathrm{Faradaic}}$, can be written as

$$\eta^{o}_{298,\text{thermal}} = \frac{\Delta H^{o}_{\text{HHV},298}}{2FV_{\text{OP}}} = \frac{V^{o}_{\text{HHV},298}}{V^{o}_{\text{OP},298}}$$

This efficiency has gained widespread acceptance, despite the fact that recovering the entire increased heating value of hydrogen in an energy application would be impossible. In theory, $\eta^{o}{}_{298,\text{thermal}}$ can be more than one. It is the reciprocal of a fuel cell's reversible thermodynamic efficiency. Typical industrial electrolyzer $\eta^{o}{}_{T,\text{thermal}}$ values vary from 60 to 80% for a certain current density (usually 1 A cm^2) and at 363 K and 0.1 MPa [30].

References

1. Abbasi KR, Shahbaz M, Zhang J, Irfan M, Alvarado R (2022) Analyze the environmental sustainability factors of China: the role of fossil fuel energy and renewable energy. Renew Energy. https://doi.org/10.1016/j.renene.2022.01.066
2. Qyyum MA, Dickson R, Ali Shah SF, Niaz H, Khan A, Liu JJ, Lee M (2021) Availability, versatility, and viability of feedstocks for hydrogen production: product space perspective. Renew Sustain Energy Rev. https://doi.org/10.1016/j.rser.2021.110843
3. Veziroğlu TN, Şahin S (2008) 21st century's energy: hydrogen energy system. Energy Convers Manag. https://doi.org/10.1016/j.enconman.2007.08.015
4. Zhiznin SZ, Timokhov VM, Gusev AL (2020) Economic aspects of nuclear and hydrogen energy in the world and Russia. Int J Hydrogen Energy. https://doi.org/10.1016/j.ijhydene.2020.08.260
5. Antonini C, Treyer K, Streb A, van der Spek M, Bauer C, Mazzotti M (2020) Hydrogen production from natural gas and biomethane with carbon capture and storage—a techno-environmental analysis. Sustain Energy Fuels. https://doi.org/10.1039/d0se00222d
6. Midilli A, Kucuk H, Topal ME, Akbulut U, Dincer I (2021) A comprehensive review on hydrogen production from coal gasification: challenges and opportunities. Int J Hydrogen Energy. https://doi.org/10.1016/j.ijhydene.2021.05.088
7. Cao L, Yu IKM, Xiong X, Tsang DCW, Zhang S, Clark JH, Hu C, Ng YH, Shang J, Ok YS (2020) Biorenewable hydrogen production through biomass gasification: a review and future prospects. Environ Res. https://doi.org/10.1016/j.envres.2020.109547
8. Mojaver M, Azdast T, Hasanzadeh R (2021) Assessments of key features and Taguchi analysis on hydrogen rich syngas production via gasification of polyethylene, polypropylene, polycarbonate and polyethylene terephthalate wastes. Int J Hydrogen Energy. https://doi.org/10.1016/j.ijhydene.2021.06.161
9. ul Haq T, Haik Y (2022) Strategies of anode design for seawater electrolysis: recent development and future perspective. Small Sci 2(9):2200030. https://doi.org/10.1002/smsc.202200030
10. ul Haq T, Mansour S, Haik Y (2021) Electronic and structural modification of Mn_3O_4 nanosheets for selective and sustained seawater oxidation. ACS Appl Mater Interfaces. https://doi.org/10.1021/acsami.1c24304
11. Li Y, Sun Y, Qin Y, Zhang W, Wang L, Luo M, Yang H, Guo S (2020) Recent advances on water-splitting electrocatalysis mediated by noble-metal-based nanostructured materials. Adv Energy Mater 10(11):1–20. https://doi.org/10.1002/aenm.201903120
12. Dau H, Limberg C, Reier T, Risch M, Roggan S, Strasser P (2010) The mechanism of water oxidation: from electrolysis via homogeneous to biological catalysis. ChemCatChem 2(7):724–761. https://doi.org/10.1002/cctc.201000126
13. ul Haq T, Haik Y (2021) S doped Cu_2O–CuO nanoneedles array: free standing oxygen evolution electrode with high efficiency and corrosion resistance for sea water splitting. Catal Today. https://doi.org/10.1016/j.cattod.2021.09.015

14. Zhang S, Zhang X, Rui Y, Wang R, Li X (2021) Recent advances in non-precious metal electrocatalysts for pH-universal hydrogen evolution reaction. Green Energy Environ. https://doi.org/10.1016/j.gee.2020.10.013
15. Yan Y, Xia BY, Zhao B, Wang X (2016) A review on noble-metal-free bifunctional heterogeneous catalysts for overall electrochemical water splitting. J Mater Chem A 4(45):17587–17603. https://doi.org/10.1039/C6TA08075H
16. Zhou Z, Pei Z, Wei L, Zhao S, Jian X, Chen Y (2020) Electrocatalytic hydrogen evolution under neutral pH conditions: current understandings, recent advances, and future prospects. Energy Environ Sci. https://doi.org/10.1039/d0ee01856b
17. Lasia A (2019) Mechanism and kinetics of the hydrogen evolution reaction. Int J Hydrogen Energy. https://doi.org/10.1016/j.ijhydene.2019.05.183
18. Zhang K, Zou R (2021) Advanced transition metal-based OER electrocatalysts: current status, opportunities, and challenges. Small. https://doi.org/10.1002/smll.202100129
19. Govind Rajan A, Martirez JMP, Carter EA (2020) Why do we use the materials and operating conditions we use for heterogeneous (photo)electrochemical water splitting? ACS Catal. https://doi.org/10.1021/acscatal.0c01862
20. Darband GB, Aliofkhazraei M, Shanmugam S (2019) Recent advances in methods and technologies for enhancing bubble detachment during electrochemical water splitting. Renew Sustain Energy Rev. https://doi.org/10.1016/j.rser.2019.109300
21. Li X, Cheng Z, Wang X (2021) Understanding the mechanism of the oxygen evolution reaction with consideration of spin. Electrochem Energy Rev. https://doi.org/10.1007/s41918-020-00084-1
22. Reier T, Nong HN, Teschner D, Schlögl R, Strasser P (2017) Electrocatalytic oxygen evolution reaction in acidic environments—reaction mechanisms and catalysts. Adv Energy Mater 7(1). https://doi.org/10.1002/aenm.201601275
23. Mefford JT, Rong X, Abakumov AM, Hardin WG, Dai S, Kolpak AM, Johnston KP, Stevenson KJ (2016) Water electrolysis on $La_{1-x}Sr_xCoO_{3-\delta}$ perovskite electrocatalysts. Nat Commun. https://doi.org/10.1038/ncomms11053
24. Li X, Wang H, Cui Z, Li Y, Xin S, Zhou J, Long Y, Jin C, Goodenough JB (2019) Exceptional oxygen evolution reactivities on $CaCoO_3$ and $SrCoO_3$. Sci Adv. https://doi.org/10.1126/sciadv.aav6262
25. Arima T, Tokura Y, Torrance JB (1993) Variation of optical gaps in perovskite-type 3d transition-metal oxides. Phys Rev B. https://doi.org/10.1103/PhysRevB.48.17006
26. Santos E, Schmickler W (2006) d-Band catalysis in electrochemistry. ChemPhysChem. https://doi.org/10.1002/cphc.200600441
27. Zhu J, Hu L, Zhao P, Lee LYS, Wong KY (2019) Recent advances in electrocatalytic hydrogen evolution using nanoparticles. Chem Rev. https://doi.org/10.1021/acs.chemrev.9b00248
28. Li J (2022) Oxygen evolution reaction in energy conversion and storage: design strategies under and beyond the energy scaling relationship. Nano-Micro Lett 14. https://doi.org/10.1007/s40820-022-00857-x
29. Yang K, Xu P, Lin Z, Yang Y, Jiang P, Wang C, Liu S, Gong S, Hu L, Chen Q (2018) Ultrasmall Ru/Cu-doped RuO_2 complex embedded in amorphous carbon skeleton as highly active bifunctional electrocatalysts for overall water splitting. Small. https://doi.org/10.1002/smll.201803009
30. Craig MJ, Coulter G, Dolan E, Soriano-López J, Mates-Torres E, Schmitt W, García-Melchor M (2019) Universal scaling relations for the rational design of molecular water oxidation catalysts with near-zero overpotential. Nat Commun. https://doi.org/10.1038/s41467-019-12994-w

Chapter 2
Electrode Setups and Water Electrolysis Technologies

2.1 Introduction

Electrodes are usually made of solid metallic materials. It is helpful to analyze the structure of solid metals to understand the surface reactions. For example, we can consider the cluster of tightly packed lattices of atomic nuclei and electrons that are partially free to move inside the lattice. Let us focus on the cloud of free electrons and examine its formation. The electron configuration of lithium is $1s^2 2s^1$, making it the lightest alkali metal. It has one additional electron than a noble gas (e.g., helium). A combination of two lithium atoms forms the molecule Li^2. According to molecular orbital theory (MOT), two molecular orbitals are created in a Li^2 molecule from the two atomic 2s orbitals of lithium: a bonding orbital (sigma 2s) that lies between the Li nuclei and an anti-bonding orbital (sigma* 2s) on the opposite sides of the nuclei (Fig. 2.1). The Aufbau principle states that the 2s electrons of the lithium atoms involved in bond formation fill the bonding molecular orbital while the four 1s electrons remain in their atomic orbitals [1]. More Li atoms in the solution result in the formation of more molecular orbitals, including anti-bonding orbitals. The bonding orbitals' energy levels become closer during this process, eventually forming a continuous energy band for the n number of Li atoms. Binding electrons occupy just half of the valence band, one electron for each Li atom. In terms of energy, there is a fully occupied band of 1s electrons below the valence band. There is such a vast energy gap between this band and the valence band that no 1s electron can jump to the valence band. A conduction band exists above the valence band that is vacant in the ground state. In the case of Li, it is formed by empty molecular orbitals. The energy gap is small, so the thermal energy ($k_B T$) excites electrons to them, making lithium electrically conducting [2]. In general, metals are good electronic conductors because their valence electrons are delocalized throughout the entire lattice.

When alkaline earth metals such as magnesium are considered similarly, the situation alters slightly. The 3s electrons of Mg would completely occupy molecular orbitals produced from atomic 3s orbitals. So, what about alkaline earth metals that

T. u. Haq and Y. Haik, *Electrochemical Water Splitting*, Materials Horizons: From Nature to Nanomaterials, https://doi.org/10.1007/978-981-99-9860-9_2

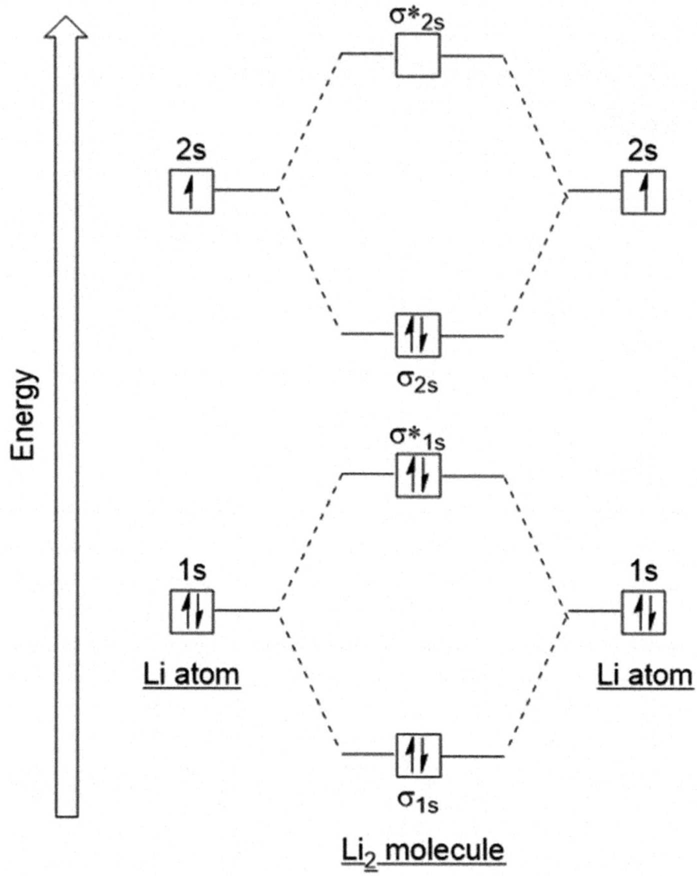

Fig. 2.1 Molecular orbital diagram of Li_2

make them such good electronic conductors? Let us proceed using Mg as an example. Molecular orbitals are also created from vacant atomic orbitals in the ground state (3p orbitals). When the atom number is high, the energy levels of these empty molecular orbitals form a continuous band. This band partially overlaps with the valence band in magnesium, allowing electron exchange between the conduction and valence bands. The materials are categorized as insulators if the energy gap is more than 5 eV because thermal energy ($k_B T = 0.025$ eV at 300 K) cannot raise an electron from the valence to the conduction band. On the contrary, when the band gap measures less than 1–2 eV, it suggests that certain electrons may be promoted to the conduction band, signifying the material's semiconductor nature. Semiconductors' conductivity increases with temperature, or thermal energy, whereas metals' conductivity decreases because the thermal motion of metal nuclei interferes with the free flow of electrons [3].

Electrons, as fermions, adhere to the Pauli exclusion principle, which stipulates that within a given system, two electrons cannot occupy identical quantum states.

As a result, electrons occupy the Fermi–Dirac distribution rather than the Boltzmann distribution [4]. The probability of finding an electron in a given energy level is

$$P(E) = \frac{1}{1 + e^{(E - E_f)/kT}}$$

where E_f is the fermi energy at a given temperature. According to the Aufbau principle, electrons are occupied in orbitals at the lower energy level at O kelvin. The highest occupied molecular orbital (HOMO) energy is known as the Fermi energy, which is also equal to an electron's electrochemical potential *t* 0 K, the probability of finding an electron on an energy level E E_f is one and zero on a level $E > E_f$. Because the electrochemical potential of an electron in a metal does not fluctuate considerably as a function of temperature, the energy at which the step function occurs is regarded to as the Fermi level. However, the Fermi–Dirac distribution no longer evolves in a step function but changes smoothly at higher temperatures. However, the distribution takes values of only one or zero around the energy where $P(E) = 0.5$, defined as the Fermi level. The Fermi level is frequently described as the lowest energy in a solid's conduction band, allowing numerical values to be assigned to the Fermi level concerning a reference point [5].

In the realm of solid-state physics, metals are often classified based on their work function. This parameter represents the energy required to extract an electron from an uncharged metal, specifically from its Fermi level. The photoelectric effect, for example, can be used to calculate the work function. The work function is a temperature-independent quantity, but it is crucial to note that it is affected by the lattice structure of a material. Metal work functions range from 2.3 eV for potassium to 5.3 eV for gold [6].

2.2 Voltage and Potential

One of the most adaptable and noticeable quantities in electrochemistry is potential. Voltage is frequently used interchangeably with potential; however, voltage is a measurable quantity, whereas possible variations can only (potentially) be measured. Voltage is the potential difference between two points, which can result from several contributions. The free energy related to the electrochemical reaction is translated to cell voltage, which is the sum of the anode and cathode contributions. Free energy demonstrates spontaneity, and thermodynamics tells us that the ΔG value is negative for thermodynamically favorable reactions [7].

Although thermodynamics cannot resolve the free energy of individual electrode reactions, a fixed potential scale is assumed. A solid electrode's potential, whether metal or semiconductor, can be considered a measure of the electron's Fermi level in the electrode. We know from electrostatics that an additional electric charge creates potential. When an electrode has a net negative charge, its potential is negative compared to an uncharged electrode or an electroneutral solution. Conversely, a positive charge implies positive potential. We can manipulate the electrical potential externally with electrochemical instruments, which means we can control the Fermi levels or the charge of electrodes. Consider some more potential-related values encountered in electrochemistry. Any material's inner phase is electroneutral. If an object has an electric charge, it spreads throughout its surface. The outer potential is the potential determined on the surface of an item or in its immediate vicinity concerning a vacuum. This is the work needed to transport an elementary charge in a vacuum from an infinite distance to the surface of a phase. A Kelvin probe can be used to test the outside potential. According to these explanations, a negatively charged object has a negative outer potential, whereas a positively charged object has a positive external potential [8].

2.3 Electrocatalysts: Function and Role in Electrode Potential

Electrocatalysis involves applying a required potential across electrodes immersed in a suitable electrolyte. The potential source adjusts the electrical density surrounding the electrodes, especially in half-cell reactions, facilitating oxidation or reduction reactions. These electrodes, acting as current collectors, are coated with catalytic substances, promoting responses at the interface between the electrode, catalyst, and electrolyte. In typical operational circumstances, the catalyst maintains an electrical connection with an electrode and an ionic connection with an electrolyte [9]. It is widely acknowledged that electrochemical reactions exclusively take place at the surface of electrode materials through the process of intermediate's adsorption and desorption, classifying them as heterogeneous catalysis. The catalyst's main function is to provide a surface for electrolyte penetration, stabilize the reaction intermediates, and facilitate electrons/ions transportation at the solid–liquid interface. Consequently, by enhancing the reaction pathways occurring on electrode surfaces, it becomes possible to lower the activation energy for various intermediates.

The onset potential signifies the reaction's feasibility under specific conditions for water splitting. Electrocatalytic redox reactions hinge on orbital energy levels at the electrode–electrolyte interface, with electron flow occurring from higher to lower energy orbitals similar to other catalytic processes. Hence, precise control of the electrode potential at the interface between the electrode and electrolyte is possible in a half-cell process. Anodic potential decreases the HOMO–LUMO orbital energy of the electrode, and electron transport from the electrolyte (source) to the catalyst becomes directed (sink). As a result, the electrolyte (H_2O) is oxidized, and electrons travel toward the source which is cathode for cycle completion by electrons conversion to protons (sink) in the water electrolysis process (Fig. 2.2). Like this, redox activities at the electrodes' surface become feasible in providing needed cell potential. In achieving a low-potential and efficient process, the electrocatalyst's

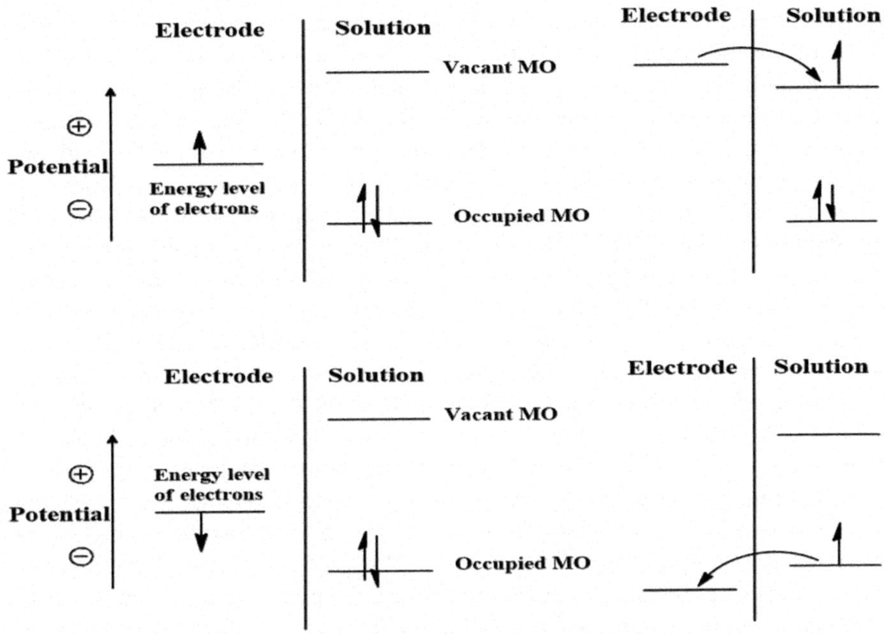

Fig. 2.2 Molecular orbital energy diagram illustrates the energy levels of electrolyte as well as electrode, with arrowheads symbolizing electron transportation between them

nature and intrinsic capacity play a pivotal role by aligning with the Sabatier principle (SP). This principle dictates an optimal bond energy between intermediates and the catalyst, enhancing chemisorption energy and reducing the energy needed for product formation. Effective implementation of the SP involves maintaining an equilibrium: excessive contact leads to unfavorable intermediate desorption, limiting active site accessibility, while weak contact hampers intermediate stabilization and electron exchange. To address this, manipulating the structural and electrical aspects of electrode materials emerges as a strategic avenue for tailoring the innate catalytic potential of electrocatalysts. This enhances their ability to interact efficiently and activate diverse intermediates [10].

2.3.1 *Working Electrode*

The working electrode controls the voltage and measures the current. In many physical electrochemistry studies, the working electrode is an "inert" material such as glassy carbon, platinum, or gold. In these circumstances, the working electrode acts as a surface for the electrochemical process. To get reliable electrochemical data, the working electrode must be configured appropriately, suitably prepared, and clean.

We start by explaining the proper settings for the ink-casting procedure. It is recommended to use a rotating disk electrode (RDE) to correctly assess the kinetics of the OER and HER because it provides well-defined O_2 or H_2 transport for mass transport limited reactions, which may be isolated from kinetic currents. The most common form is a Teflon shrouded RDE, initially invented by Frumkin et al. and popularized for ink-casted electrodes by Schmidt et al., which uses a conductive flat glassy carbon electrode as the current collector [11, 12]. The powder materials are deposited to the glassy carbon in a thin homogeneous layer using the drop-casting process. Other popular substrates with no well-defined geometric area (e.g., carbon cloth, Ni foam) are beneficial for functional performance in real devices but are challenging to measure the specific activity and mass of the electrocatalyst. A second concentric working electrode, such as a Pt or Au ring that encircles the disk, can be added to the spinning shaft in addition to a single RDE. The use of rotating ring-disk electrodes (RRDEs) enables the detection of products emitted at the disk due to a redox reaction at the ring. With the proper configuration and the preparation of an ink-casted functioning, researchers must consider catalyst loading, Nafion loading, and conductive additive addition. The catalyst loading should not be too high or too low. When the catalyst loading is too low, the ink-casted layer cannot fully cover the RDE, and the mass transfer is not clearly defined. Excessive catalyst loading causes aggregation and mass transport loss within the thick catalyst layer, which worsens kinetic current extraction. In general, the thickness of this ink-casted layer must be of the order of 1 m, which can be determined by SEM or calculated mathematically. Second, the electrode preparation for freestanding catalyst particles requires the inclusion of a conductive additive that is catalytically inert and has no effect on the original structure of the tested catalyst [13].

High-surface-area carbons are popular candidates, but they erode before OER begins and may affect electrode stability during OER. Second, the conductive component disperses the catalyst particles and enables smooth electron transit, preventing agglomeration. Third, the Nafion loading should be kept to a minimum level [14]. An extensive Nafion loading mass results in a thick Nafion film, which generates increased mass transport resistance within the Nafion layer and impedes the kinetic current estimations. To make an excellent ink-casted working electrode, we recommend using a catalyst loading that is neither too low nor too high, that a suitable conductive additive is used to distribute the catalyst particles and with minimum catalyst loading [15]. Following the preparation process, efforts must be made to ensure that the catalyst's surface is properly clean. The catalyst is frequently unclean because it adsorbs some "contaminants," such as surfactants used in nanoparticle manufacturing pathways.

Electrochemical CV cycling, which eliminates surface "contaminants" via sequential oxidation/reduction of catalysts, is a typical surface-cleaning process (particularly for Pt-based catalysts). Furthermore, we recommend exposing catalysts to CO by purging the electrolyte; thermal treatment; solvent washing; UV-Ozone treatment; and CV cycling to mainly remove the surfactants involved in wet chemical synthesis [16]. We emphasize that surface-cleaning procedures such as cycling and CO addition might cause irreversible structural changes. Hence, the activity observed may not represent the original structure of the catalyst. These structural alterations are more likely to occur on catalysts with metastable structures or chemistry.

2.3.2 Reference Electrode

The reference electrode is used to measure the potential of the working electrode. As long as no current passes through it, a reference electrode should have a constant electrochemical potential. The commonly used reference electrodes are saturated calomel electrodes (SCE) and silver/silver chloride electrodes (Ag/AgCl), and Hg/Hg–O electrodes. The working electrode is monitored concerning a reference electrode, which must maintain a stable and precisely defined potential to accurately measure or regulate the voltage applied to catalysts. We explain the RE selection, validate the need for reporting potential concerning RHE, and detail the methods for converting measured potential to RHE.

The reasonable selection of a reference electrode is crucial since it serves as the foundation for accurate potential measurement. The most dependable method is to utilize a commercially produced RHE as the reference electrode. Alternatively, all other reference electrode potentials should be translated to the RHE scale. The saturated calomel electrode (SCE), which employs the reversible redox between $Hg_2Cl_2 + 2e^- \rightarrow 2Hg + 2Cl^-$ in saturated KCl solution, and the Ag/AgCl reference electrode: $AgCl + e^- \rightarrow Ag + Cl^-$ in KCl solution with designated concentration, are two of the most widely used reference electrodes. We recommend that readers examine the concentration of KCl solution, Cl leaching, and interference from alkaline electrolytes when using SCE and Ag/AgCl reference electrodes. First, the reference potential is determined by the concentration of filled KCl solution inside these two electrodes. As an illustration, the potential of an Ag/AgCl electrode measures 0.197 V in comparison to a standard hydrogen electrode (SHE) with saturated KCl, while it is 0.288 V versus a SHE with 0.1 m KCl [17]. Second, Cl leaches out from the reference electrode in an alkaline and acidic environment, diffuses toward the working electrode adsorb, and interferes with the activity measurement. It is advised in these cases to utilize reference electrode chemistries based on mercury sulfate (Hg/Hg_2SO_4), mercury oxide (Hg/HgO), or silver sulfate (Ag/Ag_2SO_4). A more sophisticated cell design with membranes that separate electrode compartments can also be used [18]. Third, in alkaline electrolytes, the Hg/HgO reference electrode is preferable to Cl-containing reference chemistries (e.g., SCE or Ag/AgCl) because the OH-diffusion alters the pH of KCl solution inside the reference electrode, triggers the oxide formation, and shifts the reference potential to the new value, e.g., a second redox couple

$$2Ag + 2OH^- \rightarrow Ag_2O + H_2O + 2e^-$$

As indicated by the preceding effects of KCl solution, Cl leaching, and alkaline electrolyte, the choice of reference electrodes must be decided with caution based on the electrolyte. The standard practice is to convert all measured potentials to the RHE scale, regardless of the reference electrode employed. When using the SHE scales, alterations in the electrolyte's pH impact the potential window in which the reaction takes place. For example, the equilibrium potential of hydrogen electrocatalysis (i.e., HER) in acidic electrolytes is 0 V versus SHE at pH 0, but it moves to 0.83 V versus SHE at pH 14 in alkaline electrolytes. Potentials with reference to the another electrodes, like SCE and Ag/AgCl, suffer from the same issue, as the reaction potential varies with solution pH. Because the pH dependence of potential versus SHE complicates the comparison of experiments using different electrolytes, it is advised that all potentials be linked to RHE, which is expressed as

$$E_{\mathrm{RHE}} = E_{\mathrm{SHE}} + \mathrm{pH} * 2.303\frac{RT}{F}$$

where E_{SHE} represents the potential of the chosen reference electrode, pH denotes the electrolyte's pH level, R stands for the universal gas constant (8.3144598 J mol^{-1} K^{-1}), T represents the absolute temperature, and F is the Faraday constant (96,485 C/mol) [19]. Under ambient conditions, $2.303\frac{RT}{F}$ value is roughly 59 mV. It is highly recommended to calibrate the reference electrode prior to electrochemical experiment. Researchers discovered that the experiment without calibration produces erroneous potentials by several tens of mV. Possible causes of this inaccuracy include uncertainty in pH determination (significantly above pH 13), and liquid junction potentials, which can be difficult to estimate due to differences in electrolyte composition between the reference electrode compartment and working electrode compartment [20].

2.3.3 Counterelectrode

The auxiliary or counterelectrode facilitates the application of input potential to the working electrode. These electrodes serve to complete the circuit and allow charge to flow. As a result, they must be made of inert materials like carbon or platinum, and their size must be significantly more significant than the working electrode to avoid current limits. The phrase "counterelectrode" is best applied to two-electrode investigations, and the term "auxiliary electrode" is reversed for three-electrode research. Providing instructions on how to select a suitable counterelectrode is crucial to ensure that it does not interfere with the reaction at the working electrode. First, the counterelectrode must rapidly supply or sink electrons without limiting the response at the working electrode. As a result of this requirement, many researchers utilize Pt wires, foils, or meshes to balance the charge at the working electrode since they can withstand significant currents for oxygen as well as hydrogen evolution. The projected area of counterelectrode should surpass compared to working electrode, thereby guaranteeing that the counterelectrode doesn't impose limitations on the reaction rate at the working electrode. Second, there must be the suitable position of the counterelectrode to provide a homogenous electric field between the counter and working electrodes. The counter and working electrode shape determine the most appropriate counterelectrode placement. Suppose two plates/disks of equal size are sufficiently separated, a desired homogeneous field of current lines forms between them, as does a plate/disk and a wire. Third, the counterelectrode should not interfere with the detection of activity. Pt is a reasonably strong counterelectrode, but it may compromise the working electrode efficiency due to contaminations containing the electrocatalysts to be tested [21]. In HER experiments, for example, the Pt counterelectrode is at an oxidizing potential, allowing Pt to dissolve in alkaline and acidic conditions during Pt/Pt-O_x redox reaction. The dissolved Pt in the electrolyte gets electrodeposited onto the working electrode, altering the electrochemical performance, particularly of platinum group metal-free catalysts, where even a tiny quantity of Pt deposition on the working electrode might result in incorrect activity reports [22]. Thus, the apparent high activity, which increases during long-term stability tests, could be caused by counter-electrode artifacts. To avoid worries about the Pt counterelectrode, it is suggested to separate the working and counterelectrode compartments by a proton exchange Nafion membrane or replace the Pt electrode with a carbon electrode, such as a graphite rod or glassy carbon [22]. However, at high oxidizing potentials, carbon (mainly carbon with a high surface area) emits CO and CO_2, and impurities (even trace amounts) may leach into the electrolyte [21]. Therefore, a reliable counterelectrode should be able to sustain larger currents than the working electrode (counterelectrode with high geometric area comparison to working electrode), provide a uniform electric field, and avoid the working electrode contaminations.

2.4 Electrode Setups: From 2 to 3 Electrode Systems

The widely used and straightforward technique for evaluating a water electrolysis cell non-destructively involves performing measurements directly within the cell, eliminating the need for external probes like reference electrodes. This approach generates polarization plots, depicting the relationship between current and cell voltage under nearly steady conditions. These conditions correspond to gradual system stimulation, accommodating minor deviations in measured currents due to capacitive effects, which can be significant for electrodes with large surface areas, as often encountered in practical setups. To construct a polarization plot, one can either apply a current to the cell and determine its stabilized voltage (a process that may require some time) or employ a fixed voltage to measure the cell's produced current after attaining stability. This procedure typically involves chronopotentiometry or chronoamperometric steps to stabilize the signal before transitioning to the next quasi-steady operational point. Subsequently, extended chronopotentiometry can be employed to assess the sustained performance of the cell over time. Lastly, linear or cyclic polarization plots emerge as crucial tools for water electrolysis analysis; however, they often need to be executed optimally, lacking the requisite gradual pace to mitigate double-layer capacitive effects [23].

Besides two-electrode cell measurements used in industries, laboratory researchers commonly use three-electrode cell measurements. Here, they control the potentials of the working electrode (WE) and counterelectrode (CE) individually relative to a well-chosen reference electrode (RE), as illustrated in Fig. 2.3 [24]. The reference electrode (RE) needs optimal positioning and a suitable material to ensure noise-free measurements. It should not hinder the working electrode's (WE) functionality by avoiding surface coverage that disrupts current pathways. Maintaining proximity to the WE minimizes Ohmic losses and prevents electrolyte contamination, especially avoiding references with chlorine (Cl) content, which can adversely affect electrocatalysts utilized in water electrolysis [25]. Employing a reference electrode (RE) enables isolation of the working electrode's (WE) electrochemical signal from that of the counterelectrode (CE) and facilitates real-time Ohmic drop compensation. However, challenges arise during bubble evolution when electrolyte conductivity fluctuates, complicating precise Ohmic correction. Three-electrode cell measurements are applicable when a membrane or glass frit partitions the CE compartment, shielding the working electrode from substantial CE influence, particularly in scenarios where CE generates by-products during operation.

Ensuring the selection of an appropriate counterelectrode (CE) remains crucial, considering both its material composition and surface area. This pertains to averting potential limitations in the potential discrepancy between the working electrode (WE) and CE due to potentiostat compliance, especially under scenarios involving substantial currents and non-negligible Ohmic drops. The CE must be judiciously chosen to maintain unhindered current flow through the WE while circumventing electrolyte contamination from either primary or side reactions.

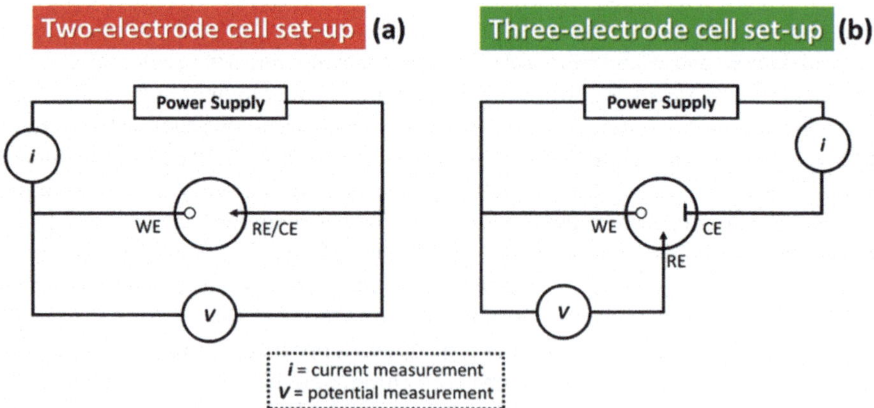

Fig. 2.3 **a** Two- and **b** three-electrode cell configurations for a Li-ion battery during the charging procedure. Reproduced with permission from Ref. [24]. © 2019, MDPI

When a metallic CE is employed, the prospect of its dissolution exists, facilitating the accumulation of deposits on the WE's surface. This occurrence becomes more pronounced when the working electrode engages in reduction reactions (e.g., HER) while the counterelectrode participates in oxidation (e.g., OER), potentially leading to metal dissolution competition. Yet, the advantage of "activating" an electrode with minimal materials, as illustrated by the dissolution of a Pt CE to induce a controlled deposition of Pt on the WE surface, is noteworthy, although instances of this phenomenon occurring uncontrolled are documented in the literature. In the realm of three-electrode cells, polarization plots can be conveniently recorded, commonly within the framework of Rotating Disk Electrode (RDE) designs, facilitating the manipulation of mass transport rates and enabling quasi-steady conditions. Nevertheless, challenges arise in water electrolysis processes due to the inevitable formation of H_2 or O_2 bubbles, particularly problematic at higher current densities. To tackle this, advanced RDE configurations have been proposed in the literature, permitting more precise measurements under elevated geometric activity levels [26]. Polarization plots, generated within the framework of Rotating Disk Electrode (RDE) or modified RDE setups, can delineate "Tafel slopes," which serve as widely employed indicators of the catalytic activity exhibited by specific materials. These plots also facilitate the determination of supplementary activity markers, including onset potential and overvoltage at a designated current density. These measurements are often conducted after compensations for Ohmic-drop effects and mass transport limitations. The latter correction is generally unnecessary for water electrolysis reactions, given that the reactant, water, acts as the solvent. This holds, provided any generated gas bubbles are effectively expelled from the electrode surface. The RDE setup, in most instances, effectively addresses this concern, although scenarios involving porous electrodes, wherein gas molecules can become trapped within the pores, could pose exceptions. For quantifying Electrochemically Active Surface Area (ECSA), the three-electrode configuration is the preferred choice due to its ability to segregate the behavior of the working electrode (WE) [27].

The distinction between a two- and a three-electrode system lies in the presence of an additional electrode known as the reference electrode in three-electrode setup. Its sole purpose is to maintain a constant potential at the working electrode. In contrast, the voltage of the counterelectrode floats. The rest is the same as with a two-electrode system. So, when we need to change our three-electrode system into a two-electrode one, we need to finish the role of the reference electrode by connecting it to the counterelectrode (Table 2.1).

2.5 Water Electrolysis Technologies

Electrocatalytic water electrolysis is the most crucial primary process to generate hydrogen gas (H_2 molecule), and its importance will grow dramatically as renewable energy demands increase. Water electrolyzers are classified into three categories

Table 2.1 Summary of two- and three-electrode configuration

	Two-electrode setup	Three-electrode setup
Advantages	• Simple setup with fewer components	• Independent control of working and reference electrodes
	• Easy to use for routine experiments	• Accurate and stable measurements
	• Cost-effective	• Can compensate for reference electrode drift
		• Can study electrocatalysis and other complex reactions
Disadvantages	• Can't accurately measure potential and current simultaneously	• More complex setup with additional components
	• No independent control of reference electrode	• More expensive than the two-electrode setup
	• Poor accuracy due to reference electrode drift	• May require more skill and experience to use effectively
		• Limited to certain types of experiments
Example applications	Basic electrochemical experiments, corrosion testing	Electrocatalysis, impedance spectroscopy, bioelectrochemistry
Common use cases	Teaching and training, routine measurements	Research and development, specialized applications
Equipment requirements	Electrochemical cell, two electrodes	Electrochemical cell, three electrodes, potentiostat

based on the electrolytes, working temperatures, separators, and pressures with the titles [28]:

1. Alkaline water electrolyzer (AWE)
2. Proton exchange membrane water electrolyzer (PEMWE)
3. Solid oxide electrolysis cell.

2.5.1 Alkaline Water Electrolyzer (AWE)

An alkaline water electrolyzer has basic architecture of any electrochemical system: two electrodes divided by a porous separator to conduct protons from the anode toward the cathode saturated with an alkali electrolyte, commonly KOH (Fig. 2.4) [29]. The inherent simplicity of this design has enabled the rapid and extensive adoption of industrial alkaline water electrolyzer (AWE) cells on a global scale. Comprising two metallic electrodes immersed in an aqueous electrolyte (typically 25–40 wt.% NaOH or KOH solutions), these cells operate within a temperature range of 70–90 °C to enhance electrical conductivity [30].

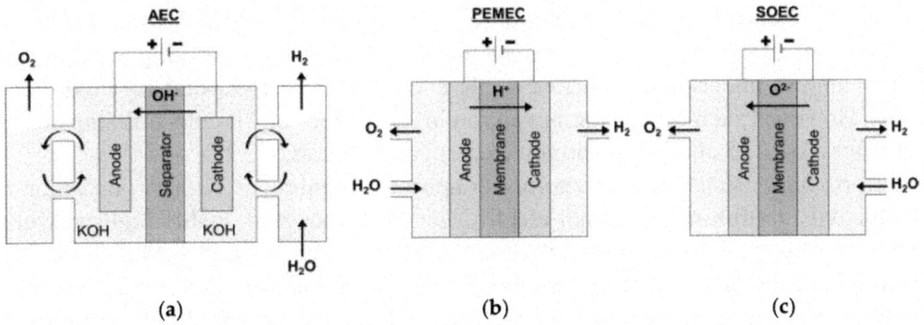

Fig. 2.4 Schematic illustration of **a** an AWE cell, **b** a PEM electrolysis cell, and **c** a SOEC. Reproduced with permission from Ref. [29]. © 2021, MDPI

The specific conductivity of KOH at 25 °C is 0.184 S cm^{-1}. Water reduction occurs at the cathode in an AWE (at pH 14).

$$2H_2O + 2e^- \rightarrow H_2 + 2OH^- \qquad E^o_{red} = -0.828 \text{ V versus SHE}$$

While the hydroxyl anions are oxidized at anode

$$2OH^- \rightarrow \frac{1}{2}O_2 + H_2O + 2e^- \qquad E^o_{ox} = +0.401 \text{ V versus SHE}$$

The AWE technology has various advantages, most of which are connected to the alkaline solution, which allows the use of transition metals (TM) catalysts without affecting durability and performance in operation. In AWE cells, nickel-based electrode materials, cobalt, or plain stainless steels are commonly employed [31]. In the context of industrial alkaline water electrolyzers, various designs are employed. Cells can be interconnected in parallel (monopolar assembly) or series (bipolar assembly) arrangements. Anodes (or cathodes) are parallel-connected in the monopolar setup on copper (or aluminum) conduction bars to mitigate Ohmic losses and ensure consistent current distribution. Conversely, series configurations involve endplates at the assembly's extremities for current collection, with neighboring unit cells' anodes and cathodes electrically interlinked. Considering the advantages and disadvantages of these mono and bipolar assemblies, it is recommended that bipolar arrangement be more efficient in energy.

A core drawback inherent to alkaline water electrolyzer (AWE) cells, irrespective of their configuration, arises from the O_2 evolution at the anode and H_2 at the cathode. These gas bubbles initially form within the liquid electrolyte, altering its ionic conductivity and increasing the cell's Ohmic resistance and operational expenditure. A second concern emerges due to the separator's porous nature, enabling the potential intermixing of bubbles at the anode and cathode. This mismatch in mass transport could undermine operational safety and gas purity. Furthermore, AWE cells necessitate considerable time—often hours—to attain a state of equilibrium concerning electrolyte flow, temperature, current density, and bubble generation. Consequently, operating AWE cells in transient states is challenging, hampering their seamless integration with renewable energy sources like solar or wind electricity. However, such limitations are absent in water electrolysis cells utilizing dense separators, such as alkaline or proton exchange membranes [32].

Efforts have been made to explore nonporous membranes with strong anionic conductivity to diminish internal electrolyzer resistance and enable high-pressure operation effectively. This approach resembles the setup used in PEMWE, involving creating porous catalyst layers on both polymer membrane sides to produce a membrane electrode assembly (MEA). The OH^- membrane should qualify for the following:

1. Excellent mechanical and thermal stability in aqueous medium and during electrolysis.
2. Electrically insulator.
3. High ionic conductivity to ensure efficient OH-migrations from cathode to anode.
4. Extremely low permeability to gases to mitigate gas crossover between the two electrode compartments.
5. Low cost.

The alkaline environment in AEMWE provides a wide range of catalyst material options due to its less corrosive nature, potentially allowing transition metals utilization for HER and OER process [33]. The ability to employ less expensive metal-based catalysts in AEMWEs is the reason why research aggressively tackles the obstacles preventing AEMWE commercialization.

2.5.2 *Proton Exchange Membrane Water Electrolyzer (PEMWE)*

The PEMWE method is the most efficient technique for electrolysis of water, with a significant reliance on the ion exchange membrane. The essential membrane-electrode assembly (MEA) takes form as the cathode as well as anode sandwich a proton-conductive polymer electrolyte (e.g., Nafions), as depicted in Fig. 2.4b. This MEA operates within pure water, with the anode's cell voltage (V cell) prompting the initiation of O_2 evolution [34].

$$H_2O \rightarrow \frac{1}{2}O_2 + 2H^+ + 2e^- \qquad E^o_{ox} = +1.23\,\text{V versus SHE}$$

The HER occurs at cathode.

$$2H^+ + 2e^- \rightarrow H_2 \qquad E^o_{red} = 0.00\,\text{V versus SHE}$$

Notably, there is no net electrolyte consumption; only water is consumed. The concentration of the ions remains constant as long as water input and consumption rate are comparable. Mobile proton species are kept contained by the highly acidic polymer membrane during electrolysis. As a result, noble metal catalysts resistant to such acidity are required at both the anode and cathode. Because of their excellent corrosion resistance, mechanical stability, and high ionic conductivity, perfluorinated sulfonic acid copolymer membranes are used in modern PEMWEs. Nafions, manufactured by DuPont de Nemours and Company, is the most frequently used membrane material (USA) [35]. Nafion membranes possess the qualities of thinness, elasticity, and transparency. When interacting with water, however, the membrane may undergo swelling and ion exchange group dissociation, thereby enabling proton migration between anodic and cathodic compartments. In contrast to alkali solutions, perfluorinated sulfonic acid membranes exhibit higher resistance (11–12 Ω cm at 20 °C and 5–6 Ω cm at 80–90 °C). To mitigate ohmic losses, ultrathin membranes spanning 100–300 μm are preferred, although their employment amplifies gas permeability, potentially compromising system efficiency.

Owing to the absence of liquid electrolytes in PEMWE, electrodes conform to a zero-gap configuration tightly pressed against the membrane. Effective interface contact is ensured by situating catalysts atop the ion-conductive membrane, culminating in a catalyst-coated membrane (CCM). This CCM is subsequently sandwiched between porous current collectors while neighboring electrolysis cells are aggregated and separated through metallic bipolar plates. Attaining high-performance PEMWE mandates precise alignment and firm contact within the porous transport layer (PTL) [36]. The Hydrogen Evolution Reaction (HER) electrode can be designed to have a similar shape to the anodes in Proton Exchange Membrane Fuel Cells (PEMFCs), where the Hydrogen Oxidation Reaction (HOR) occurs. Effective options include Pt-based catalysts affixed onto high-roughness carbon support and a conventional gas diffusion layer (GDL) encompassing a microporous layer. On the OER side, the challenge is intricate, given the instability of carbon. Consequently, titanium-based Porous Transport Layers (PTLs) are commonly employed as the porous current collector [37].

When in an oxide state, ruthenium (Ru) has significant catalytic activity in the OER. However, it should be emphasized that Ru-based electrodes might be unstable in acidic environments. Iridium (Ir) is the most often employed anode catalyst, with mass loading from 1.0 to 2.0 mg cm^{-2}. At the same time, platinum (Pt) or palladium (Pd) is the primary cathode catalyst, with anode current collectors made of porous titanium (Ti) and cathode current collectors made of carbon.

The main advantages of PEMWEs over other water electrolyzers are as follows [38]:

(1) High geometric activity and specific activity.
(2) High energy conversion efficiency.
(3) A high rate of gas production and desorption with high purity.
(4) A wide dynamic range (excellent for usage with intermittent renewable energy).

The key disadvantages are:

(1) the significant initial investment required.
(2) high-temperature electrolyzers are needed.

2.5.3 Solid Oxide Electrolysis Cell (SOEC)

Solid oxide electrolyzer cells (SOECs) integrate oxide-ion conducting ceramics, performing dual roles as the cell separator and solid electrolyte (Fig. 2.4c). Operational temperatures for SOECs typically span 800–1000 °C. These cells predominantly employ yttrium and scandium-stabilized zirconia ("YSZ") as the electrolyte, comprising a manganite-coated stabilized zirconia solid electrolyte and stainless-steel bipolar plates. In the SOEC setup, the cathode diminishes steam, enabling oxide ions' migration from the cathode to the anode through an ionic diffusion mechanism. Ultra-thin ceramic membranes (approximately 30–150 μm thick) are employed to curtail ohmic losses. Notably, the steam cathode is often fashioned from porous nickel material [39]. On the contrary, the air anode is fabricated using porous perovskite materials like lanthanum strontium manganite ("LSM"). Ongoing research involves exploring diverse catalyst blends, including rare-earth nickelates, lanthanum strontium cobalt ferrite (LSCF), and samarium-doped ceria. Computational models, accompanied by imaging data and impedance analysis, have delved into the complex landscape of heterogeneous electrocatalysis within these systems.

These studies have revealed that the oxygen surface diffusion and desorption processes on the LSM surface from the YSZ triple-phase boundary might constitute the rate-limiting step. Tinkering with the surface structure of the composite functional layer at the cathode–electrolyte interface has emerged as a promising strategy for optimizing this phenomenon. Despite operating at elevated temperatures, challenges related to multicomponent gas diffusion within porous electrodes can influence reaction rates, particularly at the steam electrode. The advancement of SOEC technologies has been underscored by their ability to perform with high geometric activity and efficiency, achieving notable outcomes such as 3.6 A/cm^2 at 1.48 V and 950 °C. Additionally, the intrinsic benefits of operating at high temperatures enable efficient and reversible electrochemical processes, facilitating the dual functionality of a single SOEC unit as both a fuel cell and an electrolysis cell [40].

Understanding and managing electrochemical deterioration and thermomechanical stability are significant research challenges to fulfill the demands of green hydrogen production from renewable technology. Each type of electrolysis has advantages and disadvantages, but all three methods are attractive and prospective for renewable energy applications. For example, alkaline electrolysis is well established, and its key benefits include economic viability due to the low cost of transition metal-based electrodes and their long-term durability. On the contrary, the acidic conditions in PEM electrolyzers impede the kinetics of redox processes, demanding the employment of costly noble metal catalysts and bipolar plate materials. Additionally, the high expense of polymeric membranes is the primary impediment to the practical viability of PEM in the near term.

In principle, alkaline electrolyzers can function at higher pressures, triggering the deterioration process. On the other hand, the use of a liquid electrolyte limits operation at differential pressures to prevent electrolyte and gas permeability through the diaphragm. On the other hand, processes at different pressures on the two sides of the electrolyzer would facilitate hydrogen storage because hydrogen would be produced at high pressure, eliminating the need for additional compressing systems. At the same time, water could be supplied (and oxygen produced) at near atmospheric pressure.

The solid electrolyte in SOEs and PEMs allows for a more compact design and operating at differential pressures is practical and beneficial. Furthermore, the solid form of the electrolyte makes PEMs and SOEs more dynamic systems with a faster response when a varied power load is applied, as opposed to liquid alkaline electrolyzers with slow diffusion rates. The limited ionic conductivity of liquid electrolytes in alkaline electrolyzers suppresses the efficiency. Additionally, various kinds of losses like ohmic loss and the presence of the diaphragm, might further impede OH transport, resulting in low geometric activity in this technology. Ohmic losses are more deficient in SOEs because of improved ion conduction at high operating temperatures and in PEM electrolyzers because of the widespread availability of polymeric membranes with excellent protonic conductivity. In alkaline electrolysis, the diaphragm significantly prevents ionic conduction.

On the other hand, the high working temperature of SOEs is the primary feature that distinguishes this technology from low-temperature electrolysis. The thermodynamic study demonstrated that the temperature increment only slightly enhances the total energy requirement of the process, ΔH, whereas the entropy factor ($T\Delta S$) contribution. Still, it can also be the source of decreasing efficiency due to undesired gas permeation. Apart from impeding overall performance, it also raises many safety concerns. There is also a probability of gas crossover in PEM electrolyzers', but the polymeric membranes significantly minimized the crossover volume where the membrane thickness can reduce gas permeability [41].

Although ceramic electrolytes prevent gas permeation, safety concerns arise in the SOE. Effective sealing between the cathodic and anodic chambers at high operating temperatures is hard to achieve over time, and the possibility of unexpected cracking is considerable. Although the more elevated temperature slightly enhanced the total energy demand of the process, ΔH, the entropy factor $T\Delta S$ increased significantly and thus contributed to the lower ΔG value cutdown of the required input voltage and the cost of the hydrogen production.

$$\Delta G = \Delta H - T\Delta S$$

SOE stacks are anticipated to operate at 1.3 V and consume 3 kWh per typical m^3 of H_2 with an average geometric activity of 7000 A/m^2 and an inlet steam temperature of 800 °C, whereas commercial alkaline electrolysis requires 4.5 kWh. These estimations, however, do not account for heating energy circulation and losses. In this regard, the net efficiency of SOEs is 40–60% (taking heating energy demands into account), which is lower than the net efficiency of low-temperature electrolyzers (59–70 and 65–82% for alkaline and PEM electrolyzers, respectively, depending on hydrogen output) [42]. This picture may vary if the required heat is generated in a renewable manner (e.g., by heat from the sun or a nuclear power plant). In the case of integrated systems, the required heat can also be supplied by waste heat from exothermic activities. In both circumstances, SOEs can run at about 100% efficiency. Another intriguing property of SOEs (in the case of O_2 conducting electrolytes), connected to their high-temperature functioning, is the ability to coelectrolyze H_2O/CO_2 mixtures for the formation of syngas (composed of CO and H_2). The Fischer–Tropsch synthesis can convert syngas to methanol or higher hydrocarbons [43]. Coelectrolysis can thus contribute to the storage of renewable energy and the reduction of CO_2 emissions. CO_2/H_2O coelectrolysis is a proven technology that can be a significant step in CO_2 recycling scenarios if suitable stable electrocatalysts are produced. However, although the operation of SOEs at high temperatures has intrinsic advantages, it is also a cause of degradation and lack of stability, which must be addressed before solid oxide electrolysis can be commercialized on a wide scale. Finally, a new trend in alkaline electrolysis has evolved in recent years based on polymeric membranes having anionic (OH) conductivity, also known as anion exchange membranes (AEM). This novel approach, known as solid alkaline electrolysis, seems promising because it constitutes the advantages of PEM as well as alkaline electrolysis. Because the OER kinetics is better in basic media than in acidic media (PEM electrolysis), transition metal-based composites may be used in this technology. At the same time, the solid form of the membrane can provide mechanical integrity for operation under differential pressures within a compact and scalable cell design. However, various difficulties (such as the high cost and low ionic conductivity of polymeric membranes and their susceptibility to degradation) must be addressed before this technology can be used (Table 2.2).

2.6 Stability of Precious and Non-precious Metals in Different Medium

Undesired oxidation and subsequent dissolution of the catalyst surface impede the development of TM-based electrocatalysts, given that the surface characteristics of the composites often dictate the performance. The vulnerability of TMs to oxidation and corrosion can significantly reduce the catalyst efficiency by limiting the duration for which they are effective. Alkaline conditions generally increase the stability of TMs by forming passivation structures such as oxides and hydroxides. Cherevko group established a dataset of different parameters to evaluate the metals dissolution rates during reduction and oxidation reactions in an electrocatalytic reaction [44]. There is a direct relation between cohesive energy and metal–metal bonding

Table 2.2 Advantages and disadvantage of different electrolyzers technologies

Advantages	Disadvantages
Alkaline water electrolyzer (AWE)	
Alkaline water electrolyzers are highly efficient with an energy efficiency of 70–80%	Alkaline water electrolyzers require high purity water, as impurities can cause electrode degradation and reduce the system's efficiency
Alkaline water electrolyzers have low operating costs, as they do not require expensive catalysts or membranes	Alkaline water electrolyzers have a slow response time because of the diffusion-limited ions transportation in the electrolyte
Alkaline water electrolyzers can produce high-purity hydrogen with minimal impurities	Alkaline water electrolyzers have a limited lifespan due to electrode degradation and the corrosive nature of the electrolyte
Alkaline water electrolyzers can operate at relatively low temperatures and pressures, reducing the risk of explosions and the need for expensive materials	Alkaline water electrolyzers are not suitable for use with intermittent sustainable resources of energy like; solar or wind power, as they cannot be easily turned on and off to match the variable energy supply
Alkaline water electrolyzers are relatively simple and robust, requiring little maintenance	Alkaline water electrolyzers have a low power density and require large surface areas of electrodes to achieve high hydrogen production rates
Proton exchange membrane water electrolyzer	
Proton exchange membrane (PEM) water electrolyzers have a fast response time and may work at high current densities, making them suitable for intermittent renewable sources of energy like wind power or solar one	PEM water electrolyzers require high-purity water and are sensitive to impurities in the electrolyte, which can cause membrane degradation and reduce the efficiency of the system
PEM water electrolyzers have a high-power density and can achieve high hydrogen production rates with a relatively small electrode surface area	PEM water electrolyzers are too much costly to manufacture compared to alkaline water electrolyzers, as they require expensive catalysts and membranes
PEM water electrolyzers produce high-purity hydrogen with minimal impurities	PEM water electrolyzers operate at higher temperatures and pressures than alkaline water electrolyzers, increasing the risk of explosions and requiring more expensive materials
PEM water electrolyzers have a longer lifespan than alkaline water electrolyzers due to the use of non-corrosive materials and the absence of electrode degradation	PEM water electrolyzers require more maintenance than alkaline water electrolyzers, as the membrane and catalysts can degrade over time
PEM water electrolyzers are compact and lightweight, making them suitable for portable applications such as fuel cell vehicles	PEM water electrolyzers have a lower energy efficiency than alkaline water electrolyzers, typically around 60–70%
Solid oxide electrolysis cell	
Solid oxide electrolysis cells (SOECs) have a high operating temperature (800–1000 °C) and can operate with a wide range of fuels, including natural gas and biogas	SOECs have a slow response time due to the high operating temperature and the diffusion-limited transport of ions in the electrolyte

(continued)

Table 2.2 (continued)

Advantages	Disadvantages
SOECs have a high energy efficiency, typically around 80–85%	SOECs are expensive to manufacture and require high-temperature materials and complex manufacturing processes
SOECs can produce both hydrogen and syngas (a mixture of hydrogen and carbon monoxide) with high purity	SOECs require a high-purity oxygen source, which can be expensive to produce
SOECs have a long lifespan and are resistant to electrode degradation and corrosion due to the high operating temperature and the use of non-corrosive materials	SOECs require a substantial input of thermal energy to reach and maintain the high operating temperature
SOECs have a high power density and can achieve high hydrogen production rates with a relatively small electrode surface area	SOECs are not suitable for portable applications due to the high operating temperature and weight
SOECs can operate with a variety of renewable sources of energy, such as solar, wind, and biomass, making them an attractive option for sustainable hydrogen production	SOECs require a complex balance of heat and mass transfer to maintain optimal operating conditions

energy. Initially, when d-band filling increases, so does the first oxidation potential or the nobility of metal. The nobility gradually enhances from 3 to 5d-shells. The tendency to form oxides, akin to nobility, rises as the d-shell filling increases but experiences a slight decrease with the growing number of d-shells. Conversely, the bond strength between metal atoms weakens as the d-band filling increases but strengthens as each successive d-shell is occupied. They have oxidized all metals at a constant input voltage (overpotential = 200 mV) to match these thermodynamic data to experimentally observe transient dissolution rates during oxidation and reduction of the metals and their electrochemically generated oxides. All electrodes were well polished just before the measurements began, which is essential to reduce the number of native oxides generated in the air. In each metal, equal overpotentials inhibit constant oxidative stress. As a result, the measured dissolution is a good measure of stability comparable among metals and directly related to a standard description. The metals studied here were predicted to generate matching thermodynamically stable oxides. They experimented with all metals in an acidic environment. However, only noble metal data form a stable oxide at acidic pH [45].

With potential, there is an exponential increase in active dissolution. All examined metals experience temporary dissolution during oxide formation at $\eta = 200$ mV in alkaline solutions. Within each d-shell, the degree of transitory dissolution rises. Only Ni and Ru show no detectable dissolution upon oxidation during these studies. Ni has previously exhibited exceptional dissolution stability during redox reactions in alkaline electrolytes, which has been attributed to the creation of $Ni(OH)_2$ via a dissolution precipitation process. The oxide is generated in this case by the precipitation of poorly soluble Ni species.

The researchers investigated that there is a linear relation between metal dissolution rate and applied potential [44]. The behavior alters more noticeably with the decrease of the oxide layers. For 3d metals, no dissolution is observed. It is vital to highlight the distinction between the protocols used for 3d and 4d/5d items. To limit bubble formation during the hydrogen evolution reaction, a constant potential of 0.05 V RHE was adopted to eliminate oxides of 4d and 5d elements (HER). This produces a substantial reductive overpotential of $E > 300$ mV. Because 0.05 V RHE is identical to their redox potentials, the same approach cannot be utilized for Fe, Ni, and Co. As a result, a steady reductive current of @0.3 mA cm^{-2} was used to decrease these 3d elements oxides, although the resulting low overpotentials ($E =$ 200 mV) may not be sufficient to reduce the surface ultimately.

The dissolution rate is linked to the pace and degree of oxide production via the stated mechanism, which both depend on the ease of M–M bond breaking. Breaking M–M bonds is easier with low bond strength, resulting in a greater oxide production rate and high dissolution via a more vital M–O interaction. The passivation effect of the oxide layer increases as it grows, whereas oxide production and M–M bond breakdown decrease. The passivation is evident in the steady reduction in the dissolution rate over the applied 300 s oxidative potential [46].

When the oxidic passivation layer is reduced, many oxygen species depart the surface, producing undercoordinated metal sites. These metal cations can be liberated from the surface when solvated by the electrolyte. The extent of cathodic dissolution should be determined by the amount of oxide on the electrode (which determines undercoordinated metal sites number) and the interaction of metal cations with the solvent. First and foremost, the total dissolved amount during reduction is proportional to oxide layer thickness. The authors proposed that the thickness of native oxides on noble metals is a function of the amount of time metals are exposed to air and that it determines the degree of dissolution during their reduction. Cathodic dissolution was observed to increase with time. When the authors compare the amount of dissolution of different metals at a given time (especially at shorter intervals) and consider the descriptors, they see an apparent increase in dissolution with adsorption energy. This is because metals with high adsorption energy create thicker oxides, resulting in increased dissolution.

In summary, the inclusion of oxygen atoms into the crystal lattice during oxidation causes M–M bonds to break. If this requires minimal energy, the dissolution tendency increases due to the formation and dissolution of produced undercoordinated metal sites. Dissolution during reduction is determined by the amount of grown oxide (controlled by both M–M and M–O interactions) and the rate of oxide reduction (mostly M–O).

References

1. Guo H, Peng Q, Chen XK, Gu Q, Dong S, Evans EW, Gillett AJ, Ai X, Zhang M, Credgington D et al (2019) High stability and luminescence efficiency in donor-acceptor neutral radicals not following the Aufbau principle. Nat Mater. https://doi.org/10.1038/s41563-019-0433-1
2. Kornblum L (2019) Conductive oxide interfaces for field effect devices. Adv Mater Interfaces. https://doi.org/10.1002/admi.201900480
3. Fu YS, Li J, Li J (2019) Metal/semiconductor nanocomposites for photocatalysis: fundamentals, structures, applications and properties. Nanomaterials. https://doi.org/10.3390/nano9030359
4. Justice P, Marshman E, Singh C (2020) Student understanding of fermi energy, the Fermi–Dirac distribution and total electronic energy of a free electron gas. Eur J Phys. https://doi.org/10.1088/1361-6404/ab537c
5. Kim SC, Arun AS, Ahsen ME, Vogel R, Stolovitzky G (2021) The Fermi–Dirac distribution provides a calibrated probabilistic output for binary classifiers. Proc Natl Acad Sci U S A. https://doi.org/10.1073/pnas.2100761118
6. Lang ND, Kohn W (1971) Theory of metal surfaces: work function. Phys Rev B. https://doi.org/10.1103/PhysRevB.3.1215
7. Lampinen MJ, Fomino M (1993) Analysis of free energy and entropy changes for half-cell reactions. J Electrochem Soc. https://doi.org/10.1149/1.2221123
8. Frankel GS, Stratmann M, Rohwerder M, Michalik A, Maier B, Dora J, Wicinski M (2007) Potential control under thin aqueous layers using a Kelvin probe. Corros Sci. https://doi.org/10.1016/j.corsci.2006.10.017
9. Lim A, Kim H-J, Henkensmeier D, Jong Yoo S, Young Kim J, Young Lee S, Sung YE, Jang JH, Park HS (2019) A study on electrode fabrication and operation variables affecting the performance of anion exchange membrane water electrolysis. J Ind Eng Chem 2019. https://doi.org/10.1016/j.jiec.2019.04.007
10. Ooka H, Huang J, Exner KS (2021) The Sabatier principle in electrocatalysis: basics, limitations, and extensions. Front Energy Res. https://doi.org/10.3389/fenrg.2021.654460
11. Dalton F (2016) ECS classics: historical origins of the rotating ring-disk electrode. Electrochem Soc Interface. https://doi.org/10.1149/2.F03163if
12. Schmidt TJ, Gasteiger HA, Stäb GD, Urban PM, Kolb DM, Behm RJ (1998) Characterization of high-surface-area electrocatalysts using a rotating disk electrode configuration. J Electrochem Soc. https://doi.org/10.1149/1.1838642
13. Mayrhofer KJJ, Strmcnik D, Blizanac BB, Stamenkovic V, Arenz M, Markovic NM (2008) Measurement of oxygen reduction activities via the rotating disc electrode method: from Pt model surfaces to carbon-supported high surface area catalysts. Electrochim Acta. https://doi.org/10.1016/j.electacta.2007.11.057
14. Patel AM, Nørskov JK, Persson KA, Montoya JH (2019) Efficient Pourbaix diagrams of many-element compounds. Phys Chem Chem Phys. https://doi.org/10.1039/c9cp04799a
15. Muthurasu A, Tiwari AP, Chhetri K, Dahal B, Kim HY (2021) Construction of iron doped cobalt-vanadate-cobalt oxide with metal-organic framework oriented nanoflakes for portable rechargeable zinc-air batteries powered total water splitting. Nano Energy. https://doi.org/10.1016/j.nanoen.2021.106238
16. Li D, Wang C, Tripkovic D, Sun S, Markovic NM, Stamenkovic VR (2012) Surfactant removal for colloidal nanoparticles from solution synthesis: the effect on catalytic performance. ACS Catal. https://doi.org/10.1021/cs300219j
17. Pavlishchuk VV, Addison AW (2000) Conversion constants for redox potentials measured versus different reference electrodes in acetonitrile solutions at 25 °C. Inorg Chim Acta. https://doi.org/10.1016/S0020-1693(99)00407-7
18. Stevens MB, Enman LJ, Batchellor AS, Cosby MR, Vise AE, Trang CDM, Boettcher SW (2017) Measurement techniques for the study of thin film heterogeneous water oxidation electrocatalysts. Chem Mater 29(1):120–140. https://doi.org/10.1021/acs.chemmater.6b02796

19. Vidal-Iglesias FJ, Solla-Gullón J, Rodes A, Herrero E, Aldaz A (2012) Understanding the Nernst equation and other electrochemical concepts: an easy experimental approach for students. J Chem Educ. https://doi.org/10.1021/ed2007179
20. Licht S (1985) PH measurement in concentrated alkaline solutions. Anal Chem. https://doi.org/10.1021/ac50001a045
21. Chen R, Yang C, Cai W, Wang HY, Miao J, Zhang L, Chen S, Liu B (2017) Use of platinum as the counter electrode to study the activity of nonprecious metal catalysts for the hydrogen evolution reaction. ACS Energy Lett. https://doi.org/10.1021/acsenergylett.7b00219
22. Chatenet M, Benziger J, Inaba M, Kjelstrup S, Zawodzinski T, Raccichini R (2020) Good practice guide for papers on fuel cells and electrolysis cells. J Power Sources. https://doi.org/10.1016/j.jpowsour.2019.227635
23. Li X, Zhao L, Yu J, Liu X, Zhang X, Liu H, Zhou W (2020) Water splitting: from electrode to green energy system. Nano-Micro Lett. https://doi.org/10.1007/s40820-020-00469-3
24. Raccichini R, Amores M, Hinds G (2019) Critical review of the use of reference electrodes in Li-ion batteries: a diagnostic perspective. Batteries. https://doi.org/10.3390/batteries5010012
25. ul Haq T, Haik Y (2022) Strategies of anode design for seawater electrolysis: recent development and future perspective. Small Sci 2(9):2200030. https://doi.org/10.1002/smsc.202200030
26. Fernandez-Delgado O, Puente-Santiago AR, Cano M, Giner-Casares JJ, Metta-Magaña AJ, Echegoyen L (2020) Facile synthesis of C_{60}-nano materials and their application in high-performance water splitting electrocatalysis. Sustain Energy Fuels. https://doi.org/10.1039/d0se00399a
27. Wei C, Rao RR, Peng J, Huang B, Stephens IEL, Risch M, Xu ZJ, Shao-Horn Y (2019) Recommended practices and benchmark activity for hydrogen and oxygen electrocatalysis in water splitting and fuel cells. Adv Mater 31(31):1–24. https://doi.org/10.1002/adma.201806296
28. Grigoriev SA, Fateev VN, Bessarabov DG, Millet P (2020) Current status, research trends, and challenges in water electrolysis science and technology. Int J Hydrogen Energy. https://doi.org/10.1016/j.ijhydene.2020.03.109
29. Santos AL, Cebola MJ, Santos DMF (2021) Towards the hydrogen economy—a review of the parameters that influence the efficiency of alkaline water electrolyzers. Energies. https://doi.org/10.3390/en14113193
30. Manabe A, Kashiwase M, Hashimoto T, Hayashida T, Kato A, Hirao K, Shimomura I, Nagashima I (2013) Basic study of alkaline water electrolysis. Electrochim Acta. https://doi.org/10.1016/j.electacta.2012.12.105
31. Zeng K, Zhang D (2010) Recent progress in alkaline water electrolysis for hydrogen production and applications. Prog Energy Combust Sci. https://doi.org/10.1016/j.pecs.2009.11.002
32. Rashid MM, Al Mesfer MK, Naseem H, Danish M (2015) Hydrogen production by water electrolysis: a review of alkaline water electrolysis, PEM water electrolysis and high temperature water electrolysis. Int J Eng Adv Technol
33. Wang J, Gao Y, Kong H, Kim J, Choi S, Ciucci F, Hao Y, Yang S, Shao Z, Lim J (2020) Non-precious-metal catalysts for alkaline water electrolysis: operando characterizations, theoretical calculations, and recent advances. Chem Soc Rev. https://doi.org/10.1039/d0cs00575d
34. Chen Z, Guo L, Pan L, Yan T, He Z, Li Y, Shi C, Huang ZF, Zhang X, Zou JJ (2022) Advances in oxygen evolution electrocatalysts for proton exchange membrane water electrolyzers. Adv Energy Mater. https://doi.org/10.1002/aenm.202103670
35. Kasperson RE (2020) Corporate culture and technology transfer. In: Social contours of risk. https://doi.org/10.4324/9781849772563-18
36. Kulkarni D, Huynh A, Satjaritanun P, O'Brien M, Shimpalee S, Parkinson D, Shevchenko P, DeCarlo F, Danilovic N, Ayers KE et al (2022) Elucidating effects of catalyst loadings and porous transport layer morphologies on operation of proton exchange membrane water electrolyzers. Appl Catal B Environ. https://doi.org/10.1016/j.apcatb.2022.121213
37. Kang Z, Alia SM, Young JL, Bender G (2020) Effects of various parameters of different porous transport layers in proton exchange membrane water electrolysis. Electrochim Acta. https://doi.org/10.1016/j.electacta.2020.136641

38. Van Pham C, Escalera-López D, Mayrhofer K, Cherevko S, Thiele S (2021) Essentials of high performance water electrolyzers—from catalyst layer materials to electrode engineering. Adv Energy Mater. https://doi.org/10.1002/aenm.202101998
39. Nechache A, Hody S (2021) Alternative and innovative solid oxide electrolysis cell materials: a short review. Renew Sustain Energy Rev. https://doi.org/10.1016/j.rser.2021.111322
40. Lei L, Zhang J, Yuan Z, Liu J, Ni M, Chen F (2019) Progress report on proton conducting solid oxide electrolysis cells. Adv Funct Mater. https://doi.org/10.1002/adfm.201903805
41. Wang Y, Li W, Ma L, Li W, Liu X (2020) Degradation of solid oxide electrolysis cells: phenomena, mechanisms, and emerging mitigation strategies—a review. J Mater Sci Technol. https://doi.org/10.1016/j.jmst.2019.07.026
42. Park BK, Zhang Q, Voorhees PW, Barnett SA (2019) Conditions for stable operation of solid oxide electrolysis cells: oxygen electrode effects. Energy Environ Sci. https://doi.org/10.1039/c9ee01664c
43. Pan X, Jiao F, Miao D, Bao X (2021) Oxide-zeolite-based composite catalyst concept that enables syngas chemistry beyond Fischer–Tropsch synthesis. Chem Rev. https://doi.org/10.1021/acs.chemrev.0c01012
44. Speck FD, Zagalskaya A, Alexandrov V, Cherevko S (2021) Periodicity in the electrochemical dissolution of transition metals. Angew Chem Int Ed. https://doi.org/10.1002/anie.202100337
45. Hughes JP, Clipsham J, Chavushoglu H, Rowley-Neale SJ, Banks CE (2021) Polymer electrolyte electrolysis: a review of the activity and stability of non-precious metal hydrogen evolution reaction and oxygen evolution reaction catalysts. Renew Sustain Energy Rev. https://doi.org/10.1016/j.rser.2021.110709
46. Latimer WM (1939) The oxidation states of the elements and their potentials in aqueous solutions. Soil Sci. https://doi.org/10.1097/00010694-193910000-00009

Chapter 3
Emerging Techniques for the Synthesis of Self-supported Electrocatalysts

3.1 Role of Electrocatalyst

A catalyst is a chemical substance that aids in accelerating a chemical reaction without being consumed in the process. In other words, it helps lower the activation barrier needed to overcome the reactant molecule. This means that reactions can happen with faster kinetics and at lower temperatures than they would otherwise. As a result, reactants interact with one another to form products. However, this process is frequently delayed because the reactants must overcome an energy barrier to create the products. Catalysts contribute to expediting the reaction process by reducing the activation energy. During a chemical reaction, the bonds between atoms in molecules are broken, reorganized, and reformed, recombining the atoms to form new molecules. Catalysts increase the efficiency of this process by decreasing the activation energy, to overcome the energy barrier for a chemical process to occur. Consequently, catalysts make it easier for atoms to break and create chemical bonds, resulting in the formation of novel combinations and compounds. Using catalysts accelerates and improves the energy efficiency of chemical reactions. Catalysts also possess an essential trait known as selectivity, which allows them to steer a reaction to enhance the amount of the desired product while decreasing the number of undesired byproducts. In other words, a catalyst creates a new active site for a reaction without being consumed [1].

On a molecular scale, catalytic mechanisms comprise the direct interactions between active sites and reactant molecules, resulting in the production and breaking of chemical bonds. These events occur at or close to where the reactants coordinate with the catalyst. The effects may entail interactions between reacting and non-reacting molecules (including solvent molecules) or between reacting molecules and atoms placed at specific distances from the binding site, where they define empty spaces. Many energetic descriptors that have been utilized successfully to build linear free energy connections between the active sites infrastructure and the composition and their reactivity as catalysts are insufficient to predict the kinetics of catalytic

T. u. Haq and Y. Haik, *Electrochemical Water Splitting*, Materials Horizons: From Nature to Nanomaterials, https://doi.org/10.1007/978-981-99-9860-9_3

events accurately. The configuration of additional atoms and directing groups that define the limits of confined spaces or voids next to the binding site can significantly affect catalytic action [2]. Their presence can influence catalytic activity by:

1. Introducing electronic effects on the binding site.
2. Creating the chemical potentials of reactants and intermediates by controlling diffusional access to the binding site.
3. Influencing the stabilities of specific transition states based on their shape, size, and polarity.
4. Enabling kinetic coupling between different reactions, as well as coupling between reactions and transport processes.

By preventing the entry of contaminants into the binding site, the expanded environment that constitutes the active site can also boost the catalyst's resistance to deactivation.

The synthetic catalysts are commonly divided into two different categories depending on the contact phase, i.e., homogenous and heterogenous catalysis. When the catalyst and reactants exist in the same phase, it's considered homogeneous. In the most prevalent type of homogeneous catalysts, metal complexes in solution catalyze a reagent also in solution. For example, Wilkinson's catalyst ($(PPh_3)_3RhCl$) is used in alkene hydrogenation. Because the catalyst and reactant molecules are in the same phase, homogeneous catalysis provides the benefit of a high level of interaction between them. However, a limitation of homogeneous catalysts is that they are frequently unrecoverable after the complete reaction. That's why using homogeneous catalysis in large-scale industrial production processes is less efficient and commercially unsustainable [3].

A heterogeneous catalyst is one whose phase is distinct from the reactants. The most widely accepted type is a solid that catalyzes a surface-adsorbing liquid or gas. One of the primary advantages of utilizing a heterogeneous catalyst is that it may be easily separated from a reaction mixture, for instance, by filtration. Therefore, it is simple and efficient to recover costly catalysts, a factor that is particularly significant in industrial manufacturing processes. The available surface area of the catalyst is a constraint of heterogeneous catalysis. When the surface of a catalyst is entirely saturated with reactant molecules (i.e., no more reactants can fit on the surface), the reaction cycle cannot continue until some of the product molecules leave the surface, making room for additional reactant molecules. Because of this, the adsorption stage is frequently the rate-limiting step in heterogeneously catalyzed reactions [4].

A suitable catalyst can reduce the water redox reaction's activation energy barrier and increase the reaction's kinetics. Recent investigations demonstrated that 3d, 4d, and 5d transition metal-based catalysts with relatively high oxidation states are more suitable for OER. These materials' surface, composition, and crystal/electronic structure change dynamically on applying high voltage. The material is inactive at open circuit potential (OCP), transforming into an active electrocatalyst during operation. Several operando characterizations have uncovered the concurrent reconstruction process of electrocatalysts during water electrolysis. Conventionally, the externally applied bias is the only driving power source for either the water electrolysis reaction or the catalyst reconstruction. The chemical driving force, however, plays diverse roles in surface reconstruction, reaction rate/path, and chemical redox interactions between post-catalyst and water molecules. Chemical driving force derives from the spatial gradient of potential energy (containing electrostatic energy and chemical potential energy) of charged/uncharged species, which can be categorized into three distinct situations. When the electrolyte interacts with the electrocatalyst surface before applying a potential, complex modifications, including surface reconstruction, occur at the catalyst–electrolyte interface. These changes are triggered by a chemical driving force related to formation enthalpies [5].

In general, the applied potential can alter the chemical potential of catalysts by the accumulation of charges on their frontier orbitals (HOMO–LUMO). The second type of chemical driving force derives from the diverse chemical potentials of catalysts, which are anticipated to reduce the reaction's activation energy and enhances its rate. Notably, the charge accumulation described here differs from that in an electrical double layer, even though both affect the electrochemical reaction rate. The applied bias causes electrocatalysis by chemically and electrostatically charging the HOMO–LUMO orbitals of the catalyst and the electric double layer, respectively. After the external potential is removed, the post-catalyst typically has a distinct chemical reactivity and redox potential compared to the pre-catalyst [6]. Due to the redox potential difference at the electrode–electrolyte interface, the post-catalyst can electrochemically split water into H_2 and O_2 gases.

1. Charge redistribution causes the electrochemical potentials of an electrolyte and a catalyst to rebalance dynamically when they come into contact. The catalyst surface will be positively charged if the electrolyte's pH level is below the point of zero charges (PZC). At the catalyst-electrolyte interface, changes such as charge transport, ion dissolution, ion adsorption, ion exchange, and hydrolysis of surface groups are frequently present throughout the charge equilibrium process. The Pourbaix diagram, which is essentially determined by the enthalpies of formation for a particular species as a function of pH and voltage, can be used to forecast the catalyst's stable state. The stability of a catalytic material can be quantitatively evaluated by the corresponding Pourbaix decomposition Gibbs free energy difference (ΔG), which is defined by the potential chemical difference between the catalytic material and the stable chemical species in the Pourbaix diagram, given the applied potential, i.e., OCP, and pH of the electrolyte. When in contact with the electrolyte, a stable catalytic material needs to have $\Delta G = 0$. The catalyst will diverge from its initial structural state upon contact with the electrolyte if the free energy is higher. Note that no external electrochemical bias is applied; all changes at the interface happen on their own. In this sense, chemical rather than electrochemical forces are what are causing these modifications. Under OCP, the parameter ΔG can be utilized to calculate the chemical driving force [7].
2. Electrocatalysis can generally be driven by applying an external electrochemical potential, decomposed into an electrostatic energy and chemical potential.

The chemical driving force for reactions between uncharged species is defined by the interfacial chemical potential difference, which controls the direction of diffusive transport. In contrast, electrostatic energy represents the electrostatic driving force for charged species. The electrostatic driving force is thought to dominate electrocatalysis by charging an electrical double layer and modifying the electrostatic energy of electrons/ions. Pioneer investigations have recently found that charge accumulation by designing the frontier orbitals of active sites also generates electrocatalysis, in which the chemical driving force dominates. The constantly changing chemical potential from charge accumulation on the active sites with varying valence states correlates to a dynamic chemical driving force affecting water splitting performance under operating conditions [8].

3. Catalysts can be reduced or oxidized during electrocatalysis under operating conditions and stay oxidized/reduced once the external potential is removed. As a result, the post-catalyst exhibits differing chemical properties from the pre-catalyst. If the post-redox catalyst's potential after oxidation is greater than that of the O_2/H_2O, the post-catalyst will chemically oxidize water to O_2 while being reduced. Meanwhile, after the reduction process, the post-catalyst has a lower redox potential than the H_2O/H_2. In that case, the post-catalyst is more likely to reduce water into H_2, with itself eventually oxidized chemically. The difference in the post-catalyst's redox potentials and the electrolyte's water generates the chemical driving force in this scenario [9].

3.2 Self-supported Electrocatalyst

Nanostructuring is a popular method for improving the efficiency of electrocatalysts. However, actual applications of nanostructured catalysts face several challenges. On the one hand, nanoparticle fabrication on the current collector needs a polymeric binder like Nafion, which inevitably raises resistance, buries active centers, and restricts mass movement. Furthermore, due to the limited electrical conductivity of electrocatalysts, conductive additives such as nanocarbon compounds are frequently employed [10]. Under OER conditions, the carbon additive is easily oxidized and etched at high voltage, reducing electrode performance [11].

Nonetheless, because of the weak bonding between the substrate and the catalytically active phase, the mass loading of active sites frequently falls below 1 mg/cm^2, leading to a constrained quantity of catalytically active sites. In addition, during extended periods of electrocatalysis, particularly under high current conditions, the coated catalysts may become susceptible to leaching from the substrate. As a result, the powder-based electrodes only perform well at a low geometric activity (100 mA cm^2) for dozens of hours. They cannot match the requirements of real industry electrolyzers, which typically operate at high current densities above 500 mA/cm^2 for thousands of hours [12].

Self-supported electrodes with catalytically active phases grown directly on a current collector are desirable than typical powdery electrocatalysts because they

Table 3.1 Comparison between self-supported and powder-based electrocatalyst

Comparison	Self-supported electrocatalyst	Powder-based electrocatalyst
Preparation method	Grown on a substrate	Powder synthesis
Porosity	High	Low
Surface area	Large	Small
Stability	Good	Susceptible to agglomeration
Active site density	High	Low
Cost	High	Low
Mass transfer	Good	Poor
Mechanical strength	High	Low
Electrode performance	Better	Lower
Catalyst loading	High	Low
Reactant concentration	Low	High
Catalyst activity	Good	Lower
Catalyst efficiency	High	Low
Durability	Good	Low
Scalability	High	Limited

have the subsequent advantages. First, the electrocatalyst in-situ development on conductive substrates eliminates the need for post-coating and adding binder and conducting agent, simplifying and lowering the cost of electrode preparation. Second, the substrate may attach and disseminate electrocatalysts, leading to high active component loading and hence a high density of catalytic sites. Third, without the inclusion of a binder, the interaction between the substrate and active areas is strong enough and works well at a higher geometric activity. The smooth contact allows quick charge transfer, while the tight integration avoids catalyst shedding. Finally, the fabrication of self-supported electrode is easier to realize surface hydrophilic/hydrophobic engineering by purposely altering the morphology and microstructure. The hydrophilic features of the electrode could improve bubble dissociation, facilitate electrocatalyst/electrolyte interaction, and boost ions and mass transfer [13]. These benefits favor increasing the catalytic performance and electrodes sustainability for practical high-current–density electrolysis (Table 3.1).

3.3 Comparative Study of Different Synthesis Techniques for Self-supported Electrocatalysts

During the synthesis process, the substrate's catalytically active phase is generated in situ to prepare a self-supported electrode. Common synthetic strategies are electrodeposition, hydro/solvothermal, vapor deposition, freeze drying, vacuum filtering, and combining diverse processes. The target catalyst component and selective substrate material determine the method used because of their low-cost and distinctive physicochemical features; transition metal elements, Ni, Co, Fe, Cu, An, W, Wo, and Ti, now the most widely explored for self-supported transition-metal-based composites. In addition, metal, carbon, and various other conductive materials are commonly used as substrates, including carbon cloth (CC), carbon fiber papers (CFP), graphite plate, metal mesh (e.g., Ti/Cu mesh), metal foam (e.g., Ni/Cu foam), metal plate/foil (e.g., Ti plate and W/Mo/Cu foil), fluorine-doped tin oxide (FTO), and so on [14, 15]. Graphene and carbon nanotubes, which have mechanical strength and flexibility, can be linked with nanocatalyst to generate freestanding film electrodes post-treatment. Furthermore, some monolithic substrate-free alloy materials can serve directly as self-supported electrocatalysts.

3.3.1 Electrodeposition

The electrodeposition technique is a viable technique for the development of self-supported electrodes with benefits including [16]:

1. The ability to easily control the amount and size of the deposits by adjusting the electrolyte concentration and plating parameters (deposition time, voltage, etc.).
2. A quick and straightforward method for controlled synthesis of uniformly dispersed electrocatalysts without additional capping and reducing agents.
3. Simplicity and high scalability at ambient conditions.

During deposition, the needed metal is deposited on the working electrode by applying an external electric field after submerging two metal electrodes in the desired electrolyte solution (cathode) (Fig. 3.1). Supercapacitor materials, nano-magnetic materials, thick and thin coatings of metals, and other materials are all synthesized using the electrodeposition process [17, 18]. The thickness of the metal films can be controlled by manipulating the electrode potential and current density. A configuration of three electrodes, such as working, counter, and reference electrodes submerged in the electrolyte-containing ions, is used to synthesize semiconducting materials. The cathode and anode electrodes typically provide a milliWatt range DC power supply. The positive ions migrate toward the cathode and discharge and chemically react to produce the desired substance. The electrical current in the electrolyte solution flows as ions. The externally applied unidirectional current source across the electrode system is the direct current (DC) power source. When an appropriate external power source is employed, anions (negative ions) migrate toward the anode, while cations (positive ions) move toward the cathode. A specific sort of substance in the electrolyte solution affects the deposition process [17]. The following are the key component of the electrodeposition setup (discussed in detail in Chap. 2).

Working Electrode: The material's superior electron transfer capabilities subject to the substrate while demonstrating higher activation energy to transfer electrons in

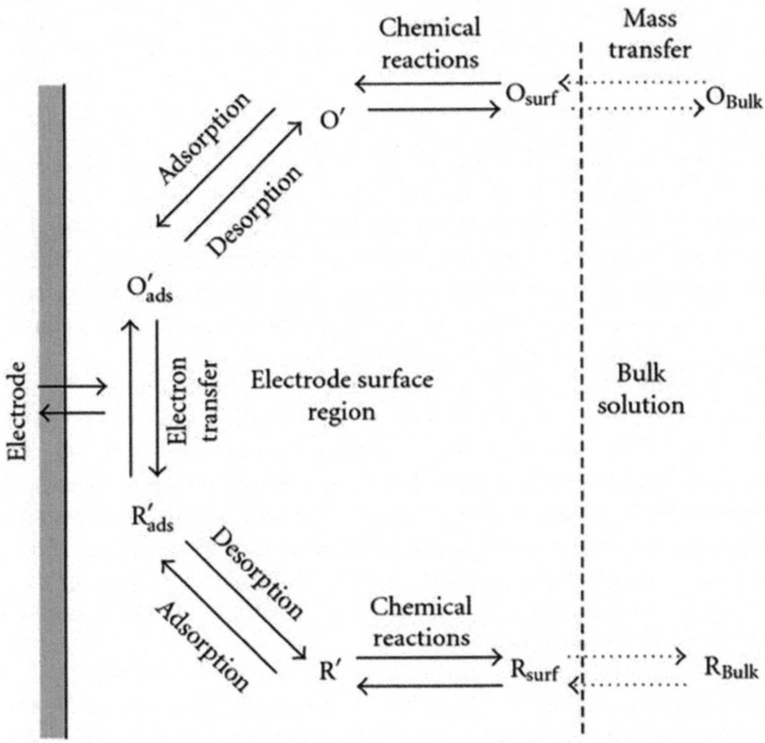

Fig. 3.1 Electrode reaction pathway. Reproduced with permission from Ref. [19]. © 2011, Hindawi

the performing reaction influence the choice of working electrode materials. Anode materials most typically used are graphite, lead oxide, and platinum. The anodic OER and cathodic HER are water's two most essential processes.

Reference Electrode: The reference electrode's electrode potential is known and utilized as a reference to manage and measure the potential of the electrochemical arrangement. A redox system with constant (buffered or saturated) concentrations of each redox reaction participant is typically used to achieve excellent stability of the reference electrode potential. Furthermore, the current across the reference electrode is kept near zero by maintaining the circuit current closed using a counterelectrode and a large input impedance (approximately > 100 *G*). Generally, one of the most difficult challenges is determining the best reference electrode for the experimental settings. According to recent findings, an "ideal RE" should have the following properties: (i) have a stable potential; (ii) fulfill the requirements of a charge transfer forced by the measuring device without attempting to change its potential (be nonpolarizable); (iii) after errant polarization it has the potential to sustain the fixed reference potential; (iv) obey the Nernst equation for some species in a solution; Saturated calomel electrodes are examples of electrodes with these features for application in aqueous solutions (SCE). Because of its simplicity, silence, and resilience, the SCE appears to be the "preferred RE" among scientists. It operates at temperatures as high as 90 °C, is filled with KCl, NaCl, and LiCl solutions, and quickly equilibrates when refilled [17].

Counterelectrode: The counterelectrode completes the electrochemical cell's circuit. Cathode materials commonly utilized include platinum, lead, mercury, cadmium, and graphite, which have thermodynamic potential in HER [8]. As the counterelectrode (CE) is not directly participating in the electrocatalytic reaction and the charge transfer occurs between the counter and working electrode, it's important for the counterelectrode to have a larger surface area than the working electrode. This ensures that the rate of electrochemical reaction processes remains unaffected, preventing any potential slowing down of the reaction [20].

Electrodeposition has long been used to improve various materials' surface characteristics, design, and ornamental purposes. In addition, electrodeposition is an efficient method for producing many nanomaterials. In recent years, the electrodeposition method has been essential for fabricating nanostructures as electrocatalysts for various energy conversion activities. In this process, little adjustments in the electrochemical deposition settings can affect the surface structure of electrocatalysts produced on conducting surfaces. The main advantage of the electrodeposition method is that no capping or conductive agents are used, which simplifies the procedure for their applicability.

The electrodeposition process involves the following procedure [18].

The electrolyte is first used to dissolve the reactants. Subsequently, by carefully adjusting the applied cell voltage, it becomes possible to continually deposit the oxidized or reduced products onto the surface of either the working or counterelectrode. Finally, the working electrode potential is observed using a reference electrode. The cell potential and deposition current, which can be adjusted as a function of time during the reaction, are two significant parameters that govern the course of the response. By maximizing different electrodeposition parameters (such as time, voltage, current, additives, temperature, concentration, and pH), the surface structure of the designed materials can be controlled. The nucleation rate of the crystal can be precisely regulated in a current-controlled synthesis process to produce deposits with a controlled shape and good adherence. However, a potentially controlled synthesis typically does not know the potential to be used. Hence, broad cell potential linear voltammetry is required to choose a suitable deposition potential. The deposition potential is found in the voltage range at which electrolyte decomposition typically occurs. Potentiostatic synthesis between the HER and OER potentials is possible for aqueous electrolytes. We should pick a solvent (such as ionic liquids, organic solvents, or molten salts) with a broad potential window if the reaction potential is outside the electrolyte's window. Due to the poor rate of reactant molecule diffusion from the bulk solution to surface of electrode in the situation of constant potential, the cell current typically degrades quickly. The electrolyte can be stirred, the electrode can be rolled, or the deposition mode can be switched to pulse potential to ease this [21].

The observed current is a precise measurement of the combined nucleation rate and development of the mature nuclei or crystallites because the continuous growth of the nuclei of electrodepositing species can only occur through a Faraday process. Many current–potential–time relationships can be investigated, and every experiment can be precisely replicated thousands of times to provide the results' mean values and statistical variation. Hence, experimental research of nucleation using electrochemical methods is possible using either deterministic or stochastic formulations. The observed response of a conducting substrate to various applied wave shapes has served as the foundation for the experimental investigation of electrochemical nucleation. Both cyclic and linear sweep voltammetry consistently and primarily qualitatively demonstrate the occurrence of nucleation. The most popular single (and double) potential step potentiostatic approach and the current step method are practical tools for quantifying the nucleation process [22].

Circular Voltammetry. The mechanics of redox and the transport characteristics of a system in a solution can be investigated using the highly versatile electrochemical technique known as cyclic voltammetry. The size of the measured Faradaic current can tell us how quickly various processes taking place on the working electrode surface are progressing as a whole. The slowest stage determines the total rate, as it does for any multistep process. In general, rates of processes like the following influence the electrode reaction rate:

(1) mass transfer, e.g., of oxygen from the bulk solution to the electrode surface
(2) transfer of electron at the electrode surface
(3) chemical reactions before or after the electron transfer, and these reactions may be homogeneous processes (e.g., protonation or dimerization) or heterogeneous ones (e.g., catalytic decomposition) on the electrode surface
(4) other surface reactions, such as adsorption, desorption, or crystallization (electrodeposition).

Simple electrodeposition processes are not controlled by chemical or surface reactions, and either mass transport or charge transfer reactions dominate the kinetics. However, (a) the homogeneous reactions can be regarded as being at equilibrium, and (b) the Nernst equation relates the surface concentrations of species engaged in the faradaic process to the electrode potential, for example, if an electrode process only involves fast heterogeneous charge transfer kinetics and mobile, reversible homogeneous reactions [21].

The mass transfer rate, mt, which determines how quickly the electrochemical ions are delivered toward surface, determines electrocatalyst net rate reaction.

$$V_{\mathrm{mt}} = i/n * F * A.$$

As the main species adhere to thermodynamic laws at the electrode surface, these electrode reactions are frequently referred to as reversible or Nernstian. The slow, irreversible electron transfer will gain control of the kinetics if the overall response is not mass transport [18]. While metal oxides often develop at the anode when an electric current flows through a metal-salt solution, most metals can be electrodeposited directly at the cathode. In electrochemical oxidation, a metal ion in a lower oxidation state can undergo an anodic oxidation reaction to reach a higher oxidation state. The metal oxide or hydroxide is quickly produced by hydrolysis of the higher oxidation state. Metal hydroxides can be electroplated at the cathode using a two-step procedure instead of the direct electrodeposition of metals and metal oxides. Initially, the reduction of the solute produced many OH ions close to the cathode surface. The solution's metal ions are then sedimented and deposited alongside OH as a hydroxide. The strategy is to raise the working electrode's surface pH. The following oxyacid anion reduction processes are typically recognized to produce hydroxide ions [21].

$$ClO_3^- + 3H_2O + 6e^- \rightarrow Cl^- + 6OH^- \quad E^0 = 1.890\,V$$

$$IO_3^- + 3H_2O + 6e^- \rightarrow I^- + 6OH^- \quad E^0 = 1.088\,V$$

$$NO_3^- + 3H_2O + 6e^- \rightarrow NO_2^- + 2OH^- \quad E^0 = 0.838\,V$$

The standard potentials of the three reactions mentioned above are higher for most metals than for metal cation reduction. Therefore, electrolyte pH and the suitable applied potential can be selected per the metal Pourbaix diagram to regulate the different types of deposits.

When chlorate is employed as the deposited electrolyte in an aqueous solution, hydroxyl groups are produced through chlorate reduction, which raises the pH in the area closer to surface of cathode surface. As a result, nickel hydroxide is produced during the 0–1.23 V potential range. This possible range can efficiently prevent the detrimental effects of oxygen evolution (hydroxyl consumption) and convert Ni^{2+} to its metallic form. Similarly, nickel hydroxide will be formed in the applied potential ranges of 0–1.09 V and 0–0.84 V, respectively, for deposition electrolytes containing iodate or nitrate. As a result, the cathodic reduction raises the pH level in the area close to the electrode and kinetically promotes the deposition of metal hydroxides. As the hydroxide is further dehydrated or oxidized in air, metal oxides may sometimes develop. Interestingly, coprecipitation of multimetallic hydroxides is also favored by cathodic electrosynthesis. Cathodic electrodeposition is a versatile and effective method for creating various metal hydroxides on conducting surfaces. Most of the periodic table's metal (hydro)oxides have been formed through electrodeposition until this point. Due to their exciting electrocatalytic capabilities, nanostructured transition metal (hydro)oxides have gained increasing interest in recent years. However, the catalysis efficiency and selectivity depend on the shape, size, and even the interparticle distance addition to their chemical composition. A variety of nanostructured catalysts are created through the process of electrodeposition. The synthetic variables of electrodeposition are simple to control compared to other methods. Moreover, electrodeposition enables interfacial modification to enhance electrode stability and precise control over the electrode's size, shape, composition, and structure. As a result, numerous nanostructured (hydro)oxides with various morphologies have been created via electrodeposition, including low dimensional 0D, 1D, 2D, and 3D nanostructures [23].

Yu's group used a two-step electrodeposition procedure to develop the Ni-FeO$_x$H$_y$-PN electrocatalyst (Fig. 3.2) [24]. It is vital to enhance the roughness to expose more active sites because the commercial metal mesh with a smooth surface has a negligible active surface area. A porous layer was produced on the nickel mesh surface using the bubble template approach to electrodeposit Ni nanoparticles. In a typical synthesis, an ultra-high cathodic current density was used, and HER on the mesh was used to generate H_2 bubbles. Concurrently, Ni^{2+} ions were converted to nickel nanoparticles on the mesh surface as they moved toward the cathode under the influence of the applied cathodic current. A porous deposition layer was created with the help of dynamic H_2 production and Ni electrodeposition, with the generated H_2 bubbles acting as a dynamic template. The as-prepared 3D porous nickel electrode has microscopic channels for releasing gas bubbles and encourages the diffusion of electrolytes, in contrast to the pure metal mesh with a flat surface. The hierarchical structure of the Ni-FeO$_x$H$_y$-PN electrocatalyst, which results from the combination of the porous substrate and the very thin Ni-FeO$_x$H$_y$ nanosheets, improves the active surface area and facilitates gas diffusion [24].

Recently, ul Haq et al. used an electrodeposition strategy to fabricate Au nanocluster-coupled Gd–CoB nanoflakes embedded in TiO_2 nanosheets, demonstrating excellent activity for unpurified seawater electrolysis [8]. The generation

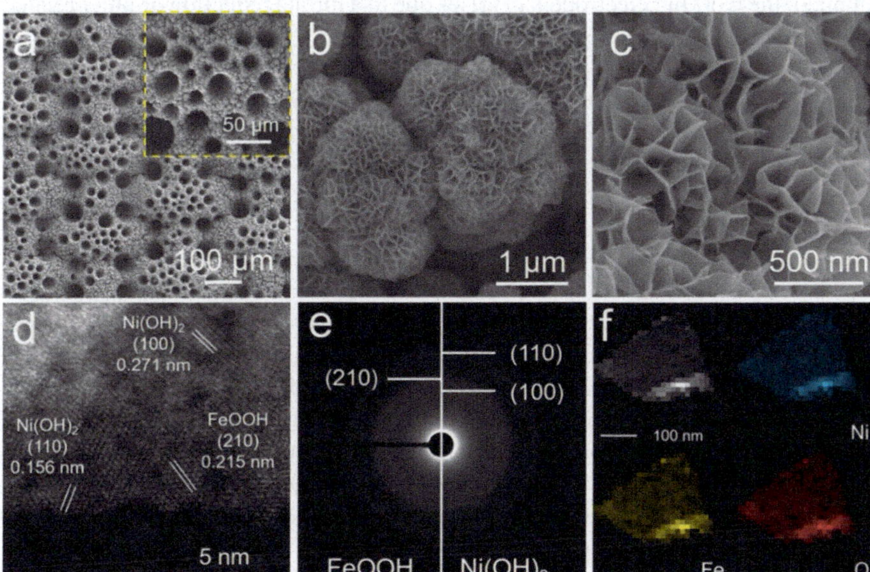

Fig. 3.2 FESEM image of **a** porous network; **b**, **c** NiFeO$_x$H$_y$-PN; **d** HRTEM image and **e** SAED pattern of NiFeO$_x$H$_y$ nanosheet; and **f** elemental mapping images of Ni, Fe and O. Reproduced with permission from Ref. [24]. © 2020, Elsevier

of nanoflakes of Gd–Co_3O_4 from solution of metal salt was made possible using a constant current density technique. The underlying TiO_2 nano possesses a greater specific surface area and offers dense nucleation centers for the growth of bi-metallic nanostructure. For diffusion-controlled development, urea was utilized as a homogeneous precipitation agent. Urea is broken down into NH^{4+}, $CO_3{}^{2-}$, and OH^- during the cathodic deposition process. Initially, a cathodic potential value increases exponentially endorsed the nucleation of Co ($Co^{2+}(aq)+2e^- \rightarrow Co(s) E^o\,(V) = -0.28$). Initially, Gd^{3+} ions don't participate in the electrochemical reaction due to its low reduction potential, ($Gd^{3+}(aq) + 3e^- \rightarrow Gd(s) E^o\,(V) = -2.279$) and maintaining the similar composition across the electrolyte. Though, the continuous increase in the negative potential confirms the successful bimetallic nanoflakes deposition. This elevates the pH in the vicinity of the electrode surface, promoting the deposition or precipitation of metals. It was observed that many independent variables, such as the applied potential, the concentration of the precursor, and the reaction duration, significantly affect the surface structure of the nanoflakes. Several Gd–Co_2B samples were produced in the subsequent chemical reduction step by soaking Gd–Co_3O_4 precursors in aqueous solution of $NaBH_4$ at room temperature for 30 min. The oxygen defects made by the released H^- ions from $NaBH_4$ and their robust reducing capacity may change the Gd–Co bimetallic layer structure.

Cathodically electrodepositing hydroxide catalyst films from a 0.1 M (total metal) aqueous solution of transition metal nitrate salt has been done by Boettcher and colleagues [17]. The most probable deposition process is that when a cathodic current is applied, the working electrode's pH rises due to a nitrate reduction or other related reactions.

$$NO_3^- + 7H_2O + 8e^- \rightarrow NH_4^+ + 8OH^-$$

The insoluble metal hydroxide precipitates onto the electrode surface when the pH rises:

$$M^{n+} + nOH^- \rightarrow M(OH)_n$$

where n is the metal ion's oxidation state, M, the magnitude of the applied cathodic geometric current, and the duration for which it is applied might change properties such as nanostructure shape, coverage density, and thickness. The deposition method is probably impacted by the current density being used. For instance, Merrill et al. report that metallic NiO production and subsequent quick oxidation by $NO_3{}^-$ in the solution were the predominant routes for $Ni(OH)_2$ deposition under specific circumstances. A two-electrode cell without a reference electrode was utilized for deposition. For first-row transition metals Mn, Ni, Co, and Fe this electrodeposition technique works well because it typically prevents the residue of the metallic form, which is present in cathodic electrodeposition techniques that use ions that are difficult to reduce to consume protons electrochemically. For films containing iron, however, $FeCl_2$ is employed as the metal source and $NaNO_3$ as the nitrate source instead of directly using Mn, Co, and Ni nitrate salts. The authors have discovered that $Fe(NO)_3$ precipitates uniformly throughout the solution over time or when a cathodic current is applied, most likely due to the homogeneous production of insoluble FeOOH. To avoid oxidation and precipitation, Fe-containing deposition solutions are purged with N_2 both before and after adding $FeCl_2$.

Another method is anodic electrodeposition. The Boettcher group anodically deposited $Ni(Fe)O_xH_y$ (0–20% Fe) from a pH 9, 0.4 mM (total metals) $Ni(NO_3)_2$/$FeCl_2$ solution [17]. An anodic deposition is advantageous because the current driving the deposition directly causes the soluble metal cation to oxidize and transform into an insoluble phase. Since each cation has a different reduction potential, it isn't easy to link the deposition charge directly to the film loading in multimetal cation systems, but it is still possible in theory. Moreover, it has been asserted that catalysts that have been anodically deposited are more connected and function better when overloaded. It is also possible to produce high connectivity at high loads using a pulsed cathodic deposition.

Depending on the question that needs to be answered, one of the synthetic approaches is often preferred because each has benefits and drawbacks. The actual use of spin casting (or comparable procedures) is that the film's stoichiometry may be carefully regulated because it matches that of the precursor solution. For research on catalysts containing several metals, precise composition control is crucial. There is remarkable repeatability from sample to sample, even on various substrates, because film thickness depends on the precursor concentration, solvent, and spin speed. In general, the films are more uniform than their electrodeposited counterparts. Yet thin films (20 nm or fewer) are the only ones that can be created using the spin casting technique. When thickness increases, low-temperature nitrate removal becomes more challenging. Surfactants must be removed at temperatures between 250 and 300 °C when they are required to wet the substrate. This complicates research on the active substance since, during such heating, oxides frequently formed rather than the hydroxide phases.

Thicker films and the direct targeting of (oxy)hydroxide phases benefit from electrodeposition. A more homogeneous composition with thickness, a more compact morphology, and films that do not experience a drop-in intrinsic activity with increasing loading is produced for thick films using pulse electrodeposition instead of continuous deposition. The amount of each metal cation depends on its local solubility at a specific pH and its diffusion to the electrode surface, making composition control for mixed-metal films more challenging for electrodeposition than for spin-casting. The stoichiometry of the solution is typically different from that of the film deposited. Regular calibration of the desired film composition to the deposition circumstances involves direct film composition and loading analysis, such as elemental analysis. Higher-quality salts are also necessary due to the deposition process' reliance on cation solubility rather than solution concentration. If they can be incorporated into the target components more quickly than low-level contaminants, they may even become concentrated in the electrodeposited films [17].

3.3.2 Hydro/solvothermal Synthesis

Hydro/solvothermal synthesis is a method that can be performed at high ambient temperature and pressure ($T > 100$ °C, $P > 1$ atm) in the presence of water or organic solvent. If water is used in the process, then the method is referred to as a hydrothermal technique, and if an organic solvent is utilized, it is referred to as a solvothermal technique. Both the temperature and the pressure provided make it possible for the chemical precursors (metal chlorides and nitrates) in the reaction medium to dissolve, making it possible for the desired products to crystallize. In general, the preparation of metal oxide nanoparticles can be accomplished by using these approaches in a single step. The autoclaves are typically utilized as the reaction vessel in this synthesis, made from high pressure-resistant materials, such as titanium or stainless steel. They include a polytetrafluoroethylene (PTFE) lid on the interior to avoid corrosion. Stainless steel is also an option. The hydro/solvothermal method is efficient for producing nanocomposites because it allows for precise control over a wide variety of reaction parameters, including solvent type, reaction time, stirring, pressure, and temperature. These parameters all have an impact on the characteristics of the nanocomposites that are produced. To be more precise, the autoclave-based hydrothermal synthesis method uses the dissolution–recrystallization of sparingly soluble or insoluble components to create inorganic compounds. This uses an aqueous solution as the reaction medium and generates autogenous vapor pressure, a liquid-phase chemical that results in a relatively high pressure and temperature reaction environment. The qualities of the products can be adjusted with great convenience by adjusting the variable parameters [25].

Hydrothermal synthesis has the following figure of merits:

1. Various starting materials and additional substances (such as surfactants, morphology-controlled agents, and cosolvents) exist.
2. It is easy with a simple operation process, and a mild reaction process reduces the activation barrier for nucleation and favors the nanostructure's burst nucleation and controlled growth.
3. The diffusion-controlled growth process in hydrothermal synthesis accelerates the homogenous particle distribution with high-phase purity and crystallinity.

When it comes to solvothermal synthesis, the conditions for the synthesis are practically the same, except that an organic solution is utilized as the primary medium rather than water. In addition to the many advantages offered by hydrothermal synthesis, solvents also have the following effects: (1) they restrict development and flexibility, which is necessary for regulating morphologies and size; (2) they are reducing agents; (3) they attach to portions of the lattice planes; and (4) they have a chelating capability that captures the cation and promotes stability; they diminish anti-site detection by preventing aggregation [26]. The problem of solute precipitation arises in the process of hydrothermal synthesis. In most cases, it involves three different procedures. The first step is the creation of a solution that is supersaturated. After that, a nucleation procedure will take place. The next step in the process is the growth of the crystal's nucleus. Variations in the supersaturated solution ultimately determine whether or not the fluid phase will transform into the powder phase. There is a functional relationship between the supersaturation ratio, *S*, and the Gibbs free energy, the rate of the nucleation, and the critical radius. The degree of supersaturation, denoted by the symbol "*S*," is the factor that most strongly influences the nucleation and growth processes. This term is defined as follows:

$$S = \frac{C}{C_s}$$

C represents the solution's concentration, and C_s represents the saturation concentration.

It is generally believed that the phase can be distinguished from the state of a metastable solution through nucleation and growth processes. The total sum of the nuclei's bulk enthalpy and surface energy is referred to as the modified Gibbs free energy of the sealed system. Thus,

$$\begin{aligned}\Delta G &= \Delta G_s + \Delta G_v \\ &= 4\pi r^2 \sigma + \frac{3}{4}\pi r^3 \Delta G_v\end{aligned}$$

The radius is referred to as r, and sigma is the surface tension in the solution.

Nucleation won't take place until the radius is over a threshold size. The critical radium r_c can be obtained in accordance with setting to $\frac{d\Delta G}{dr} = 0$. The critical radius of the nuclei can be determined from the equation;

$$r_c = 2\sigma v_m / KT \ln S$$

r_c refers to the smallest possible nucleus radius at which nuclei can form spontaneously and remain stable in a supersaturated solution. The nucleation process will move along quickly once S is large enough to allow the nuclei to reach the critical radius [27].

3.3.3 Supercritical Hydro/solvothermal Process

The supercritical fluid is a physical entity that is elevated in pressure and temperature beyond its critical point. When the temperature and pressure of the material are increased to levels higher than its critical values, there is no phase separation. Even though there is only one phase present, the density of the fluid can be adjusted in several different ways by adjusting the pressure and temperature of the system [28]. The dielectric constant of water demonstrates a substantial reduction from 80 MPa close to room temperature to nearly 0 MPa when it is in its critical state. When heated over its critical temperature, the plate almost does not change. In addition, the ionic dissociation constant of supercritical water is relatively low, and there is only a trace amount of remaining hydrogen bonding present. Supercritical water has the properties of a non-polar dense gas and a low-polarity organic solvent. There is a switch between supercritical water and liquid water in this process. Water has a high solubility for inorganic salts but a low solubility for organic compounds in the liquid state. On the other hand, supercritical water has a low polarity, which prevents it from precipitating inorganic compounds. This allows it to dissolve organic compounds quickly. The lower dielectric constant explains the reduced solubility of inorganic salts in supercritical water. Supercritical hydro/solvothermal processing, which occurs when the hydro/solvothermal system operates at a pressure and temperature above its critical value, has several benefits. Producing homogenous particles with high crystallinity and a narrow grain size distribution; having a high degree of supersaturation; being a homogeneous reaction, particularly for the addition of organics; and having a viable, faster, and continuous procedure make it simple to scale up production. Further benefits can be gained by using supercritical synthesis with safe and inexpensive solvents and surface modifiers instead of supercritical hydrothermal: In addition, supercritical solvothermal synthesis (SSS) using nontoxic and low-cost solvents and surface modification provides further benefits over supercritical hydrothermal synthesis (SHS):

(1) In solvothermal synthesis, conditions are mild, and the atmosphere is reductive, whereas in hydrothermal, conditions are extreme, and the atmosphere is oxidizing.
(2) Line and filter clogging during continuous synthesis caused by unreacted and precipitated surface modifiers can be avoided by using a supercritical solvothermal procedure.
(3) Organic solvent use can improve dispersibility, especially at high concentrations, and modify specific crystal orientation and shape in situ.

Similarly, to supercritical hydrothermal, solvothermal has distinctive qualities that make it possible to manufacture commercially through a quick and continuous process [27].

Wang and coworkers reported the development of $Ni_{12}P_5$ nanowires using a hydrothermal approach [19]. Based on the physio-chemical characterizations, authors have proposed the following mechanism for the growth of nanowires. At first, powders of red phosphorus are submerged in an aqueous solution containing Ni^{2+}. This is done because red phosphorus is insoluble in a solution of nickel acetate at room temperature. P on the surface of the red phosphorus continues to spread outward as hydrothermal synthesis continues. This forms a $Ni_{12}P_5$ nucleus by combining P with Ni^{2+} ions in solution (Fig. 3.3). As the reaction proceeded in these specific conditions, it consumed a significant amount of red P while the number of $Ni_{12}P_5$ crystals expanded. When three hours had passed since the beginning of the reaction, the majority of the red phosphorus bulk had been consumed and was now coated with agglomerated $Ni_{12}P_5$ nanowires. As a cationic surfactant, CTAB possesses strong coordination properties with anions and good dispersibility in solution. During the hydrothermal process, the phosphorus (P) that is present on the surface of the red phosphorus moves steadily in a direction that is beneficial for creating the $Ni_{12}P_5$ nanowire structure. This movement is caused by the attraction of a cation that CTAB has ionized. Similarly, Liu et al. have outlined a straightforward and scalable method for producing a hybrid network consisting of mesoporous nanosheets composed of Co_3O_4 and $Co_3(PO_4)_2$ [29]. The porous nanosheet structure with oxygen vacancies and $Co_3(PO_4)_2$ that received electrons from Co_3O_4 demonstrate exceptional performance for oxidizing water. As per the author's account of the synthesis procedure, guanosine 5-monophosphate (GMP) was initially decomposed into phosphate ions, pentose and guanine under hydrothermal conditions.

$$C_{10}H_{12}O_8N_5P^{2-} + 2H_2O \rightarrow C_5H_{10}O_5 + C_5H_5N_5O + PO_4^{3-} + H^+$$

Pentose's dehydration, followed by the carbonization of organic waste, created the carbon phase.

$$C_5H_{10}O_5 \rightarrow 5C + 5H_2O$$

Subsequently, the hydrophilic surface of the carbon phase initiates the adsorption of Co^{2+} ions through mechanisms involving coordination or electrostatic interaction. Ultimately, guanine hydrolysis serves as a pH buffer, leading to an increase in the concentration of ${CO_3}^{2-}$ and OH^- ions during the process.

$$C_5H_5N_5O + 5H_2O \rightarrow 5NH_3 + CO_2 + 4CO$$

$$NH_3 + H_2O \rightarrow NH_4^+ + OH^-$$

$$CO_2 + H_2O \rightarrow CO_3^{2-} + 2H^+$$

Cobalt hydroxide and $Co_3(PO_4)_2$ are the results of the Co ions' reactions with the ${CO_3}^{2-}$ and OH^- ions and phosphate ions.

$$Co^{2+} + OH^- + 0.5CO_3^{2-} + 0.11H_2O \rightarrow Co(OH)_1(CO_3)_{0.5}0.11H_2O$$

$$Co^{2+} + 2PO_4^{3-} \rightarrow CO_3(PO_4)_2$$

The acidity of Co_3O_4 has been enhanced as a result of its electrical interaction with $Co_3(PO_4)_2$. This interaction destabilizes Lewis bases through acid–base interactions, reducing the potential energy barrier for O–H bond cleavage. Furthermore, the inherent reactivity of active sites is heightened, and the resistance to charge transfer is diminished as a result of electronic interactions taking place at the interface of the 2D nanosheets.

Wang et al. used a straightforward solvothermal approach with dodecyl amine as the solvent to create a CdS with a triangular-like form, nanorods, and multipods with

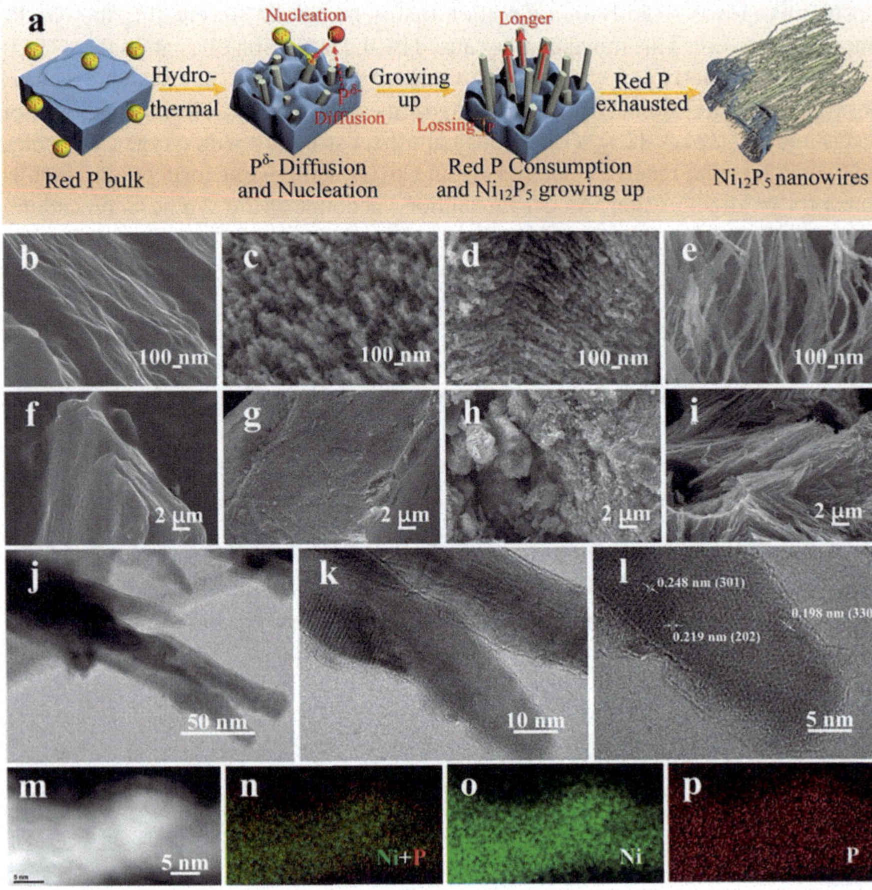

Fig. 3.3 **a** Systematic representation for the development of $Ni_{12}P_5$ nanowires. SEM images recorded at different time interval: **b, f** zero hour; **c, g** 1 h; **d, h** 3 h; **e, i** 7 h. **j, k** The TEM images. **l** HRTEM images. **m–p** STEM images of the $Ni_{12}P_5$ nanowires with corresponding elemental distribution maps of Ni and P. Reproduced with permission from Ref. [19]. © 2020, Elsevier

adjustable arm diameter and length [30]. The precursor concentration, stoichiometry ratio, reaction temperature, and duration significantly impacted the morphology, phase composition, and crystallinity of CdS. Due to its impact on creating shape-determinant CdS nuclei, the cadmium concentration controls CdS morphology. Using 1.5, 3, and 6 mmol of Cd, respectively, nanorods, multipods, and CdS triangular nanocrystals were produced. When the sulfur content is increased while the cadmium concentration is kept at three mmol, the arm diameter in CdS multipods grows from 10 to 60 nm. With CdS multipods, the solvothermal time could regulate the arms' length (Table 3.2).

Table 3.2 Comparison between hydrothermal and solvothermal synthesis

Comparison	Hydrothermal synthesis	Solvothermal synthesis
Solvent	Water	Organic solvents
Temperature	High	High
Pressure	High	High
Reaction time	Long	Short
Reactor type	Autoclave	Sealed reactor
Energy efficiency	Low	High
Control over morphology	Good	Good
Control over particle size	Limited	Good
Crystallinity	High	High
Purity of product	High	High
Cost	Low	High
Toxicity of reactants	Low	High
Scale-up capability	Limited	Good
Surface area of products	High	High
Homogeneity of products	Good	Good

3.3.4 *Chemical Vapor Deposition for the Development of Self-supported Electrocatalyst*

The term "chemical vapor deposition," or "CVD," refers to a procedure of depositing material in which the deposition is accomplished through chemical reactions. In contrast to the method known as physical vapor deposition (PVD), which uses actual physical force as the driving force behind the deposition process, chemical vapor deposition (CVD) is in and of itself an entirely different method. In chemical vapor deposition or CVD, reactive precursors are introduced into a tightly sealed chamber, where they react with the surface of a substrate, the gas phase, or both the gas phase and the surface simultaneously. The end products of gas-phase reactions are either the desired material, which sticks to the substrate, or reactive intermediate intermediates, which trigger chemical interactions between the substrate and the product. Both of these outcomes are possible depending on the nature of the gas-phase reactions. Film growth on the substrate is the consequence of both of these occurrences as a result. The type of reactor used in CVD can be either a batch-wafer reactor or a single-wafer reactor [31]. It is simpler to control the procedure for smaller batch sizes, but it needs significantly longer throughput time, and batch reactors are therefore preferred if the process can accommodate them. Compound and highly pure elemental films, such as epitaxial silicon films that would be extremely challenging or expensive to create using PVD techniques, are mostly grown via CVD. The growth of silicon oxides and nitrides via CVD is another widely used application. Thermal constraints of previously used materials would prevent these materials from growing thermally.

Moreover, CVD is the best approach if deep trenches or holes need to be conformally filled, such as in the case of transistor interlayer connections. By tailoring the reaction kinetics to the requirements of the finished product, it is feasible to manage the conformality and uniformity of the deposited thin layer in CVD. Nevertheless, CVD cannot achieve atomic scale conformality [32].

The basic steps of the conventional process flow for thermal CVD are as follows:

- Introducing a known inert gas, such as argon or nitrogen, to the chamber.
- Increasing the chamber's temperature and pressure until the desired results are achieved.
- The reactive precursor is introduced into the chamber.
- There are chemical reactions and depositions.
- Flushing a known inert gas through the chamber.

According to the Arrhenius equation, the CVD reaction rate is exponentially proportional to the reactive components' temperature because temperature strongly influences chemical reactions [33].

$$K = Ae^{\frac{-E_a}{RT}}$$

Plasma-enhanced CVD (PECVD) has a different process flow than thermal CVD because the needed energy in PECVD is produced by plasma rather than by the surrounding temperature [34]. Since the minimum quantity of E_a is necessary for chemical reactions, this energy must be generated somehow. In most cases, energy is created by raising the temperature of the entire surrounding area, heating a specific substrate area with a laser, or applying an electrical discharge to the gaseous media inside a chamber. While the temperature of solitary particles in plasma may reach very high total energies in plasma-enhanced CVD (PECVD), the ambient and substrate temperatures nevertheless stay relatively modest because PECVD occurs under lower pressure (= fewer particles colliding with surfaces and producing heat). Because of the ability for chemical reactions to occur, films can be deposited at lower temperatures. A material will evaporate more quickly at a given temperature the more volatile it is. It is preferred that byproducts be volatile to desorb from the substrate surface after the reaction has finished. However, volatility is also required whether liquid or solid precursors are utilized to facilitate deposition at lower temperatures. These are just a handful of the primary justifications for using metal-organic precursors in CVD, sometimes known as metal-organic CVD (MOCVD).

Although gaseous precursors are utilized in most CVD processes, liquid or solid precursors are also viable options. Both liquid and solid precursors are brought into the chamber with the assistance of carrier gas. However, while liquid precursors keep their liquid phase and are brought into the chamber as vapors, solid precursors are reduced to their elemental state [35].

The same film may typically be generated using a variety of various precursors. Many other factors come into play when deciding which precursor is the right option, including the following:

- To what temperature must the deposition process be carried out successfully?
- Is there a potential that the precursor will generate undesired reactions with the material on the substrate, or is the chamber material appropriate for the precursor?
- What are the products that are produced? How difficult is it to take them off? Are there any reactions between them and the substrate?
- How much does the precursor cost?
- The precursor's level of toxicity

The CVD deposition rate is variable and depends on both the speed of the chemical reaction and the accessibility of the chemicals. The deposition rate is then controlled by mass transport: this can be thought of as the reaction beginning instantly when the reagent diffuses into the reaction site. If the system has sufficient energy for the reaction to take place but does not have the reagents, then the deposition rate is controlled by mass transport. However, if the location at which a chemical reaction occurs is crowded with a sufficient number of reagents, then the reaction rate will control the deposition rate. This can be considered an adequate number of reagents, but more energy is needed for the reaction to occur more quickly.

Mass transport or the reaction rate always controls the chemical vapor deposition (CVD) process. However, in most cases, it is preferable to have some control over the response rate during the process. This is because having such control results in a growing film that is more conformal and uniform. The temperature and pressure are the important variable that control the deposition rate in CVD. The system's temperature increases surface diffusion as the temperature is converted into kinetic energy. This encourages the distribution of attached particles to locate the energetically best sites, such as vacancies, step, and kink sites, where the binding energy of adatoms is greater than that of a planar surface. More conformality and homogeneity, as well as fewer voids and other defects, are the natural outcomes of this process. Low-temperature deposition produces less crystalline, and occasionally even amorphous, films. The rate at which the precursors for film growth emerge is higher than the surface diffusion rate due to inhibited atom diffusion. As a result, grain development occurs locally close to the deposition sites as opposed to at the thermodynamically favorable places. It develops with smaller grains and a less directed structure.

The CVD chamber's base pressure can range from atmospheric pressure (AP) (1 atm = 760 torr) to ultra-high vacuum (UHV) before precursors are introduced (10^{-6}–10^{-9} torr). To eliminate all the remaining gases inside the chamber that could affect chemical processes or the quality of the film, an ultra-high vacuum might be required. During deposition, the pressure can range from air to 10^{-3} torr, with the UHV system producing the latter. While APCVD typically functions in a zone with limited mass transit, switching to a lower-pressure system can move the operating point to a site with limited reaction time. The operational end of the system cannot be in the mass-transport limitation zone if batch processing is needed since this would result in unequal layer thicknesses not just between the wafers but also wafer-wise.

He et al. have used low-temperature chemical vapor deposition to create the core–shell nanostructure of Ni-NC@Ni with monodispersed Ni–N species [36]. When heated to a temperature of more than 160 °C, urea breaks down into carbon-containing molecules and ammonia. NiO was converted into metallic Ni nanoparticles due to this reductive atmosphere. Simultaneously, C and N compounds cover the surface of the metallic Ni and are catalyzed to form an N-rich carbon (NC) coating. This NC layer has many defect sites due to many defect sites within the NC layer. The strong Lewis acid–base interaction of Ni–N coordination caused the atoms of Ni on metallic Ni surface to migrate into the NC layer and occupy the defects within the layer [34]. This migration has been driven by the Ni–N coordination. During the transformation process, the NiO nanosheets serve as the self-template. The NC coating applied to the Ni nanoparticles functions as a binder to join the distinct core–shell nanoparticles. This results in the sheet-like shape of the Ni-NC@Ni material. Notably, the Ni particles could give a suitable Ni source to the NC framework that is closely covered, making it more straightforward for the carbon layer to collect more Ni atoms and, as a result, obtain high-loading SACs.

By CVD growth and subsequent pyrolysis, various x-CVD/Fe-N-C-kat catalysts with different Fe contents in ZnO substrates (x: Fe content in ZnO) have been created. Usually, Fe-doped ZnO substrates were made by combining $Fe(NO)_3 \cdot 9H_2O$ and $Zn(Ac)_2 \cdot 2H_2O$ using the sol–gel method [37]. However, creating ZnO templates does not constrain the scalability of the CVD synthesis. A range of tried-and-true techniques can quickly and easily create ZnO with tunable particle sizes and nanostructures. Then, two different alumina combustion boats were used in a furnace with two temperature zones to hold the created Fe–ZnO nanosheets and 2-Melm, respectively. Upstream and downstream zones are heated to temperatures 280 and 350 °C, respectively. As a result, the gaseous 2-Melm that has evaporated flows with argon and settles on Fe–ZnO nanosheets. According to the suggested reaction mechanism, the gas–solid interaction results in the creation of the intermediate $Fe–Zn(Melm)_2$:

$$2\text{Melm (g)} + \text{ZnO (s)} \rightarrow \text{Zn(Melm)}_2\text{ (s)} + \text{H}_2\text{O (g)}$$

The crystalline structure of the intermediates is greatly influenced by the downstream CVD growth temperature (Table 3.3).

Table 3.3 Comparison between electrodeposition and CVD for the development of thin films

Comparison	Chemical vapor deposition (CVD)	Electrodeposition
Method of deposition	Gas phase	Liquid phase
Substrate type	Heat-resistant	Conductive
Deposition rate	High	Low
Film thickness control	High	Low
Uniformity	Good	Good
Control over morphology	Limited	Good
Film composition control	High	Low
Adhesion strength	High	Moderate
Cost	High	Low
Scalability	High	Moderate
Process complexity	High	Low
Surface area coverage	High	Low
Deposited material purity	High	High
Materials depositioned	Inorganic materials	Metals and alloys
Control over film thickness	Good	Limited

References

1. Zhou Y, Fan HJ (2021) Progress and challenge of amorphous catalysts for electrochemical water splitting. ACS Mater Lett. https://doi.org/10.1021/acsmaterialslett.0c00502
2. Liu K, Zhong H, Meng F, Zhang X, Yan J, Jiang Q (2017) Recent advances in metal-nitrogen-carbon catalysts for electrochemical water splitting. Mater Chem Front. https://doi.org/10.1039/c7qm00119c
3. Thalji RK, Ahrendt KA, Bergman RG, Ellman JA (2001) Annulation of aromatic imines via directed C–H activation with Wilkinson's catalyst. J Am Chem Soc. https://doi.org/10.1021/ja016642j
4. Argyle MD, Bartholomew CH (2015) Heterogeneous catalyst deactivation and regeneration: a review. Catalysts. https://doi.org/10.3390/catal5010145
5. Cornils B (2004) Handbook of commercial catalysts. Heterogeneous catalysts. By Howard F. Rase. Angew Chem Int Ed. https://doi.org/10.1002/anie.200385146
6. Chen G, Wei C, Zhu Y, Huang H (2023) Emerging chemical driving force in electrocatalytic water splitting. EcoMat 5(2):1–11. https://doi.org/10.1002/eom2.12294
7. Appel AM, Helm ML (2014) Determining the overpotential for a molecular electrocatalyst. ACS Catal. https://doi.org/10.1021/cs401013v
8. ul Haq T, Pasha M, Tong Y, Mansour SA, Haik Y (2022) Au nanocluster coupling with Gd–Co_2B nanoflakes embedded in reduced TiO_2 nanosheets: seawater electrolysis at low cell voltage with high selectivity and corrosion resistance. Appl Catal B Environ. https://doi.org/10.1016/j.apcatb.2021.120836
9. ul Haq T, Haik Y (2021) S doped Cu_2O–CuO nanoneedles array: free standing oxygen evolution electrode with high efficiency and corrosion resistance for seawater splitting. Catal Today. https://doi.org/10.1016/j.cattod.2021.09.015
10. ul Haq T, Mansour SA, Munir A, Haik Y (2020) Gold-supported gadolinium doped CoB amorphous sheet: a new benchmark electrocatalyst for water oxidation with high turnover frequency. Adv Funct Mater 1910309, 1–11. https://doi.org/10.1002/adfm.201910309

11. ul Haq T, Haik Y, Hussain I, Rehman H, Al-Ansari TA (2020) Gd-doped Ni-oxychloride nanoclusters: new nanoscale electrocatalysts for high-performance water oxidation through surface and structural modification. https://doi.org/10.1021/acsami.0c17216
12. Liu J, Wang Z, Su K, Xv D, Zhao D, Li J, Tong H, Qian D, Yang C, Lu Z (2019) Self-supported hierarchical IrO_2@NiO nanoflake arrays as an efficient and durable catalyst for electrochemical oxygen evolution. ACS Appl Mater Interfaces 11(29):25854–25862. https://doi.org/10.1021/acsami.9b05785
13. Yang H, Driess M, Menezes PW (2021) Self-supported electrocatalysts for practical water electrolysis. Adv Energy Mater. https://doi.org/10.1002/aenm.202102074
14. ul Haq T, Bicer Y, Munir A, Mansour SA, Haik Y (2020) Surface assembling of highly interconnected and vertically aligned porous nanosheets of Gd−CoB on TiO_2 nanoflowers for durable methanol oxidation reaction. ChemCatChem **12**(13):3585–3597. https://doi.org/10.1002/cctc.202000392
15. Sun H, Yan Z, Liu F, Xu W, Cheng F, Chen J (2019) Self-supported transition-metal-based electrocatalysts for hydrogen and oxygen evolution. Adv Mater 1806326, 1–18. https://doi.org/10.1002/adma.201806326
16. Bernal Lopez M, Ustarroz J (2021) Electrodeposition of nanostructured catalysts for electrochemical energy conversion: current trends and innovative strategies. Curr Opin Electrochem. https://doi.org/10.1016/j.coelec.2021.100688
17. Stevens MB, Enman LJ, Batchellor AS, Cosby MR, Vise AE, Trang CDM, Boettcher SW (2017) Measurement techniques for the study of thin film heterogeneous water oxidation electrocatalysts. Chem Mater 29(1):120–140. https://doi.org/10.1021/acs.chemmater.6b02796
18. Mallik A, Ray BC (2011) Evolution of principle and practice of electrodeposited thin film: a review on effect of temperature and sonication. Int J Electrochem 2011:1–16. https://doi.org/10.4061/2011/568023
19. Gan Y, Wang C, Chen X, Liang P, Wan H, Liu X, Tan Q, Wu H, Rao H, Wang H et al (2020) High conductivity $Ni_{12}P_5$ nanowires as high-rate electrode material for battery-supercapacitor hybrid devices. Chem Eng J. https://doi.org/10.1016/j.cej.2019.123661
20. Kale MB, Borse RA, Gomaa Abdelkader Mohamed A, Wang Y (2021) Electrocatalysts by electrodeposition: recent advances, synthesis methods, and applications in energy conversion. Adv Funct Mater 31(25):1–24. https://doi.org/10.1002/adfm.202101313
21. Yan Z, Liu H, Hao Z, Yu M, Chen X, Chen J (2020) Electrodeposition of (hydro)oxides for an oxygen evolution electrode. Chem Sci. https://doi.org/10.1039/d0sc01532f
22. ul Haq T, Mansour S, Haik Y (2022) Electronic and structural modification of Mn_3O_4 nanosheets for selective and sustained seawater oxidation. ACS Appl Mater Interfaces. https://doi.org/10.1021/acsami.1c24304
23. Yan H, Yang Y, Fu Z, Yang B, Xia L, Fu S, Li F (2005) Fabrication of 2D and 3D ordered porous ZnO films using 3D opal templates by electrodeposition. Electrochem Commun. https://doi.org/10.1016/j.elecom.2005.08.011
24. Zhong B, Kuang P, Wang L, Yu J (2021) Hierarchical porous nickel supported $NiFeO_xH_y$ nanosheets for efficient and robust oxygen evolution electrocatalyst under industrial condition. Appl Catal B Environ. https://doi.org/10.1016/j.apcatb.2021.120668
25. Chen W, Du L, Wu C (2020) Hydrothermal synthesis of MOFs. In: Metal-organic frameworks for biomedical applications. https://doi.org/10.1016/B978-0-12-816984-1.00009-3
26. Machmudah S, Ceaser MR, Alwajdy MF, Widiyastuti W, Winardi S, Wahyudiono W, Kanda H, Goto M (2019) Hydrothermal and solvothermal synthesis of cerium-zirconium oxides for catalyst applications. Int J Technol. https://doi.org/10.14716/ijtech.v10i3.2930
27. Ye N, Yan T, Jiang Z, Wu W, Fang T (2018) A review: conventional and supercritical hydro/solvothermal synthesis of ultrafine particles as cathode in lithium battery. Ceram Int. https://doi.org/10.1016/j.ceramint.2017.12.236
28. Baloch HA, Siddiqui MTH, Nizamuddin S, Riaz S, Haris M, Mubarak NM, Griffin GJ, Srinivasan MP (2021) Effect of solvent on hydro-solvothermal co liquefaction of sugarcane bagasse and polyethylene for bio-oil production in ethanol-water system. Process Saf Environ Prot. https://doi.org/10.1016/j.psep.2021.02.015

29. Liu B, Peng HQ, Ho CN, Xue H, Wu S, Ng TW, Lee CS, Zhang W (2017) Mesoporous nanosheet networked hybrids of cobalt oxide and cobalt phosphate for efficient electrochemical and photoelectrochemical oxygen evolution. Small. https://doi.org/10.1002/smll.201701875
30. Wang X, Feng Z, Fan D, Fan F, Li C (2010) Shape-controlled synthesis of CdS nanostructures via a solvothermal method. Cryst Growth Des. https://doi.org/10.1021/cg101166t
31. Smith JD (1996) The chemistry of metal CVD. J Organomet Chem. https://doi.org/10.1016/0022-328x(96)06312-7
32. Vahlas C, Caussat B, Serp P, Angelopoulos GN (2006) Principles and applications of CVD powder technology. Mater Sci Eng R Rep. https://doi.org/10.1016/j.mser.2006.05.001
33. Gracia KC, Llanas-Cornejo D, Husi H (2017) CVD and oxidative stress. J Clin Med. https://doi.org/10.3390/jcm6020022
34. Yi K, Liu D, Chen X, Yang J, Wei D, Liu Y, Wei D (2021) Plasma-enhanced chemical vapor deposition of two-dimensional materials for applications. Acc Chem Res. https://doi.org/10.1021/acs.accounts.0c00757
35. Raiford JA, Oyakhire ST, Bent SF (2020) Applications of atomic layer deposition and chemical vapor deposition for perovskite solar cells. Energy Environ Sci. https://doi.org/10.1039/d0ee00385a
36. He Y, Li Y, Zhang J, Wang S, Huang D, Yang G, Yi X, Lin H, Han X, Hu W et al (2020) Low-temperature strategy toward Ni-NC@Ni core-shell nanostructure with single-Ni sites for efficient CO_2 electroreduction. Nano Energy. https://doi.org/10.1016/j.nanoen.2020.105010
37. Liu S, Wang M, Yang X, Shi Q, Qiao Z, Lucero M, Ma Q, More KL, Cullen DA, Feng Z et al (2020) Chemical vapor deposition for atomically dispersed and nitrogen coordinated single metal site catalysts. Angew Chem Int Ed. https://doi.org/10.1002/anie.202009331

Chapter 4
Electrochemical Methods for Measuring Water Splitting Efficiency

4.1 Electrochemical Methods

Electrocatalysts play a crucial role in a wide range of electrochemical processes, from fuel cells to water electrolysis to electrochemical sensors. Evaluating the performance of an electrocatalyst is a complex task that requires the measurement of multiple parameters, such as activity, selectivity, stability, and durability. To simplify this process, researchers often use figure of merit (FOM) metrics that combine multiple parameters into a single value. FOMs provide an overall assessment of the electrocatalyst's performance and enable researchers to compare the performance of different catalysts. In this regard, FOMs are an essential tool for the development and optimization of electrocatalysts for various applications. In this chapter, we will discuss different figure of merits in detail for electrocatalyst evaluation.

4.1.1 Cyclic Voltammogram

A cyclic voltammogram is a graph that depicts the changes in electrical current as a function of applied voltage in an electrochemically reactive system. Cyclic voltammetry (CV) is a frequently used electrochemistry technique that offers information about a compound's redox characteristics, the kinetics of electron transfer processes, and the stability of reaction intermediates. Cyclic voltammetry involves applying a voltage to an electrochemical cell and measuring the current that results. The potential is modulated throughout a predetermined range, and the current is measured as a function of the potential. The potential range can be scanned in either a positive or negative direction, or both (hence the name "cyclic"). A potentiostat is used to monitor the resulting current when the potential is scanned at a certain rate. A working electrode, a reference electrode, and a counter electrode are the three electrodes that make up an electrochemical cell. The working electrode is the one where

T. u. Haq and Y. Haik, *Electrochemical Water Splitting*, Materials Horizons: From Nature to Nanomaterials, https://doi.org/10.1007/978-981-99-9860-9_4

the electrochemical reaction occurs. The working electrode's potential is measured by the reference electrode, and the electrical circuit is completed by the counter electrode. The working electrode is typically made of a conducting material, such as platinum or gold, and can be modified with the chemical of interest. A saturated calomel electrode (SCE) or a silver/silver chloride electrode (Ag/AgCl) is commonly used as the reference electrode. In most cases, the counter electrode is comprised of an inert substance such as platinum or graphite [1].

An electrochemical reaction occurs at the electrode surface when a voltage is applied to the working electrode. This reaction could be an oxidation or reduction of the compound under investigation, or it could be a reaction between the compound and the solvent or other species in the solution. The rate of the electrochemical reaction is proportional to the current flowing through the electrochemical cell. The current changes as the voltage are scanned due to changes in the rate of the electrochemical reaction. Cyclic voltammetry is an effective technique for researching the electrochemical water-splitting reaction because it offers information on the reaction kinetics, mechanism, and electrode stability. The anode in electrochemical water splitting is usually constructed of an oxide or a mixed oxide material such as iridium oxide (IrOx) or ruthenium oxide (RuOx). Typically, the cathode is comprised of a platinum-based substance. The anode and cathode of an electrochemical cell used for water splitting are separated by an electrolyte solution. An acid or base is present in the electrolyte solution, which aids in proton transfer processes [2].

The cyclic voltammogram for electrochemical water splitting depicts current changes as a function of applied voltage. The potential is scanned from negative to positive, and the current is measured as a function of it. The anode's oxidation of water to create oxygen gas is shown in the first portion of the cyclic voltammogram. The anodic peak potential is proportional to the energy required to oxidize water and is normally approximately 1.23 V in comparison to the reversible hydrogen electrode (RHE), which is the thermodynamic potential necessary to split water. The shape and position of the anodic peak can reveal information about the reaction mechanism and the anode material's stability. Following the anodic peak, the current drops until it reaches a minimum current, which corresponds to the potential at which the anode is entirely oxidized (Fig. 4.1a). The rest potential is the section of the cyclic voltammogram that demonstrates the stability of the anode material in the oxidized state. If the rest potential is too low, the anode material may become unstable and oxidize further, resulting in electrode deterioration. The cyclic voltammogram's cathodic part depicts the reduction of protons at the cathode to produce hydrogen gas. The cathodic peak potential, which is proportional to the energy required for proton reduction, is typically around − 0.41 V versus RHE. The shape and position of the cathodic peak can reveal information about the reaction process and the cathode material's stability [3].

Following the cathodic peak, the current climbs until it reaches a maximum current, which corresponds to the voltage at which the cathode is totally reduced. The rest potential is a component of the cyclic voltammogram that demonstrates the stability of the cathode material in the reduced state. If the rest potential is too high, the cathode material may become unstable and further reduce, resulting in

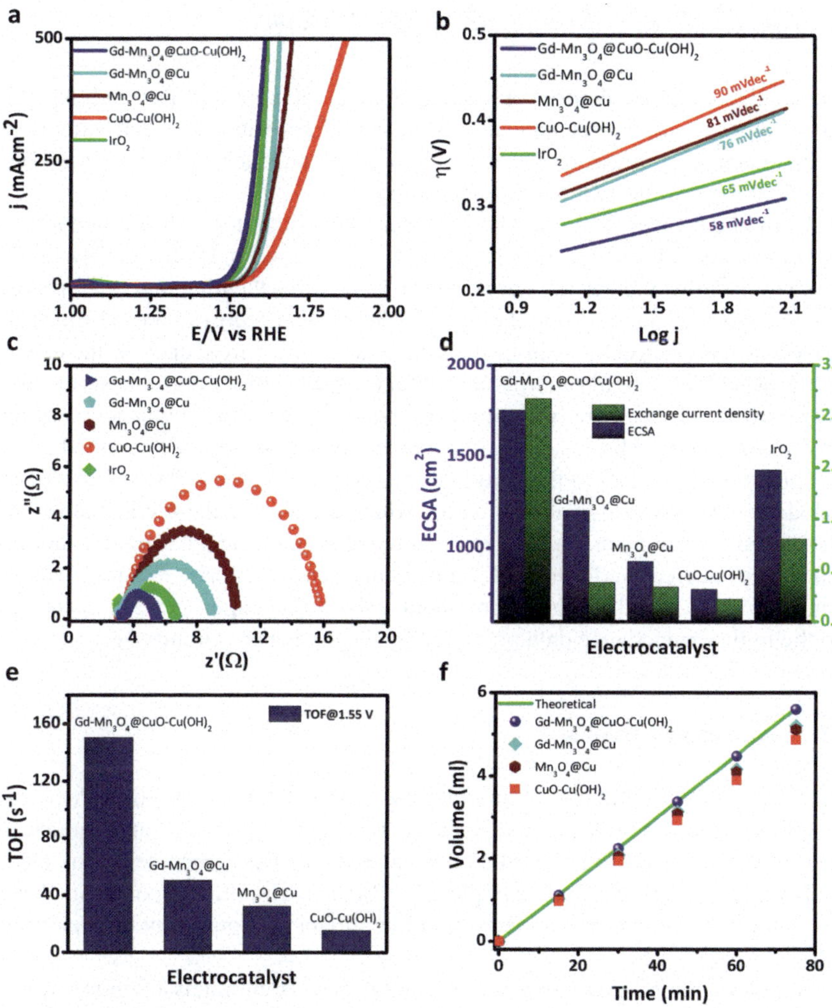

Fig. 4.1 **a** Cyclic voltammetry plot, **b** Tafel plots, **c** EIS, **d** ECSA and I_O, **e** TOF, and **f** Faradic efficiency recorded for comparative catalysts

electrode degradation. The cyclic voltammogram for electrochemical water splitting can be used to calculate reaction kinetic parameters such as exchange current density and Tafel slope. These characteristics can be utilized to optimize the performance of the electrochemical cell by providing information about the efficiency of the water-splitting reaction [3].

4.1.2 Electrochemical Impedance Spectroscopy

Impedance electrochemical spectroscopy (EIS) is a valuable tool for studying the electrical behavior of electrochemical systems. EIS evaluates an electrochemical cell's impedance as a function of frequency or time and offers information on the electrochemical reactions at the electrode surface.

An electrochemical system impedance is a complex quantity with real and imaginary components. The real component of the impedance indicates the system's resistance to the flow of electric current. At the same time, the imaginary part represents the system's reactance due to the storage of electrical energy in the system. The impedance of the electrochemical system as a function of frequency is measured in EIS by providing a small alternating voltage to the system and measuring the resulting current. The voltage is commonly sinusoidal, with the frequency varying across various values. The resulting impedance values are typically plotted on a Nyquist plot, which is a plot of the imaginary (Y-axis) versus the real (X-axis) part of the impedance (X-axis). Equivalent circuits, which are electrical circuits that represent the various components of the electrochemical system, can be used to examine the impedance data received from EIS. Equivalent circuits comprise resistors, capacitors, and inductors, and their values are obtained by fitting experimental impedance data to theoretical impedance data predicted by the equivalent circuit [4].

4.1.2.1 Equivalent Circuits

Equivalent circuits are utilized to simulate the electrochemical system's electrical activity. Equivalent circuits are made up of resistors, capacitors, and inductors arranged in various ways to reflect the physical and chemical processes at the electrode–electrolyte interface. For example, the Randles circuit, the constant phase element (CPE) circuit, and the transmission line model are equivalent circuits used to model EIS data (TLM).

The Randles circuit, also known as the Randles–Sevcik circuit, is a famous equivalent electrochemistry circuit for modeling electrode–electrolyte interface behavior. The Randles circuit comprises a resistor connected in series with a capacitor, miming the charge transfer resistance and the double-layer capacitance. The circuit was created in the 1950s by D. G. Randles and J. Sevcik.

The Randles circuit assumes that the electrochemical system is in a steady state, which means that the electrochemical reaction rate is equal to the rate of mass transit of reactants and products to and from the electrode surface. Under these conditions, the impedance of the electrochemical system can be expressed as [5]:

$$Z = \frac{R_{\mathrm{ct}}}{1 + \mathrm{i}\omega\tau_{\mathrm{d}} + R_{\mathrm{s}}}$$

where Z is the system's complex impedance, R_{ct} is the charge transfer resistance, is ω the applied angular frequency, d is the diffusion time constant, R_s is the solution resistance, and i is the imaginary unit. The charge transfer resistance represents the resistance of the electron transfer reaction at the electrode–electrolyte interface. It is proportional to the rate constant of the electrochemical reaction and the electrode's surface area. The capacitance, C_{dl}, is related to the double-layer capacitance, which results from charge separation at the interface due to ion adsorption on the electrode surface. The diffusion time constant, d, is proportional to the diffusion coefficient of the electrolyte's reactants and products and the distance between the electrode and the bulk electrolyte. A Nyquist plot, a plot of the imaginary component of the impedance (Y-axis) versus the real part of the impedance, can illustrate the Randles circuit graphically (X-axis). The Randles circuit's Nyquist plot typically shows a semicircular arc at high frequencies, corresponding to the charge transfer resistance, and a linear section at low frequencies, corresponding to the solution resistance. The Randles circuit is frequently used in electrochemical impedance spectroscopy (EIS) experiments to investigate the electrochemical behavior of various systems, including batteries, fuel cells, and corrosion processes. The Randles circuit can offer data on the electrochemical reaction's kinetics, the electrode–electrolyte interface's characteristics, and the reactant's and products' movement in the electrolyte [6].

The CPE circuit is a modified Randles circuit with a constant phase element (CPE) instead of a capacitor. A CPE's impedance varies with frequency, acting like a capacitor at low frequencies and a resistor at high frequencies. The CPE circuit simulates EIS data for more complicated electrochemical systems, such as those with surface roughness or heterogeneity. A constant phase element (CPE) is frequently used as a model in electrochemical impedance spectroscopy (EIS) to depict the behavior of specific interfaces or materials that do not behave like a simple capacitor. The CPE is a capacitor generalization that appropriately reflects various materials and systems. To depict the behavior of an electrochemical system, a CPE circuit is frequently used in conjunction with a resistor (R) [7].

The CPE circuit is represented by the following equation:

$$Z = (R(\mathrm{CPE}(\alpha, Q)))^{-1}$$

where Z is the system's complex impedance, R is its resistance, CPE(, Q) is the constant phase element, is the exponent, and Q is a constant related to the system's capacitance. The CPE circuit is frequently used to model the behavior of non-ideal capacitive materials such as porous electrodes, electrode–electrolyte interfaces, and biological membranes. The CPE circuit can also be used to model the behavior of fractal geometries or systems with various relaxation durations. In addition, the CPE circuit can provide useful information on a system's electrochemical behavior. For example, the exponent can give insight into the system's relaxation time distribution and can be related to the system's fractal dimension. The constant Q can provide information about the system's capacitance and can be connected to the surface area or roughness of the electrode.

The transmission line model (TLM) is frequently used in electrochemical impedance spectroscopy (EIS) to evaluate the behavior of planar electrode geometries. The TLM model comprises a sequence of resistors and capacitors that represent the electrode structure's different layers and the solution resistance and double-layer capacitance. The TLM model implies that the electrode structure comprises layers with varying resistances and capacitances. A simple resistor and capacitor may represent the solution resistance and double-layer capacitance in parallel. The TLM model is based on transmission line theory concepts and is frequently used to investigate the impedance of thin films, coatings, and other planar structures [8].

The equation commonly represents the TLM circuit:

$$Z = R_s + \frac{R_{dl}}{1 + j\omega R_{dl} C_{dl}} + \sum \left(\frac{R_i}{1 + j\omega R_i C_i} \right)$$

Z is the system's complex impedance, R_s is the solution resistance, R_{dl} is the double-layer resistance, C_{dl} is the double-layer capacitance, R_i is the ith layer's resistance, and C_i is the ith layer's capacitance. The total is applied to all layers of the system. The TLM model can provide useful information about the electrode structure's attributes, such as layer thickness and conductivity. The TLM model can also investigate the effect of various characteristics, such as temperature or electrolyte composition, on system impedance. The TLM model has been used in multiple electrochemical systems, such as batteries, fuel cells, sensors, and corrosion protection coatings. In addition, the TLM model is very effective for investigating the behavior of thin films and coatings, where the electrode structure's characteristics significantly influence the system's impedance.

A Nyquist plot is a graphical representation of the complex plane impedance of a system [9]. The x-axis represents the real component of the impedance, and the y-axis represents the imaginary part. The resulting plot is typically a semicircle or a loop. The plot's shape can reveal details about the physical processes taking place in the system. The type of electrochemical system determines the physical interpretation of Nyquist plots under consideration. The Randles circuit, which consists of a resistance in series with a constant phase element (CPE) and a charge transfer resistance (Fig. 4.1c), can be used to characterize the impedance of a simple electrochemical cell. The CPE represents the electrode's double-layer capacitance, which measures the electrode's charge storage capability. The resistance to charge transfer at the electrode/electrolyte contact is characterized by charge transfer resistance. The Nyquist plot of an electric double-layer capacitor (EDLC) usually comprises a semicircle in the high-frequency zone and a line in the low-frequency region (Fig. 4.1c). Because of the solution resistance and the contact resistance between the electrode and the collector, the semicircle indicates the system's resistive behavior. The line in the low-frequency region illustrates the capacitive behavior of the system as a result of the electrode's double-layer capacitance [10].

The double-layer capacitance measures the electrode's charge storage capacity and is proportional to the electrode's surface area. The line's intercept with the real axis offers an estimate of the electrode's double-layer capacitance. The Nyquist plot for a redox pair usually consists of a semicircle in the high-frequency region and a diagonal line in the low-frequency zone. The semicircle depicts the system's resistive behavior caused by solution resistance and charge transfer resistance at the electrode/electrolyte interface. The diagonal line illustrates the system's diffusion-limited behavior caused by redox species mass transport to the electrode surface. The slope of the diagonal line can be used to calculate the redox species' diffusion coefficient. The position of the semicircle in the plot can reveal details about the redox reaction's kinetics.

A semicircle in the high-frequency region and a diagonal line in the low-frequency region are common components of the Nyquist plot for water splitting. The semicircle depicts the system's resistive behavior caused by solution resistance and charge transfer resistance at the electrode/electrolyte interface. The diagonal line represents the system's diffusion-limited behavior caused by the mass transport of reactants and products to and from the electrode surface. The location of the semicircle in the plot can reveal information about the electrochemical reaction's kinetics. The charge transfer resistance at the electrode/electrolyte contact is proportional to the radius of the semicircle. A smaller semicircle radius means less charge transfer resistance and a more efficient electrochemical process. The diagonal line's slope can reveal information about the diffusion coefficients of the reactants and products. A steeper slope denotes a higher diffusion coefficient, which might result in more effective mass transfer [11].

4.1.3 Tafel Plots

When doing experiments in electrochemistry, electric currents are measured experimentally by applying voltage to the electrodes. Across the electrodes, the electric currents have a magnitude proportional to the reaction rate. As the rate of the electrocatalytic reaction depends on the potential, the rate constant similarly relies on the potential. Each elementary step forward and reverse reaction rates depend on the voltage, and electrocatalytic reactions are often composed of several elementary steps. The electrocatalytic reaction rate is dependent on the potential to function. This dependence is coupled with the potential-dependent coverage of the intermediate species, which is related to the rates at which these species are formed and consumed. In most cases, a Tafel analysis is used to compare the electrocatalytic activity of different electrocatalysts and to understand the reaction mechanism of various electrocatalysts. In this procedure, the sensitivity of the electric current response to the applied potential is evaluated, which offers information connected with the rate-determining stages. A Tafel plot is created by plotting the logarithm of the current density against the applied potential (Fig. 4.1b). The difference between the real potential and the thermodynamic potential gives rise to a curve that takes the form of a straight line with a slope that is proportionate to the overpotential. Calculating the exchange current density, which is the rate of electron transfer at the equilibrium potential, may be done with the help of the Tafel slope, which is an important parameter that represents the kinetics of the electrochemical reaction. Experiments can be performed to estimate the Tafel slope by measuring the current density as a function of the potential and then fitting the data to a linear equation. The rate of the reaction can be determined by the slope of the line, which is also proportional to the amount of energy required to activate the reaction. A common method for determining the exchange current density, also known as the rate of electron transfer at the equilibrium potential, is to make use of the slope of the Tafel plot [12].

To have a considerable magnitude of current density, applying a high overpotential (η) is necessary. This is because of the practical considerations involved. In general, having a lower overpotential (Z) in conjunction with a quicker increase in the associated current density is preferable. The well-known Butler–Volmer equation can be used to accurately describe both the current density I and the applied overpotential [13].

$$i = i_o\left[\exp\frac{\alpha_a nFE}{RT} + \exp\frac{\alpha_c nFE}{RT}\right]$$

According to the Butler–Volmer equation, when there is a high anodic overpotential, most of the current flowing through the system may be assigned to the anodic end. At the same time, the contribution from the cathodic part is almost nonexistent. Hence, the Butler–Volmer equation can be rewritten as Equation, also known as the Tafel equation [14].

$$i = i_o \exp\frac{\alpha_a nFE}{RT}$$

Exchange current density (i_0) and Tafel slope (b) can be determined by rewriting the above equation in logarithmic scale, which is the logarithm version of the Tafel equation.

From the equation, the Tafel slope (b) is defined as "how rapidly the current grows against overpotential (η)," with the value of the Tafel slope primarily dependent on the transfer coefficient [15].

$$\log(i) = \log(i_o) + \frac{\text{overpotential}}{\text{Tafel Slope}}$$

The equation demonstrated that if the Tafel slope (b) is small, then the current density can increase with a minor change in overpotential (Z) (i.e., a faster reaction rate constant), indicating good electrocatalytic kinetics. Additionally, the Tafel slope (b) can be used to get important and valuable insight into the reaction mechanism, particularly in explaining the rate-determining phase [16].

The equation that describes the relationship between the transfer coefficient (a) and the symmetry factor (b) in a process involving the transfer of a single electron reads as follows. As the overpotential is almost always substantially lower than the reorganization energy, the symmetry factor, *b*, is almost always equal to 0.50 in most cases. If this assumption is correct, then the Tafel slope for a single electron reaction will result in a 120 mV/dec[1] for the slope. This suggests that the step in the electrochemical system that determines the rate is regulated by the step that involves the transfer of a single electron [2].

$$\alpha = \frac{1}{2} + \frac{\text{overpotential}}{\text{re-organization energy}}$$

In many electrochemical systems, the situation is significantly more complex, and the systems typically consist of consecutive reaction stages. These processes may involve the transport of electrons or chemical events like association and dissociation. Equation, where n_b is the number of electrons that transfer back to the electrode before the rate-determining step, and *n* is the number of rate-determining steps in the overall reaction, shows the transfer coefficient obtained by Bockris and Reddy, for a multiple-electron reaction. The symbol n_r denotes the number of electrons involved in the rate-determining step. When the value of n_b is 1, electron transfer is the rate-determining step, whereas the value of n_b is 0, indicating that chemical reaction is the rate-determining phase [17].

$$\alpha_a = \frac{n_b}{v} + n_r\beta$$

The values of n_r and *b* are 1 and 0.5, respectively, and both n_b and *n* are equal to 0 if the first electron oxidation is the rate-determining step. According to the calculation, the transfer coefficient is 0.5, and the related Tafel slope is 120 mV dec^{-1} (resembling a single electron transfer reaction). The values of n_b and *n* are equal to 1, whereas the value of n_r is zero if the chemical reaction after the one-electron transfer reaction serves as the rate-determining step. As a result, the transfer coefficient value is one, and the Tafel slope becomes 60 mV dec^{-1}. While the third electron transfer step is the rate-determining step in some systems, such as OER, n_b, and *n* are equal to 2 and 1, respectively (n_r and *b*, are 0). This results in a transfer coefficient of 2, and the Tafel slope value is 30 mV dec^{-1}. As shown above, distinct Tafel slopes indicate different rate-determining phases. The transfer coefficient and Tafel slopes are strongly related to the number of electrons involved. For example, an excellent electrocatalyst has a lower Tafel slope, indicating that the rate-determining step is toward the conclusion of the multiple-electron transfer reaction [18].

Comparisons can be made between the Tafel slopes that have been physically seen and those that have been theoretically computed by presuming distinct rate-determining steps based on the model. Because its derivation is usually somewhat intricate, the surface coverage of the intermediate species is typically believed to be constant: 0 or 1, depending on which value is chosen. This simplification makes it easier for the electrochemist to evaluate the surface kinetics, and in many studies, Tafel slopes calculated by this method are used. This simplification reduces the number of variables that need to be considered. However, simplification leads to an imperfect representation of the actual surface dynamics, dependent on the coverage. This was previously noted, but it needs repeating: the coverage should vary with the applied potential. In addition, this assumption of the invariable range may be valid for steady-state conditions at constant potential and current conditions; nevertheless, the applicability of such an assumption for the Tafel analysis includes questionable accuracy due to its variable nature.

4.1.4 Exchange Current Density

The exchange current density is a fundamental parameter in electrochemistry that describes the rate of electrochemical reactions at an electrode surface. Specifically, it is the current density at which the rate of the forward and reverse electrochemical reactions is equal, meaning that there is no net current flow through the electrode at this point. The exchange current density is closely related to the kinetics of the electrochemical reaction. It is determined by the rate at which reactants come into contact with the electrode surface and undergo the reaction. The exchange current density can be affected by various factors, including the reactants' nature, the electrode surface's composition and structure, and the surrounding electrolyte solution. The exchange current density is often used to measure an electrode material's activity or efficiency for a particular electrochemical reaction. Higher exchange current densities generally indicate more active electrode surfaces that can promote faster electrochemical reactions. The exchange current density (i_0) can be determined experimentally using cyclic voltammetry, a technique commonly used in electrochemistry to study the kinetics of electrochemical reactions [19]. The basic steps for calculating the exchange current density are as follows:

Set up the experimental apparatus: The electrochemical cell consists of a working electrode (the electrode where the reaction occurs), a counter electrode, and a reference electrode. The working electrode is typically made of the material being studied, and it should have a large surface area to maximize the reaction rate. The counter electrode should be an inert material (e.g., platinum) that does not participate in the reaction. Finally, the reference electrode monitors the potential difference between the working and reference electrodes.

Measure the cyclic voltammogram: The cyclic voltammogram plots the current versus the applied voltage. The voltage is swept between two limits, typically starting at a negative potential and ending at a positive potential. During the first sweep, the working electrode potential is changed linearly with respect to time, and the resulting current is recorded. The voltage is swept back to the starting potential during the second sweep, and the current is recorded again.

Extract the peak current density: The peak current density is the maximum current observed in the cyclic voltammogram. This current is due to the electrochemical reaction at the electrode surface and is proportional to the rate of the reaction.

Determine the scan rate: The scan rate is the rate at which the potential changes during the cyclic voltammetry experiment. It is typically expressed in units of V/s.

Calculate the exchange current density: The exchange current density can be calculated using the following equation:

$$i_o = \left(2.69 \times 10^5\right) * A * D^{\frac{1}{2}} * n^{\frac{3}{2}} * c^{1/2}$$

where A is the surface area of the working electrode, D is the diffusion coefficient of the reactant, n is the number of electrons transferred in the reaction, and c is the concentration of the reactant. In practice, the calculation of the exchange current density can be simplified by assuming that the diffusion coefficient, number of electrons transferred, and concentration of the reactant is constant over the potential range of the cyclic voltammogram. Under these conditions, the exchange current density can be calculated as follows [20]:

$$i_o = \frac{(0.446 * n * F * v)}{A * R}$$

where F is the Faraday constant, v is the sweep rate, and R is the gas constant. This simplified equation is known as the Tafel equation.

Overall, the current exchange density is an essential parameter in electrochemical water-splitting measurement because it provides information about the kinetics and efficiency of the electrochemical reaction. By understanding and optimizing this parameter, researchers can develop more effective and efficient water-splitting systems for various applications, including renewable energy generation and storage.

4.1.5 Turnover Number and Turnover Frequency

Turnover number (TON) is a measure of the catalytic efficiency of a catalyst, which tells us how many times a single molecule of catalyst can react with a substrate to form a product. It is defined as the number of moles of substrate that are converted to product per mole of catalyst. TON is an important parameter in catalysis because it allows us to compare the efficiency of different catalysts for a given reaction. To calculate TON, we first need to know how much product is formed and how much catalyst is used. We can then use the following formula [21]:

$$\text{TON} = (\text{moles of product formed})/(\text{moles of catalyst used})$$

For example, suppose we have a reaction in which 1 mol of the substrate is converted to 1 mol of product using 0.1 mol of catalyst. If we run the reaction multiple times and find that a total of 10 mol of the product are formed, then the TON would be:

$$\text{TON} = (10\,\text{mol of product})/(0.1\,\text{mol of catalyst})$$

$$\text{TON} = 100$$

This means that for every molecule of catalyst used, 100 molecules of the product are formed. In other words, a single molecule of catalyst can catalyze the conversion of 100 molecules of the substrate to a product. TON can be used to compare the efficiency of different catalysts for the same reaction. The higher the TON, the more efficient the catalyst is because it can convert more substrate to the product with the same amount of catalyst. Turnover frequency (TOF) is the number of moles of product formed per mole of catalyst per unit of time. It is calculated as follows:

$$\text{TOF} = (\text{moles of product formed})/(\text{moles of catalyst used}) \times (\text{time})$$

TOF is a measure of how quickly the catalyst can convert substrate to product. It takes into account the amount of catalyst used and the time it takes to produce the product. The higher the TOF, the faster the reaction rate and the more efficient the catalyst. The formula to calculate TOF for OER and HER is as follows [22]:

$$\text{TOF} = \frac{i \times N_{\text{A}}}{A \times F \times n \times r}$$

where

i Current in Ampere
N_{A} Avogadro number
A Geometrical surface area of the electrode
F Faraday constant
n Number of electrons
r Surface concentration of atoms.

4.1.6 *Faradic Efficiency*

Faradic efficiency measures an electrochemical process's ability to transform electrical energy into a desired chemical product. The ratio of the amount of a desired chemical product produced by an electrochemical process to the amount of electrical charge passed through the system throughout the process is defined. Faradic efficiency is an essential metric in electrochemical engineering because it quantifies an electrochemical process's efficiency and may be used to modify process parameters and enhance process economics. Faraday's law of electrolysis describes the link between the quantity of electrical charge flowing through an electrochemical system and the amount of chemical product produced. According to this law, the amount of a substance created at an electrode is directly proportional to the amount of electrical charge transferred through the system. The proportionality constant is the Faraday constant, which is equal to the charge on one mole of electrons (96,485 C/mol). As a result, the quantity of chemical product created (in moles) is given by [23]:

$$\text{Moles of product} = \text{electrical charge (in Coulombs)}/\text{Faraday constant}$$

Faradic efficiency is therefore defined as the ratio of actual moles of product produced to theoretical moles created based on the quantity of electrical charge carried through the system. Faradic efficiency can be stated mathematically as follows:

$$\text{Faradic efficiency (\%)} = (\text{actual moles produced}/\text{theoretical moles created}) \times 100$$

A Faradic efficiency of 100% means that all the electrical charge supplied through the system was employed to create the required chemical product. A Faradic efficiency of less than 100% implies that some electrical charges were lost due to side reactions or other inefficiencies in the electrochemical process. Faradic efficiency is a critical parameter in electrochemical water splitting, as it determines the amount of hydrogen produced for electrical energy. Several factors can affect the Faradic efficiency of electrochemical water splitting, including:

1. *Electrolyte composition*: The electrolyte composition used in the electrochemical water-splitting process can significantly impact the Faradic efficiency. The presence of impurities or contaminants in the electrolyte can lead to side reactions and reduce the overall efficiency of the process. The composition of the electrolyte can affect the rate of the electrochemical reaction. For example, increasing the electrolyte's concentration can increase the solution's ionic conductivity and promote faster ion transport, leading to faster reaction rates and higher Faradic efficiency.
2. *Electrode material*: The choice of electrode materials is another critical factor that can significantly affect the Faradic efficiency of electrochemical reactions. The electrode material should have good catalytic activity for the reaction of interest and be stable under the reaction conditions. Some of how electrode materials can affect the Faradic efficiency are:
3. *Catalytic activity*: The electrocatalytic activity of the electrode material is a critical factor that can influence the Faradic efficiency. The electrode material should have a high catalytic activity for the reaction of interest, such as hydrogen evolution reaction (HER) or oxygen evolution reaction (OER), to minimize the energy input required to drive the reaction. Different materials have different catalytic activity for these reactions, and the choice of the electrode material can significantly affect the Faradic efficiency.
4. *Surface area*: The surface area of the electrode material can also affect the Faradic efficiency. Increasing the surface area of the electrode can provide more active sites for the reaction and enhance the overall reaction rate, leading to higher Faradic efficiency.

5. *Stability*: The electrode material should be stable under the reaction conditions to avoid degradation or corrosion, which can reduce the catalytic activity and the Faradic efficiency. The stability of the electrode material can depend on the composition of the electrolyte, the pH, temperature, and other factors.
6. *Selectivity*: The selectivity of the electrode material for the reaction of interest is also an essential factor that can influence the Faradic efficiency. For example, if the electrode material has poor selectivity, it may catalyze undesirable side reactions, reducing Faradic efficiency.

Morphology: The morphology of the electrode material can also influence the Faradic efficiency. For example, some materials may form nanoparticles or nanowires that can enhance the catalytic activity by providing a large surface area and efficient charge transport.

7. *Temperature*: The temperature of the electrolyte can also affect the Faradic efficiency. Higher temperatures can lead to increased reaction rates and higher Faradic efficiency, but the temperature should be controlled within a specific range to avoid side reactions and electrode degradation. Temperature can affect the Faradic efficiency in several ways, including:
8. *Reaction rate*: Temperature affects the reaction rate by increasing the kinetic energy of the reacting species. This can increase the Faradic efficiency as the reaction can proceed more quickly and effectively.
9. *Activation energy*: The activation energy required for a reaction to occur decreases with increasing temperature. This means that a lower amount of energy is needed to initiate the response, which can result in higher Faradic efficiency.
10. *Electrode potential*: The electrode potential can be affected by temperature, which can, in turn, affect the Faradic efficiency. For example, at higher temperatures, the electrode's potential may shift, leading to changes in the reaction mechanism and overall Faradic efficiency.
11. *Side reactions*: Temperature can also influence the occurrence of side reactions, which can reduce the Faradic efficiency. At higher temperatures, side reactions may become more prevalent, leading to a decrease in the overall efficiency of the reaction.
12. *pH*: The pH of the electrolyte can also affect the Faradic efficiency. The optimum pH for electrochemical water splitting depends on the electrode materials and can vary from system to system.

Reaction mechanism: The pH of the solution can alter the reaction mechanism and the rate of the electrochemical reaction, thereby affecting Faradic efficiency. For example, at a low pH, protonation or deprotonation of the reactants or intermediates may occur, leading to a change in the reaction pathway and a decrease in Faradic efficiency.

Electrode potential: The electrode potential can be influenced by the pH of the solution. The electrode's potential can shift with changes in pH, leading to a change in the reaction mechanism and thus affecting the Faradic efficiency.

The concentration of ions: The concentration of ions in the solution can affect the Faradic efficiency. At a low pH, the concentration of H^+ ions are high, which can lead to competition for electrons, resulting in lower Faradic efficiency. Conversely, at high pH, OH^- ions can compete for the same reaction sites, reducing Faradic efficiency.

Passivation: The pH can affect the formation of passivating layers on the electrode surface, which can interfere with the electrochemical reaction and result in lower Faradic efficiency.

The following formulas are used to calculate the Faradic efficiency of OER and HER [24].

$$\text{Faradaic efficiency} = V_{\text{experimental}} / V_{\text{Theoretical}}$$

While

$$V_{\text{Theoretical for } O_2} = 1/4 * Q/F * V_m$$

The number 1 means 1 mol of O_2 per mole of H_2O and 4 means 4 mol of electrons per mole of H_2O.

$$V_{\text{Theoretical for } H_2} = 1/2 * Q/F * V_m$$

The number 1 means 1 mol of O_2 per mole of H_2O and 2 means 2 mol of electrons per mole of H_2O.

Q It (amount of charged passed through electrode)
F Faraday constant (96,485 C mol^{-1})
V_m Molar volume of gas (24.1 L mol^{-1}, 293 K, 101 kPa).

4.1.7 Chronoamperometry and Chronopotentiometry

Two electrochemical techniques often used to analyze electrode stability, and kinetics are chronoamperometry and chronopotentiometry. These methods are widely employed in various domains, such as electrocatalysis, electrochemical energy storage, and corrosion research.

4.1.7.1 Chronopotentiometry

Chronopotentiometry is an electrochemical technique for studying the kinetics of electrochemical reactions. It entails passing a continuous current through an electrochemical cell and measuring the resulting potential as time passes. Understanding the chemistry of chronopotentiometry requires an examination of the electrochemical reactions that occur within the cell. The electrode and the electrolyte solution interface are at the electrochemical cell's heart. When a current is delivered to an electrode, electrons move between the electrode and the electrolyte. This electron flow might result in the creation or consumption of ions at the electrode surface. Ion creation or consumption at the electrode surface might result in an ion concentration gradient near the electrode. A potential difference between the electrode and the electrolyte can result from this concentration gradient. The electrode potential is the difference in potential. A constant current is delivered to the electrode in chronopotentiometry, and the possibility of the electrode is measured as a function of time—the potential changes as the delivered current changes the concentration of ions at the electrode surface. The kinetics of the electrochemical reaction will determine the rate of change of the potential. It is possible to learn about the kinetics of an electrochemical process by examining variations in potential over time. The reaction rate, for example, can be calculated by measuring the time it takes for the potential to reach a specific value. Furthermore, the prospect can quantify the reaction rate by determining the concentration of ions at the electrode surface [25].

4.1.7.2 Chronoamperometry

In chronoamperometry, a constant potential is applied to an electrode and the resulting current is measured as a function of time. By analyzing the changes in current over time, information about the reaction kinetics can be obtained. During chronoamperometry, the current response can be characterized by two distinct stages. The initial stage is known as the transient period, during which the current changes rapidly due to changes in concentration of the electroactive species at the electrode surface. The second stage is known as the steady-state period, during which the current stabilizes and remains constant over time [26].

Both chronoamperometry and chronopotentiometry can provide helpful information about a system's electrochemical behavior. The specific experimental setting and the information sought to determine the decision between the two strategies. In general, chronoamperometry is used to research electrode reactions, including electron transfer reactions, whereas chronopotentiometry studies surface processes such as adsorption and desorption.

4.1.8 Corrosion Experiment

Corrosion is a natural phenomenon that occurs when metals react with the environment. It can cause severe structural and mechanical damage, posing safety risks and financial losses. Electrochemistry is a potent tool for investigating corrosion causes and devising solutions to prevent or reduce the impacts of corrosion. Therefore, we will utilize electrochemical techniques to assess the corrosion resistance of various metal samples in this experiment.

4.1.8.1 Setup for the Experiment

A working electrode (the metal sample being tested), a reference electrode (typically a saturated calomel electrode), and a counter electrode will be used in the experiment (usually a platinum wire). The working electrode will be immersed in an electrolyte solution containing chloride ions to imitate a corrosive environment. A potentiostat will be used to adjust the working electrode's potential, and the resulting current will be monitored with an ammeter.

4.1.8.2 Preparation of the Sample

The metal samples will be cleaned and polished to remove any surface impurities. To maintain consistency, the samples should be of similar size and shape. Throughout the experiment, the surface area of each sample should be measured to compute the current density. Then, using a non-conductive glue such as epoxy, the samples will be attached to the working electrode holder.

4.1.8.3 Solution of Electrolytes

This experiment's electrolyte solution will contain chloride ions, a common cause of corrosion. Therefore, the solution should be made with deionized water and high-purity ingredients to avoid contamination. The concentration of chloride ions can be changed to imitate different corrosive conditions. Procedure for conducting experiments is as follows:

1. Soak the working electrode (metal sample) in the electrolyte solution for one minute.
2. Attach the potentiostat to the reference and counter electrodes.
3. Adjust the working electrode's potential so that no appreciable current is noticed (open circuit potential).
4. Wait for the system to attain a stable state.
5. Take note of the open circuit potential and current density.
6. Do a potential scan on the working electrode, typically from open circuit to higher positive potential. This will produce an anodic current, a measure of the metal dissolving rate.
7. Note the resulting current density as a function of time or potential.
8. Steps 6 and 7 should be repeated for different potential scan rates and chloride ion concentrations.

The experiment's data can be evaluated using various electrochemical techniques, including polarization curves and electrochemical impedance spectroscopy (EIS). A corrosion experiment's potentiodynamic polarization curve reveals information about the electrochemical reactions taking place at the metal's surface. Sweeping the working electrode's potential while measuring the current response yields the curve. The curve is often divided into three separate regions: anodic, cathodic, and passive. The voltage of the working electrode is swept in the positive direction at the anodic area, and the current response grows fast due to the oxidation of the metal at the surface. The following is a representation of the anodic reaction:

$$M \rightarrow M^{+} + e^{-}$$

where M represents the metal being oxidized and M^+ represents the corresponding metal ion. As the potential of the working electrode is swept further in the positive direction, the current response begins to level off in the passive region. This occurs when a protective oxide layer forms on the surface of the metal, which inhibits further oxidation. The passive region is characterized by a low current response, indicating that the metal is no longer actively corroding. In the cathodic region, the potential of the working electrode is swept in the negative direction, and the current response increases due to the reduction of species present in the electrolyte solution. The cathodic reaction can be represented as follows:

$$O_2 + 2H_2O + 4e^{-} \rightarrow 4OH^{-}$$

where O_2 represents dissolved oxygen, H_2O represents water, and OH^- represents hydroxide ions. The point at which the anodic and cathodic currents intersect on the potentiodynamic polarization curve is known as the corrosion potential [27]. At this point, the rate of oxidation of the metal is equal to the rate of reduction of species in the electrolyte solution. The difference between the corrosion potential and the open circuit potential (OCP) is known as the corrosion potential shift, and it provides information about the tendency of the metal to corrode [28].

EIS is a technique that measures the impedance of a system as a function of frequency. It can provide information about the resistance of the metal surface to corrosion, as well as the capacitance and resistance of the electrolyte solution. EIS data can be analyzed using equivalent circuit models to extract quantitative parameters, such as the corrosion rate and the polarization resistance. The corrosion resistance of different metals can vary widely depending on their composition and surface properties. For example, stainless steel is highly resistant to corrosion due to the presence of chromium oxide on the surface, which forms a protective layer. Copper, on the other hand, is susceptible to corrosion due to its high reactivity with chloride ions. The results of this experiment can be used to compare the corrosion resistance of different metals and to identify factors that influence the rate of metal dissolution. The data can also be used to evaluate the effectiveness of corrosion inhibitors or coatings.

4.1.9 Electrochemical Active Surface Area

The surface area of an electrode that is available for electrochemical reactions is referred to as its electrochemical active surface area (ECSA). The ECSA is critical in electrochemistry because it controls how much charge may be transferred between an electrode and an electrolyte during an electrochemical reaction. Several techniques, including cyclic voltammetry, electrochemical impedance spectroscopy, and charge–discharge experiments, can be used to calculate an electrode's ECSA. We will cover the significance of ECSA in electrochemistry, the many methodologies used to quantify ECSA, and the factors that influence ECSA in this study. The amount of surface area accessible for the response on the electrode determines the electrochemical reactions at an electrode. As a result, an electrode's ECSA is a significant metric in electrochemistry. During an electrochemical reaction, the ECSA is directly proportional to the amount of charge that may be transferred between the electrode and the electrolyte [29]. This indicates that a larger ECSA electrode can transfer more charge and, as a result, has higher electrochemical activity than a smaller ECSA electrode. The electrochemical reaction's kinetics are also affected by an electrode's ECSA. The electrochemical reaction rate is proportional to the electrode's ECSA. This means that a larger ECSA electrode has a faster reaction rate than a smaller ECSA electrode. Understanding and managing an electrode's ECSA is critical for optimizing electrochemical reactions.

ECSA can be measured using a variety of approaches, including cyclic voltammetry (CV), electrochemical impedance spectroscopy (EIS), and charge–discharge measurements, the most often employed techniques.

Cyclic voltammetry is a technique that includes applying a voltage ramp to an electrode and measuring the current that results. Typically, the voltage ramp is used between a positive and negative potential limit. The electrode conducts an oxidation reaction when the voltage increases and a reduction reaction when the voltage is dropped. The resultant current is proportional to the electrochemical reaction rate and is used to calculate the electrode's ECSA. The Randles–Sevcik equation, which connects peak current to ECSA, can be used to compute an electrode's ECSA:

$$i_{\mathrm{p}} = \left(2.69 \times 10^5\right) n^{3/2} A D^{1/2} C^{1/2} \upsilon^{1/2}$$

where i_{p} is the peak current, n is the number of electrons transferred in the electrochemical reaction, A is the EASA, D is the diffusion coefficient of the redox species, C is the concentration of the redox species, and v is the scan rate.

To determine the ECSA from EIS, one can follow the following steps:

- First, measure the EIS data for the electrode of interest in a suitable electrolyte solution.
- Analyze the EIS data to obtain the electrochemical impedance, Z, which is a complex number representing the electrode–electrolyte interface's resistance and capacitance.

- Fit the EIS data to an appropriate equivalent circuit model with double-layer capacitance and charge transfer resistance. The double-layer capacitance represents the capacitive behavior of the electrode–electrolyte interface, while the charge transfer resistance represents the resistance to charge transfer during the electrochemical reaction.
- Calculate the electrochemical active surface area (ECSA) using the following equation:

$$\mathrm{ECSA} = C_{\mathrm{dl}} * A * \frac{F}{n * R * T}$$

where C_{dl} is the double-layer capacitance, A is the geometric surface area of the electrode, F is Faraday's constant, n is the number of electrons transferred in the electrochemical reaction, R is the gas constant, and T is the temperature. Note that this equation assumes that the electrochemical reaction is under quasi-reversible conditions and that the double-layer capacitance is purely capacitive. Additionally, the ECSA calculated from EIS may overestimate the true ECSA, as the electrode surface may not be fully accessible to the electrolyte solution.

To determine ECSA from charge–discharge measurements, one needs to follow these steps:

- Prepare an electrochemical cell with the electrode whose ECSA is to be determined, a reference electrode, and a counter electrode.
- Charge the electrode to a specific potential in the cell, ensuring that the potential is high enough to completely oxidize any adsorbed species on the electrode surface.
- Discharge the electrode to a potential where reduction of the adsorbed species on the electrode surface is complete.
- Calculate the total charge passed during the charge–discharge cycle.
- Determine the double-layer capacitance of the electrode from the charging portion of the curve using cyclic voltammetry or electrochemical impedance spectroscopy.
- Calculate the ECSA of the electrode using the following equation:

$$\mathrm{ECSA} = \frac{Q}{2FC_{\mathrm{dl}}}$$

where Q is the charge passed during the charge–discharge cycle, F is the Faraday constant (96,485 C/mol), and C_{dl} is the double-layer capacitance of the electrode. Note that this method assumes that the electrode has a flat surface and that the adsorbed species on the surface are completely oxidized or reduced during the charge–discharge cycle. The method may not be suitable for electrodes with complex surface structures or for systems where adsorbed species are not fully removed during the charge–discharge cycle.

Figure 4.1 displays a compilation of various electrochemical techniques employed for the analysis of Gd–Mn_3O_4@CuO–$Cu(OH)_2$ and other catalysts (Table 4.1).

Table 4.1 Summary of the features of an efficient catalyst in terms of various parameters

Parameter	Description	Features of an efficient catalyst
Overpotential	The amount of extra energy required for a reaction to occur in the presence of a catalyst	Low overpotential indicates efficient catalytic activity
Electrochemical impedance spectroscopy (EIS)	A technique for measuring the electrochemical behavior of a catalyst	Efficient catalysts exhibit low impedance values in EIS experiments
Tafel slope	A measure of the rate at which current density changes with the applied voltage	Low Tafel slope indicates efficient catalytic activity
Exchange current density	The rate of the forward and reverse reactions when the net current is zero	High exchange current density indicates efficient catalytic activity
Turnover frequency (TOF)	The number of reaction events occurring at the active site per unit time	High TOF indicates efficient catalytic activity
Turnover number (TON)	The number of reaction events per catalyst site	High TON indicates efficient catalytic activity
Faradic efficiency	The fraction of electrons participating in the desired reaction	High Faradic efficiency indicates efficient catalytic activity
Mass activity	The catalytic activity per unit mass of the catalyst	High mass activity indicates efficient catalytic activity
Corrosion resistance	Ability of the catalyst to withstand corrosion in harsh reaction environments	High corrosion resistance indicates efficient catalytic activity
Electrochemically active surface area (ECSA)	The surface area of the catalyst that is active for catalytic reactions	High ECSA indicates efficient catalytic activity
Chronopotentiometry	A technique for measuring the electrochemical behavior of a catalyst	Efficient catalysts exhibit stable and reversible behavior during chronopotentiometry

References

1. Wei C, Rao RR, Peng J, Huang B, Stephens IEL, Risch M, Xu ZJ, Shao-Horn Y (2019) Recommended practices and benchmark activity for hydrogen and oxygen electrocatalysis in water splitting and fuel cells. Adv Mater 31(31):1–24. https://doi.org/10.1002/adma.201806296
2. Suen NT, Hung SF, Quan Q, Zhang N, Xu YJ, Chen HM (2017) Electrocatalysis for the oxygen evolution reaction: recent development and future perspectives. Chem Soc Rev 46(2):337–365. https://doi.org/10.1039/c6cs00328a
3. Elgrishi N, Rountree KJ, McCarthy BD, Rountree ES, Eisenhart TT, Dempsey JL (2018) A practical beginner's guide to cyclic voltammetry. J Chem Educ. https://doi.org/10.1021/acs.jchemed.7b00361

4. Srinivasan R, Fasmin F (2021) An introduction to electrochemical impedance spectroscopy. https://doi.org/10.1201/9781003127932
5. Choi W, Shin HC, Kim JM, Choi JY, Yoon WS (2020) Modeling and applications of electrochemical impedance spectroscopy (EIS) for lithium-ion batteries. J Electrochem Sci Technol. https://doi.org/10.33961/jecst.2019.00528
6. Alavi SMM, Mahdi A, Payne SJ, Howey DA (2017) Identifiability of generalized Randles circuit models. IEEE Trans Control Syst Technol. https://doi.org/10.1109/TCST.2016.2635582
7. Pandey S, Kumar D, Parkash O, Pandey L (2017) Equivalent circuit models using CPE for impedance spectroscopy of electronic ceramics. Integr Ferroelectr. https://doi.org/10.1080/10584587.2017.1376984
8. Sadiku MNO, Agba LC (1990) A simple introduction to the transmission-line modeling. IEEE Trans Circuits Syst. https://doi.org/10.1109/31.56072
9. ul Haq T, Haik Y (2021) S doped Cu_2O–CuO nanoneedles array: free standing oxygen evolution electrode with high efficiency and corrosion resistance for seawater splitting. Catal Today. https://doi.org/10.1016/j.cattod.2021.09.015
10. Lei C, Markoulidis F, Ashitaka Z, Lekakou C (2013) Reduction of porous carbon/Al contact resistance for an electric double-layer capacitor (EDLC). Electrochim Acta. https://doi.org/10.1016/j.electacta.2012.12.092
11. Anantharaj S, Noda S (2020) Appropriate use of electrochemical impedance spectroscopy in water splitting electrocatalysis. ChemElectroChem. https://doi.org/10.1002/celc.202000515
12. Khadke P, Tichter T, Boettcher T, Muench F, Ensinger W, Roth C (2021) A simple and effective method for the accurate extraction of kinetic parameters using differential Tafel plots. Sci Rep. https://doi.org/10.1038/s41598-021-87951-z
13. Dickinson EJF, Wain AJ (2020) The Butler–Volmer equation in electrochemical theory: origins, value, and practical application. J Electroanal Chem. https://doi.org/10.1016/j.jelechem.2020.114145
14. Petrii OA, Nazmutdinov RR, Bronshtein MD, Tsirlina GA (2007) Life of the Tafel equation: current understanding and prospects for the second century. Electrochim Acta. https://doi.org/10.1016/j.electacta.2006.10.014
15. Aoki K (2004) An interpretation of small values of the transfer coefficient at conducting polymers. J Electroanal Chem. https://doi.org/10.1016/j.jelechem.2004.02.019
16. ul Haq T, Haik Y, Hussain I, Rehman H, Al-Ansari TA (2020) Gd-doped Ni-oxychloride nanoclusters: new nanoscale electrocatalysts for high-performance water oxidation through surface and structural modification. ACS Appl Mater. https://doi.org/10.1021/acsami.0c17216
17. Bockris JOM, Otagawa T (1983) Mechanism of oxygen evolution on perovskites. J Phys Chem. https://doi.org/10.1021/j100238a048
18. Li Y, Sun Y, Qin Y, Zhang W, Wang L, Luo M, Yang H, Guo S (2020) Recent advances on water-splitting electrocatalysis mediated by noble-metal-based nanostructured materials. Adv Energy Mater 10(11):1–20. https://doi.org/10.1002/aenm.201903120
19. Tang D, Lu J, Zhuang L, Liu P (2010) Calculations of the exchange current density for hydrogen electrode reactions: a short review and a new equation. J Electroanal Chem. https://doi.org/10.1016/j.jelechem.2009.11.031
20. ul Haq T, Pasha M, Tong Y, Mansour SA, Haik Y (2021) Au nanocluster coupling with Gd–Co_2B nanoflakes embedded in reduced TiO_2 nanosheets: seawater electrolysis at low cell voltage with high selectivity and corrosion resistance. Appl Catal B Environ 2021(301):120836. https://doi.org/10.1016/j.apcatb.2021.120836
21. Polshettiwar V, Varma RS (2009) Nanoparticle-supported and magnetically recoverable palladium (Pd) catalyst: a selective and sustainable oxidation protocol with high turnover number. Org Biomol Chem. https://doi.org/10.1039/b817669h
22. Ardagh MA, Abdelrahman OA, Dauenhauer PJ (2019) Principles of dynamic heterogeneous catalysis: surface resonance and turnover frequency response. ACS Catal. https://doi.org/10.1021/acscatal.9b01606
23. Lv W, Zhang R, Gao P, Lei L (2014) Studies on the Faradaic efficiency for electrochemical reduction of carbon dioxide to formate on tin electrode. J Power Sources. https://doi.org/10.1016/j.jpowsour.2013.12.063

24. ul Haq T, Mansour S, Haik Y (2022) Electronic and structural modification of Mn_3O_4 nanosheets for selective and sustained seawater oxidation. ACS Appl Mater Interfaces. https://doi.org/10.1021/acsami.1c24304
25. Harris AR, Newbold C, Carter P, Cowan R, Wallace GG (2019) Using chronopotentiometry to better characterize the charge injection mechanisms of platinum electrodes used in bionic devices. Front Neurosci. https://doi.org/10.3389/fnins.2019.00380
26. Lingane PJ (1964) Chronopotentiometry and chronoamperometry with unshielded planar electrodes. Anal Chem. https://doi.org/10.1021/ac60215a010
27. Wang HR, Hou BR, Wang J, Wang Q, Li WY (2008) Effect of process conditions on microstructure and corrosion resistance of cold-sprayed Ti coatings. J Therm Spray Technol. https://doi.org/10.1007/s11666-008-9256-3
28. Doménech-Carbó A, Peiró-Ronda MA, Vives-Ferrándiz J, Duffó GS, Farina S (2021) 'Dry' electrochemistry: a non-invasive approach to the characterization of archaeological iron objects. Electrochem Commun. https://doi.org/10.1016/j.elecom.2021.106992
29. Cossar E, Houache MSE, Zhang Z, Baranova EA (2020) Comparison of electrochemical active surface area methods for various nickel nanostructures. J Electroanal Chem. https://doi.org/10.1016/j.jelechem.2020.114246

Chapter 5
Best Practices for Accurately Reporting Electrocatalytic Performance of Nanomaterials

5.1 Introduction

Electrocatalytic water splitting has recently gained substantial attention within the research community. The primary objective of this chapter is to proactively mitigate inadvertent errors from permeating future scientific records. Our review of recent literature on catalytic water electrolysis has identified recurrent pitfalls. These include the misapplication of overpotential without detailing the extent of iR drop compensation, the underestimation of the pivotal electrochemical parameter of iR uncompensated overpotential, and a general unawareness regarding the methodologies underpinning the acquisition of Tafel plots and the determination of corresponding Tafel slopes and exchange current densities.

To circumvent such errors, we present an encompassing perspective that addresses electrolyte preparation and pivotal activity parameters. These parameters encompass overpotential at a definite current density, iR-corrected overpotential at a specific current density, mass activity, Tafel slope, Faradaic efficiency (FE), specific activity, turnover frequency (TOF), double-layer capacitance (C_{dl}), and electrochemically active surface area (ECSA). Each parameter undergoes an in-depth exploration, accompanied by the delineation of recommended best practices. These guidelines encompass the execution of experiments for data acquisition and the subsequent data processing procedures before dissemination.

T. u. Haq and Y. Haik, *Electrochemical Water Splitting*, Materials Horizons: From Nature to Nanomaterials, https://doi.org/10.1007/978-981-99-9860-9_5

5.2 Electrolyte Preparations

The preparation of alkaline electrolytes is crucial to achieve accurate concentrations and minimize impurities, which can have detrimental effects on electrochemical measurements. Inaccurate alkali concentrations can lead to erroneous pH estimation, which can result in questionable reports of voltage efficiency and overpotentials of electrocatalysts. This issue can arise due to CO_2 absorption that modifies the effective molarity caused by dissolved carbonate, incorrect weighing due to the hygroscopic nature of alkali solids, and inaccurate calculations caused by improper use of compositions provided by manufacturers. CO_2 absorption can be prevented by boiling deionized water before use and sparging electrolytes with inert gases. Handling hygroscopic solids can be improved if measurements are conducted in low-humidity environments. Finally, correct solid calculations are done when accurate alkali purities are retrieved from certificates of analysis from the manufacturer. These three problems can be effectively controlled by standardizing the final electrolyte concentration [1].

Impurities can harm the reproducibility of electrochemical measurements. Glass etching products and metal impurities can deactivate electrocatalysts, while Fe impurities enhance the OER activity. However, Fe impurities can also form hydroxyl radicals that poison and degrade critical components such as cation exchange membranes. Al impurities are also responsible for significant performance degradation due to changes in the ORR mechanism, and some of these effects occur only after long-term tests. Therefore, when impurities are present in electrochemical energy devices, special attention must be given to stability examinations [2].

Alkaline electrolytes can be easily contaminated with impurities if not prepared and handled correctly. Strong bases can etch glasses, including borosilicate glass, resulting in unwanted contaminants. These can also originate from incorrect cleaning of glassware or using chemicals with low purities. Thus, it is recommended to replace glassware with plasticware whenever possible, mainly when concentrated alkaline electrolytes are used for long periods. Cleaning glassware and plasticware with strong acid solutions is also highly recommended. Researchers must carefully select acids with high purity to avoid systematic errors. The use of glass-free components in electrochemical cells is encouraged to prevent glass etching in alkaline media. It is worth noting that Fe, Ti, Mn, and other metal impurities have been found in other materials often regarded as "metal-free," such as 3D printing filaments and other polymers (Fig. 5.1). Therefore, future studies should always examine the impact of impurities, especially after prolonged testing [3].

Mullins group proposed the following protocol to prepare the Alkaline Electrolyte [4].

It is recommended to remove CO_2 from the DI water used to prepare alkaline electrolytes. The absorption of CO_2 increases the effective molarity of the solution by forming HCO_3^- and CO_3^{2-} ions, thus deteriorating the electrolytes. To remove CO_2, DI water can be boiled in a Florence flask covered with an inverted beaker and cooled under cold water or overnight. The resulting CO_2-free DI water can be stored

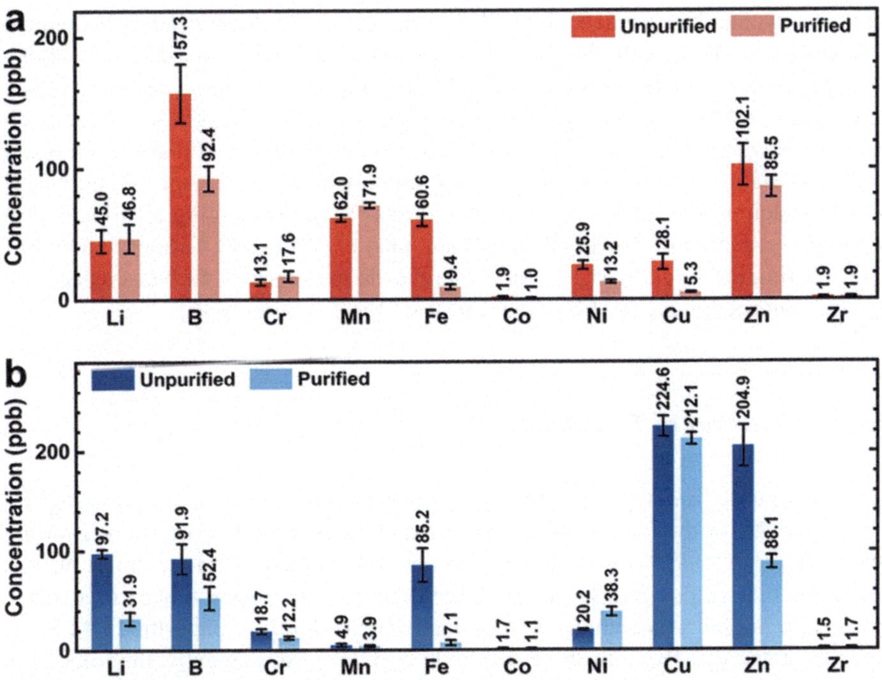

Fig. 5.1 Concentrations of primary elements in purified and unpurified alkaline electrolytes analyzed with ICP-MS **a** 1 M KOH and **b** 1 M NaOH. Reproduced with Permission from Ref. [4]. © 2023, American Chemical Society

in clean polypropylene bottles wrapped with Parafilm TM. Additionally, in the case of NaOH electrolytes, CO_2 can be removed by precipitating Na_2CO_3 from 50 wt.% NaOH (~ 19 M) solutions, allowing the Na_2CO_3 to settle down for several days and then decanting the supernatant. The remaining carbonate-free NaOH solution can then be diluted with CO_2-free DI water and standardized via titration. It should be noted that this method does not work with KOH because K_2CO_3 remains soluble.

After CO_2 removal, the electrolytes should be kept away from air exposure as much as possible. Standardizing the electrolytes with a primary standard is recommended for accurate alkali concentration determination and reproducibility comparison between operators and research groups. Sparging the electrolytes with inert gases such as N_2 or Ar is recommended to remove other gases that could interfere with electrochemical tests. For example, O_2 should be removed when electrodepositing Fe or Co to prevent oxidation of Fe^{2+} or Co^{2+} ions in the solution. In some cases, bubbling O_2 or H_2 is necessary to study gas evolution reactions, but care should be taken to remove bubbles from the surface of the electrodes [5].

KOH and NaOH are highly hygroscopic, so they should be weighed as quickly as possible and stored in environments free of moisture, such as desiccators. In addition, the purity of the alkali can vary significantly, even among different lots, so acquiring a certificate of analysis (COA) from the company website and using the absolute alkali purity when calculating the required alkali weight is recommended. Electrochemical cells and their components should be washed with acids to remove adsorbed impurities, particularly metals, before use. In addition, alkalis, scorching solutions, attack borosilicate, and soft glasses are recommended to avoid using glassware as much as possible when preparing and handling alkaline electrolytes. If plasticware is not readily available, glassware should be used quickly and only for specific purposes, such as reaching the mark on volumetric flasks. When dissolving alkali solids in water, it is essential to slowly add small amounts of solid and DI water to a beaker while stirring the solution to dissipate heat. Plastic beakers and stir bars coated with polytetrafluoroethylene (PTFE) should be used for this purpose [6].

5.2.1 Removal of Fe Impurities

Trotochaud et al. reported the following experimental procedure for removing Fe impurities [7]. In the experiment, approximately 2 g of high-purity nickel nitrate hexahydrate (> 99.99%) was added in deionized water of 4 mL within a 50 mL centrifuge tube of polypropylene. First, the tube was vigorously shaken in a vortex mixer until the salt was completely dissolved. Then, 20 mL of unpurified 1 M KOH electrolyte was dissolved, forming insoluble $Ni(OH)_2$. Subsequently, the sonication of tube was done in an ultrasonic bath for at least 10 min and further shaken in a vortex mixer. The centrifugation of tube was done for 10 min at 8000 rpm, and the resulting supernatant was slowly decanted. This step was followed by three washing cycles, where each cycle involved the addition of 20 mL of DI water and KOH electrolyte (around 2 mL) to the tube, sonicating the mixture, redispersing the solid using the vortex mixer, centrifuging, and decanting the supernatant. The residue was then rinsed thrice with 5 mL of unpurified electrolyte to remove residual DI water from the washing cycles. Next, 30 mL of unpurified electrolyte was added to the tube, and the solid underwent two redispersing cycles by shaking in the vortex mixer for 5 min and sonicating for 10 min. The precipitate disintegrated into fine particles at this point, and any large $Ni(OH)_2$ chunks were removed from the suspension. Additional cycles may have been necessary for complete disintegration. Finally, 20 mL of electrolyte was added to reach a total volume of 50 mL, and the tube was left to rest for at least 48 h until all the solid phases settled down. After resting, the centrifugation of tube was done for 10 min at 10,000 rpm, along with decantation of supernatant. To avoid carrying small $Ni(OH)_2$ particles, the supernatant was filtered with a disposable polyether sulfone membrane filter fitted to a 60 mL polypropylene syringe. The filtered electrolyte was then collected in a clean polypropylene bottle for storage.

Mullins and coworkers have recommended additional steps to remove Fe impurities from the electrolyte effectively (Fig. 5.2) [4].

Centrifuge tubes, polypropylene bottles, and plastic syringes used in the purification process should be thoroughly cleaned with 3 M HNO_3 and DI water to eliminate trace metal impurities. The nitric acid used for cleaning should be of high purity.

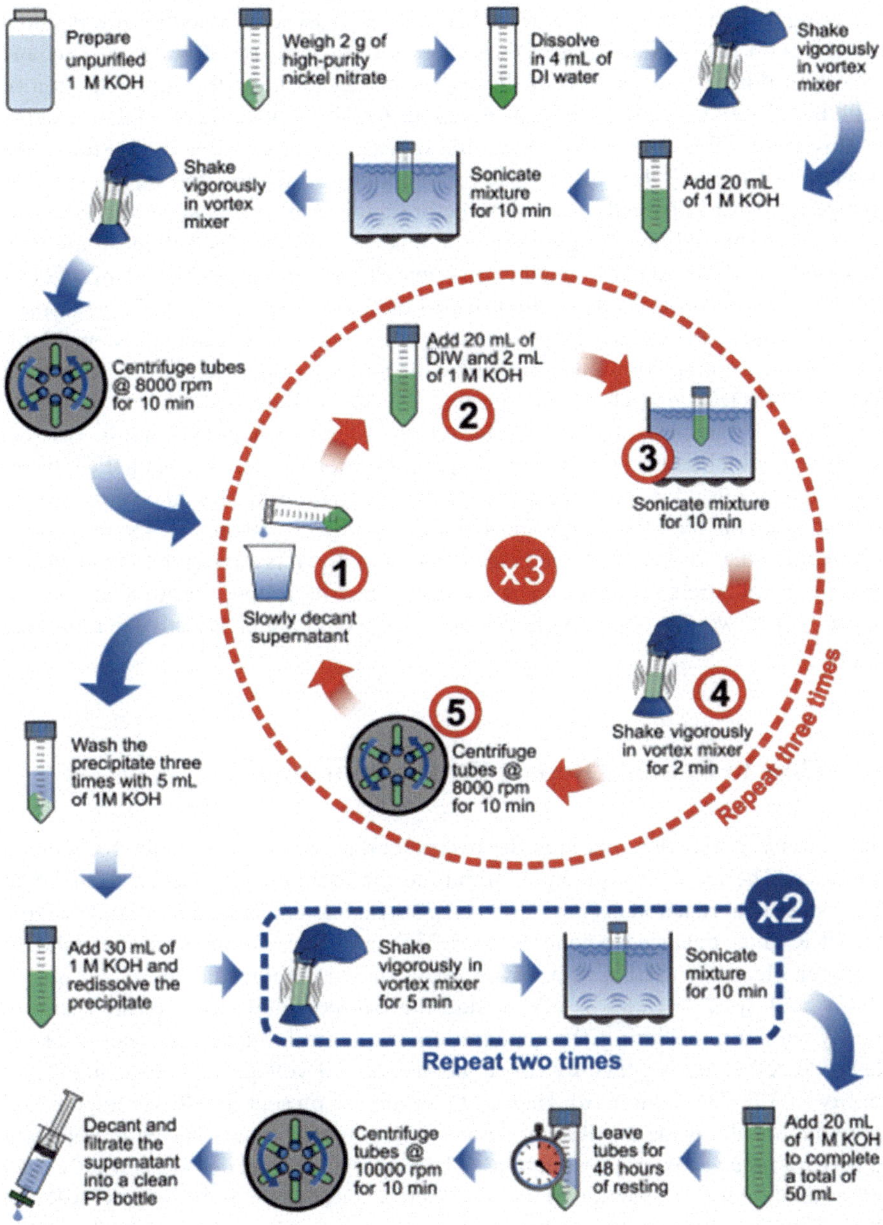

Fig. 5.2 Suggested route for alkaline electrolyte purification. Reproduced with permission from Ref. [4]. © 2023, American Chemical Society

It is advisable to avoid using other acids such as H_2SO_4 as S and Cl traces could interfere with ICP-MS analysis of the electrolyte [4]. The use of vortex mixing and sonication during redispersion of the sediment can improve the fragmentation of insoluble $Ni(OH)_2$ particles, increasing contact with the electrolyte and enhancing Fe absorption. This technique is essential during washing cycles to maximize the formation of the $Ni(OH)_2$ phase, leading to better purification and reproducibility. Precipitating washing with electrolyte before adding 50 mL ensures that water is thoroughly removed, avoiding diluting the final Fe-purified electrolyte. In the final stage of the process, redispersing the solid phase in 30 mL rather than 50 mL of electrolyte enhances solid fragmentation due to increased turbulence in the headspace and air bubbles created during shaking in the vortex mixer. After redispersion, 20 mL is added to complete 50 mL and maximize the amount of purified KOH. To prevent insoluble $Ni(OH)_2$ particles from entering the purified electrolyte, it is crucial to filter the supernatant using nanoporous syringe filters such as hydrophilic 0.1 μm PES filters, which reduce Fe, Ni, and Co concentrations to below 10 ppb. PES filters should not be reused, as air typically clogs them, and filtering should be done slowly to avoid breaking the filter. Large, clean plastic syringes (~ 50 mL) are recommended to maximize the amount of filtered electrolyte. Finally, it is advisable to scale up the mass of nickel nitrate hexahydrate based on the electrolyte concentration, as concentrations > 1 M would contain more Fe, and ~ 2 g would not be effective in removing Fe [4].

5.3 How to Reliably Report the Overpotential

The activity parameter, particularly overpotential, is crucial in electrochemical processes. The term "overpotential" refers to the extra energy needed to make an electrochemical reaction occur consistently beyond its balanced potential. For OER and HER, their balanced potentials are 1.23 V and 0 V compared to the standard hydrogen electrode (SHE), respectively. We can use the formulas $\eta_{OER} = E_{RHE} - 1.23$ V and $\eta_{HER} = E_{RHE} - 0$ V to find the overpotentials for a desired current density without considering iR compensation. In practical water electrolysis, the two-electrode cell setup is generally run at a constant cell voltage or a constant current density of 1.0–2.0 A cm^{-2} [8]. Hence, it's crucial to present the iR uncompensated overpotential alongside the iR compensated overpotential at the same defined current density for precise material evaluation. A 10 mA cm^{-2} current density is widely used as the standard for comparison of various electrocatalytic materials for both HER and OER. The iR uncompensated overpotential is particularly important at high current densities, where the difference with iR compensated overpotential can be significantly large. Therefore, it is essential to report iR uncompensated overpotential and other activity parameters for accurate characterization of electrocatalytic materials [9].

In recent years, the parameter of iR drop compensated overpotential has been given more attention than the iR drop uncompensated overpotential in electrocatalysis research, despite the latter being a more meaningful activity parameter. Understanding the notion of iR compensation in polarization curves is crucial. Some researchers contend that adjusting the iR drop in the working electrode potential with the measured solution resistance (R_s) effectively mirrors what would be observed in a two-electrode cell setup [10]. Nonetheless, this assumption needs to be revised, as the magnitude of R_s is influenced by diverse variables like the electrolyte's ionic strength and the distance between the reference and counter electrodes. In OER and HER investigations, the ionic strength ensures high ionic conductivity, separating the counter and reference electrodes as the likely cause of R_s and ensuing iR drop. Researchers have employed specialized Luggin capillary configurations for reference electrodes to address this to minimize R_s. Nevertheless, other factors, like heterojunctions and establishing connections with the electrochemical workstation, also contribute to overall resistance. Consequently, the term "solution resistance (R_s)" is often substituted with "uncompensated resistance (R_u)" to account for this [11]. Even when achieving total iR drop compensation for an electrocatalyst, the adjusted potential won't fully mirror the outcome of an analogous two-electrode assembly involving the same electrocatalyst. This is mainly due to the series resistance interposed in the R_s of a three-electrode setup persisting within the two-electrode framework, rendering complete iR drop compensation unattainable. Thus, the iR drop uncompensated overpotential at a specific current density offers a more pertinent activity parameter than the iR drop compensated overpotential at the same current density. Should researchers report the iR drop compensated overpotential, providing the percentage of iR drop compensation is imperative. The subsequent equations are applied to compute iR drop compensated overpotentials for an electrocatalyst in both OER and HER electrocatalysis [12]:

$$\text{iR drop free } \eta_{\text{OER}} = E_{\text{RHE}} - 1.23 - E_{\text{iR}}$$

$$\text{iR drop free } \eta_{\text{HER}} = E_{\text{RHE}} - 0 - E_{\text{iR}}$$

Correcting for potential drops caused by solution resistance during electrode reactions, iR drop compensation is a well-established technique in electrochemical evaluations. Two main approaches are commonly employed: (1) automated compensation by the electrochemical workstation and (2) manual calculation of potential reduction using current density and Ru value. The latter method, supported by experimental substantiation, is widely favored. The extent of iR drop compensation varies among studies due to different current densities and practical potential ranges, exerting a substantial influence on the iR drop-corrected overpotential activity metric. A study was conducted on an OER measurement using a RuO_2-modified electrode developed via the universal drop-casting method [9]. The impedance plot obtained from Nyquist analysis demonstrated the lowest impedance, situated within the high-frequency domain, corresponds to the Ru of the studied RuO_2-modified electrode.

By systematically augmenting the compensation percentage for the iR drop, it was observed that the progressive increase in iR drop compensation yields a gradual reduction in overpotentials for OER current densities across all potentials surpassing the onset overpotential. This observation signifies that manipulations in the iR drop compensation percentage can impart variability to the iR drop-exempt overpotential at a designated current density. Therefore, including the iR drop compensation percentage alongside the iR-corrected polarization curves becomes imperative, allowing readers to evaluate the electrocatalyst's catalytic characteristics comprehensively. Nonetheless, significant challenges may arise when the catalyst or catalytic material generates a exceeding current density of 100 mA cm^{-2}, coupled with a Ru value surpassing 6 Ω [9]. Selecting suitable substrate electrodes characterized by low intrinsic resistance, such as glassy carbon electrodes, carbon cloth, carbon fiber paper, and various metal foils and foams, including Ti, Cu, and Ni, holds the potential to mitigate Ru (uncompensated resistance) and diminish the influence of substrate electrodes on the electrode's Ru. Unless specifically mandated, it is imperative to avoid utilizing substrate electrodes with intrinsic resistance, notably tin-doped indium oxide and fluorine-doped tin oxide.

Observations reveal that the electrochemical workstation-assisted iR drop compensation approach yields reduced overpotentials compared to manual compensation. However, this method concurrently results in an augmented overall current density. This discrepancy underscores the importance of specifying the employed iR drop compensation technique, alongside the corresponding compensation percentage, in research accounts. Moreover, employing iR drop uncompensated overpotentials at established benchmark current densities emerges as a more meaningful activity parameter than iR drop compensated overpotentials. Notably, the consensus remains elusive concerning the preferred iR drop compensation methodology. Thus researchers must opt for an approach aligning optimally with their experimental requisites.

5.4 How to Calculate the Tafel Slope

During electrochemicatalytic analysis, the construction of Tafel plots, a pivotal tool, is facilitated by utilizing three well-established techniques: voltammetry, chronoamperometry/chronopotentiometry, and electrochemical impedance spectroscopy (EIS). These methodologies offer the means to compute crucial parameters, such as the iR drop-free overpotential and the logarithm of current density ($\log j$), essential for delineating the Tafel plot. Each technique provides accurate and reliable data of $\log j$ and iR drop-free overpotential, with the choice of approach contingent upon the specific experimental requisites [13]. Researchers often choose the method based on the complexity of the experiment, the nature of the system under investigation, and the availability of the required instrumentation.

5.4.1 Tafel Plot from Polarization Curve

The Tafel plot is a valuable tool for assessing the electrocatalytic efficacy of materials HER and OER. While cyclic voltammetry (CV) is commonly employed for extracting Tafel plots, it doesn't generate an actual steady-state polarization curve according to the Tafel equation. Recent reviews emphasize using linear sweep voltammetry (LSV) or a segment of the CV curve to derive Tafel plots. It is recommended to record iR drop compensated LSV or CV at a scan rate below 5 mV s^{-1} and manually correct for iR drop to minimize inaccuracies.

Elevated scan rates introduce substantial capacitive current to the electrocatalytic process, leading to unreliable Tafel slope, and exchange current density values. Hence, employing voltammograms obtained at very low scan rates is crucial to ensure accurate experimental results. For precise exchange current density measurements, voltammetry-based Tafel plots are not suitable. The backward sweep of the CV curve is suggested for extracting Tafel plots for catalysts exhibiting distinct active redox peaks in their voltammograms. For high-performance catalytic materials in OER and HER, a more reasonable and meaningful approach involves either employing the amperometry method or resorting to the potential step method to extract the Tafel plot, ensuring robust values for Tafel slopes and exchange current densities [14].

5.4.2 Tafel Plot from Amperometry/Potentiometry

Tafel plots serve as valuable tools for examining the kinetics of electrochemical reactions, especially in electrocatalysis. However, it's important to note that Tafel plots generated from voltammograms conducted at slower scan rates may not accurately represent accurate steady-state polarization curves [15]. This discrepancy arises because the current and overpotentials acquired continuously change with time rather than reaching a consistent state. Additionally, voltammogram-derived Tafel plots can encompass significant contributions from the capacitance of the interface to the overall gas evolution current.

It's advisable to employ static electrochemical methods such as amperometry or potentiometry to depict steady-state polarization curves accurately. These techniques grant the catalytic interface ample time to achieve a steady state, enabling the collection of corresponding potential and current density data for plotting Tafel curves. This approach yields Tafel slopes that genuinely reflect the intrinsic catalytic activity, devoid of capacitance current contributions, and obtained at true steady states, unlike slopes derived from voltammograms.

To acquire steady-state potential or current for constructing Tafel curves in HER and OER, a gradual stepwise increase in gas evolution reaction current density is suggested. This should be conducted in small intervals, such as 1 mA, during chronopotentiometry, over several intervals. The overpotential from the gas evolution reaction's equilibrium potential should be incrementally increased in smaller intervals, like 0.05 V, until the catalyst can deliver the desired current density. The obtained steady-state overpotential must be corrected for iR drop before plotting against the log current density. For high-performance catalysts capable of delivering substantial current densities exceeding 200 mA cm^{-2}, it's beneficial to focus the steady-state experiment within the initial low current density ranges, such as below 50 mA cm^{-2}. In such cases, smaller potential intervals, as tiny as 0.01 V, can yield additional data points for determining the Tafel slope region. For high-performance and high surface area catalysts, the capacitance current should ideally fall within the 1–3 mA cm^{-2} range, which could affect chronopotentiometry. Hence, utilizing the chronoamperometry technique is preferable for this type of catalyst.

5.4.3 *Tafel Slope from EIS*

Accurately determining the Tafel slope, a key indicator of electrocatalyst activity, is pivotal for performance evaluation. However, challenges arise from factors like series resistance (Ru) and inherent catalyst resistance. Conventionally, Tafel slope assessment has relied on linear sweep voltammetry (LSV) at different scan rates [16]. However, the precision of the derived Tafel slope through this technique can be compromised due to the impact of series resistance. In response, Hu et al. introduced an innovative strategy that mitigates the effects of both series and intrinsic catalyst resistance. This approach employs successive electrochemical impedance spectroscopy (EIS) measurements at varying overpotentials, with intervals adjusted based on the catalyst's activity range. Subsequently, only the logarithmic reciprocal of the charge transfer resistance (R_{ct}) is plotted against the overpotentials to yield the corresponding Tafel slope. This refined methodology elucidates the intrinsic kinetics of the catalyst material while circumventing the influence of series resistance (as depicted in Fig. 5.3) [17].

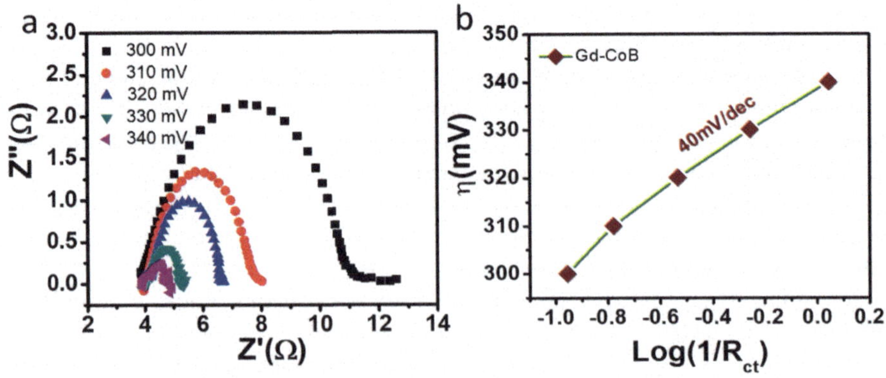

Fig. 5.3 **a** EIS spectra recorded at different DC voltage and **b** corresponding Tafel slope value derived from the EIS. Reproduced with permission from Ref. [17]. © 2020, Wiley

A comparative analysis of Tafel slopes was conducted for the Pt/C (Sigma) 20 wt.% catalyst, employing LSV, chronoamperometry (CA), and EIS techniques. Notably, substantial distinctions emerged among the Tafel slopes determined through all three methods. Specifically, the Tafel slope obtained from the Tafel curve extracted from LSV, executed at a scan rate of 5 mV s^{-1} in 0.5 M H_2SO_4 at ambient conditions, was 43 mV dec^{-1}. Conversely, the Tafel slopes derived from the Tafel curves extracted via CA and EIS techniques, operating under identical experimental parameters, are 33 mV dec^{-1} and 28 mV dec^{-1}, respectively. Remarkably, the Tafel slope of 28 mV dec^{-1} ascertained from the EIS-derived Tafel curve represents the lowest value, notably approximating the theoretical Pt value of 30 mV dec^{-1} [9]. This demonstrates the superiority of the EIS method in providing exact information on the catalytic activity of an electrocatalyst in terms of the Tafel slope.

To attain a meaningful Tafel slope, one must exclusively rely on voltammetric responses acquired at lower scan rates. However, this approach may yield inaccurate readings for high-performance catalysts that exhibit HER and OER current densities surpassing 200 mA cm^{-2}, coupled with uncompensated resistance (Ru) exceeding 6 Ω. In such instances, employing partially iR drop compensated LSV/CV curves to deduce corresponding Tafel curves is not advisable. The electrochemical impedance spectroscopy (EIS) method is strongly advocated for precise Tafel slope measurements, especially concerning high-performance catalysts. Additionally, the chronoamperometry/chronopotentiometry (CA/CP) approach presents a valuable technique to extract the Tafel slope and exchange the current density of a target catalyst. Irrespective of the chosen Tafel analysis method, the potential of the Tafel region should ideally exceed 120 mV from the HER and OER onset overpotentials, and a linear segment spanning two to three decades is necessary for accurate slope determination.

5.5 How to Properly Report TOF

The half-cell reactions involving HER and OER process are derived by electrocatalytic water splitting. However, many recent studies need more critical data, such as turnover frequency (TOF) values, that are crucial for determining the intrinsic electrocatalytic performance. In contrast to classical heterogeneous catalysis, the amount of product formed in electrocatalysis is not measured. Instead, the current density is used along with other necessary parameters to calculate TOF. The equation below is frequently employed to compute the TOF for HER and OER [18].

$$\mathrm{TOF} = \frac{j * N_{\mathrm{A}}}{A * F * n * \tau}$$

where

j Current in Ampere

N_A Avogadro number
A Geometrical surface area of the electrode
F Faraday constant
n Number of electrons transferred to generate one molecule of product
τ Surface concentration of atoms.

The quantification of turnover frequency (TOF) for the OER process, often in conjunction with the concurrent oxygen reduction reaction (ORR) at the ring electrode through a rotating ring-disk electrode (RRDE) assembly, represents an indirect method of assessment. This methodology can also be adapted for HER process by maintaining a consistent anodic potential at the ring electrode. However, careful consideration is crucial in selecting the anodic potential, as potential complications, such as competitive hydrogen underpotential deposition (HUPD), can impact the intended investigation.

An alternative strategy entails measuring the evolved quantities of H_2 or O_2, typically facilitated by techniques like water displacement or gas chromatography-mass spectrometry (GC-MS). Despite their utility, these approaches are accompanied by challenges, including potential catalyst degradation during extended studies and the potential for human errors during measurement procedures. Although TOF values provide valuable insights into intrinsic electrocatalytic activity, the existing methodologies for calculating TOF present persistent complexities and uncertainties that require further attention to ensure accurate and dependable TOF determinations. Addressing these intricacies is essential to enhance the precision and robustness of reported TOF values, thereby advancing the understanding of catalytic behaviors and materials [19].

The turnover frequency (TOF) is a fundamental metric employed in electrocatalytic investigations to assess catalyst efficiency. However, recalibration of the TOF calculation is necessary due to complexities linked to determining the precise count of active sites contributing to the catalytic process. The quantification of active sites significantly influences catalyst efficiency, underscoring its significance. Furthermore, the utilization of geometrically normalized current density to standardize current density to the catalyst's surface area adds an additional layer of intricacy to the accurate determination of TOF.

These aspects collectively exert a substantial impact on the dependability and precision of TOF determination within the realm of electrocatalytic research. To accurately compute the catalyst's turnover frequency, various electrochemical strategies are employed to pinpoint the actual count of active centers or the authentic surface area of the catalyst. These methodologies play a pivotal role in achieving a precise assessment of TOF. Prominent among these methods are hydrogen underpotential deposition (HUPD), copper underpotential deposition (Cu-UPD), lead underpotential deposition (Pb-UPD), CO stripping, and redox peak integration (as depicted in Fig. 5.4). These approaches collectively contribute to refining the accuracy of TOF calculations in the context of electrocatalytic investigations [20–22].

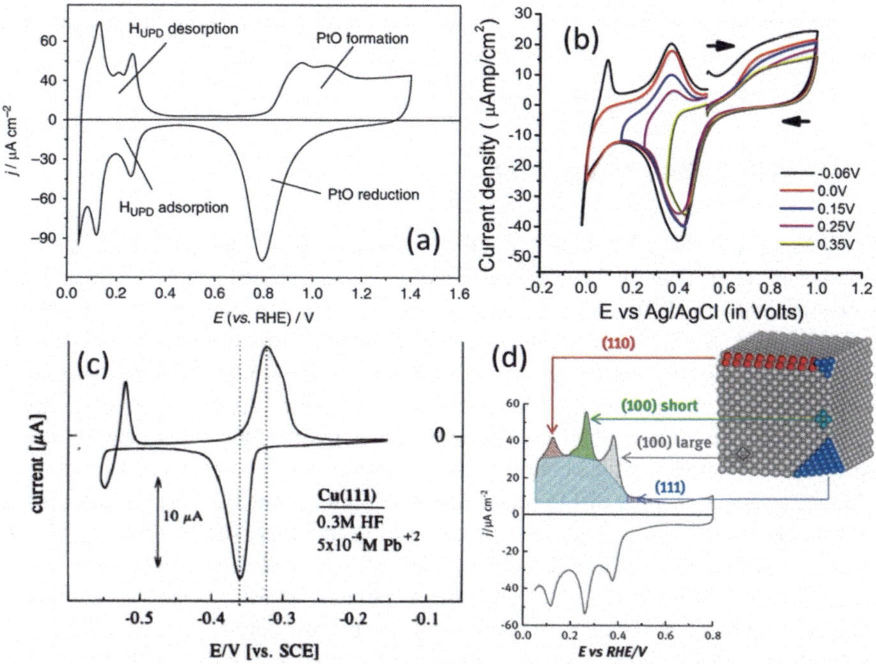

Fig. 5.4 **a** Cyclic voltammogram plot of a polycrystalline Pt electrode immersed in a H_2SO_4 solution with 0.5 M molarity obtained using a scan rate of 10 mV s^{-1}, revealing the occurrence of hydrogen underpotential deposition (HUPD) peaks. **b** Cyclic voltammetry profiles illustrating the emergence of Cu underpotential deposition (Cu-UPD) peaks on a palladium (Pd) surface. The potentials are incrementally reduced in a solution containing varying concentrations of $NaClO_4$, $HClO_4$, and $Cu(ClO_4)_2$, namely 0.500, 0.010, and 0.001 M. **c** Cyclic voltammetry response highlighting lead underpotential deposition (Pb-UPD) on a copper (Cu) (111) crystallographic plane. The experiment is carried out in a 0.3 M HF electrolyte containing 0.0005 M Pb^{2+} ions. **d** Cyclic voltammogram features corresponding to hydrogen underpotential deposition (HUPD) on different crystallographic planes of a platinum (Pt) electrode. Reproduced with Permission from Ref. [18]. © 2021, Wiley

The HUPD technique employs hydrogen to ascertain the number of active centers present on the catalyst's surface. Cu-UPD and Pb-UPD processes utilize underpotential deposition of copper or lead to estimate catalyst's surface area. The CO stripping method involves the removal of CO molecules from the catalyst surface to determine the surface area. The CO stripping method entails eliminating CO molecules from the catalyst's surface for determining its surface area. Redox peak integration consists of integrating the redox peaks in the cyclic voltammogram to select the number of active centers and surface area of the electrocatalyst. However, these electrochemical methods are materials-specific and cannot be universally applied to all electrocatalysts. Furthermore, the suitability of each method depends on the specific properties of the catalyst being studied. Therefore, it is essential to carefully select the appropriate method based on the nature of the electrocatalyst and the study's objectives.

5.5.1 Redox Peak Integration

Determining turnover frequency (TOF) via redox peak integration is a well-established electrochemical method utilized to assess the intrinsic catalytic activity of electrocatalysts. TOF provides crucial insights into the rate at which catalytic reactions occur per active site on the catalyst surface. This method is particularly significant in studying various electrochemical reactions, including OER and HER process, as it allows researchers to quantify the number of catalytic events facilitated by individual active sites. The fundamental principle underlying this technique involves the integration of the catalytic current associated with redox peaks in cyclic voltammograms (CVs) or differential pulse voltammograms (DPVs) corresponding to the catalytic reaction of interest. This integration encompasses the total charge (Q) which is transferred in the redox process, and it may be linked with number of active sites (N) contributing in the reaction using Faraday's law of electrolysis ($Q = nF$, where n is the number of electrons exchanged and F is the Faraday constant) [17, 23]. By dividing the total charge by electrons number exchanged in the catalytic reaction (n), one can estimate the number of active centers contributing to the electrochemical process. For instance, in OER studies, the redox peaks relating to the reduction as well as oxidation of surface-bound intermediates, such as metal oxides or hydroxides, are usually observed in CVs. By integrating the catalytic current associated with these peaks, researchers can quantify the number of catalytic events involving the release of oxygen molecules. Similarly, in HER studies, redox peaks corresponding to the evolution of hydrogen gas can be integrated to estimate the number of active sites that facilitate the reaction. It is important to note that the accuracy of this method relies on several factors, including the proper identification of the relevant redox peaks, the careful consideration of background currents, and the correct identification of the number of electrons exchanged in the catalytic process. Additionally, the determination of TOF via redox peak integration assumes that all active sites are equally accessible and exhibit identical catalytic behavior, which might not be the case in complex electrocatalytic systems. While this method provides a direct and quantitative measure of TOF, it is essential to consider potential limitations. The complexity of redox peaks overlaps with other electrochemical processes, and the presence of side reactions can introduce uncertainties in the calculated TOF values. Therefore, a comprehensive analysis and thorough validation are required to ensure the reliability and accuracy of TOF determination via redox peak integration [24, 25].

5.6 Double-Layer Capacitance

The electrochemically active surface area (ECSA) is determined from double-layer capacitance (C_{dl}) is a valuable electrochemical technique used to quantify the accessible electrode's surface area. ECSA plays a pivotal role in understanding the electrochemical performance of electrocatalyst, particularly in HER and OER process. The basis of such method lies in the relationship between the ECSA and (C_{dl}) of the electrode [26]. The double-layer capacitance is an electrochemical parameter that reflects the ability of the electrode's electrical double layer to store charges, which is inherently linked to electrocatalyst surface area. Increased surface area corresponds to heightened charge storage capacity within the double layer, leading to a larger C_{dl} value. To determine ECSA from C_{dl}, a reference electrode is typically used, and the electrode is subjected to cyclic voltammetry (CV) scans in an electrolyte solution that supports a reversible redox couple, such as a ferri/ferrocyanide couple. During the CV scan, the current response related to the charging and discharging of the double layer at different potentials is measured. The (C_{dl}) can be determined by creating a plot of the square root of the scan rate against the current response (I) and then extracting the slope from the graph can be calculated using the equation $C_{dl} = (2.69 \times 10^5) \times (I/V^{0.5})$, where V denotes scan rate and I is the current [27].

Subsequently, the ECSA can be calculated using the relationship between C_{dl} and ECSA, which is expressed as ECSA $= C_{dl}/C_{specific}$, where $C_{specific}$ is the specific capacitance of the material (F cm^{-2}). The specific capacitance is obtained from the material's geometric surface area, typically through techniques like the Brunauer–Emmett–Teller (BET) method for porous materials or geometric measurements for planar electrodes [28, 29].

It is important to note that the accuracy of ECSA determination from C_{dl} relies on factors like the proper identification of the double-layer region in the CV, the use of a reliable reference electrode, and the appropriate consideration of background currents. Furthermore, the method assumes a linear relationship between C_{dl} and ECSA, which may deviate in certain cases. Despite these considerations, determining ECSA from double-layer capacitance provides valuable insights into the surface availability of catalytically active sites, enhancing the understanding of electrocatalyst performance.

5.7 Mass and Specific Activity

The current density is a pivotal parameter for assessing catalysts, representing the current flow per unit electrode area. This current density is conventionally normalized by the geometric surface area of the electrocatalyst, often denoted in mA cm^{-2}. Even considering electrocatalytic surfaces with varying degrees of roughness, this normalization approach might not effectively capture the catalyst's intrinsic performance. In contemporary research, there has been a growing emphasis on evaluating catalysts based on their mass activity. This metric involves normalizing the current density by the mass of the composite loaded onto the electrode, typically presented as A g^{-1}. This mass-normalized approach aims to provide a more accurate depiction of the catalyst's effectiveness, accounting for surface roughness and loading variations, thereby enabling a more meaningful comparison of catalyst performance across different materials and configurations [30]. This parameter has garnered significance, particularly in the context of nanostructured electrocatalysts. The emergence of diverse nanostructured materials composed of various metals and their compounds has yielded highly efficient catalysts. The conventional practice of employing geometrically normalized activity proves inadequate for roughened electrocatalytic surfaces, as it assumes a flat and even structure. Yet, this assumption is valid only for certain electrocatalytic surfaces, which exhibit pronounced roughness and non-planarity. Thus, the adoption of mass-normalized activity is increasingly vital for the assessment of such catalysts, offering a more meaningful and dependable gauge of their genuine activity, especially when confronted with non-planar surfaces [31].

In the context of high-performance catalysts, typically generated through in situ growth techniques, resulting in substantial catalyst loading per unit area ranging from 1 to 3 mg cm^{-2}, the adoption of mass-normalized activity becomes even more pivotal (Fig. 5.5). Furthermore, alongside this considerable catalyst loading, the surface area exposed to the electrolyte within these hierarchically structured nanoarrays surpasses smooth and planar surfaces. Unreflective application of a geometric surface area of 1 cm^2, even when the electrode dimensions measure 1×1 cm^2, yields imprecise measurements. Mass-normalized activity emerges as the most appropriate choice in such instances, offering results with minimal discrepancies. Hence, a dual approach employing mass-normalized and geometric area normalized activity is indispensable to garner a profound understanding of catalytic attributes within non-planar and significantly roughened interfaces. Adopting this practice enables a more precise assessment of catalyst activity, particularly pertinent to electrochemical applications like OER and HER process.

5.7.1 BET Surface Area Normalized Activity

In recent times, as advanced nanostructured electrocatalysts have emerged for efficient water splitting, there is a growing notion that evaluating an interface's actual surface area should supersede the conventional reliance on its geometrical area—a parameter suited primarily for smooth and planar surfaces. The term "real surface area" pertains to the exposed sites within the nanostructured catalyst dedicated to the water-splitting process [9]. In customary materials chemistry, quantifying surface area involves applying the Brunauer–Emmett–Teller (BET) adsorption isotherm technique. This consists of assessing nitrogen gas adsorption and desorption measurements to ascertain the real accessible surface area of a material. Previously, researchers employed the BET surface area for activity normalization, labeling it

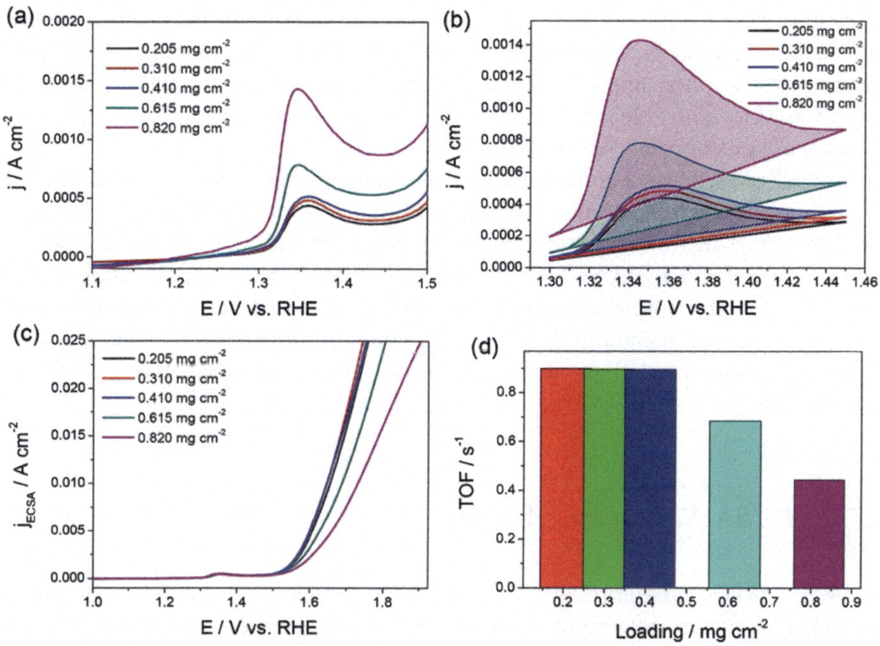

Fig. 5.5 **a** Linear sweep voltammograms (LSVs) of NiO OER electrocatalysts exhibit a distinct oxidation peak that increases with increasing loading. This peak is attributed to enhanced number of accessible centers for catalytic performance. **b** The isolated oxidation peaks of the LSVs used for charge integration and calculating the number of active sites can be observed. **c** Specific activity and **d** turnover frequency (TOF) were determined using the calculated number of sites from the charge integrated from the oxidation peaks illustrated in (**b**). Reproduced with permission from Ref. [18]. © 2021, Wiley

as a specific activity (*j*BET). However, this approach presents a significant limitation: while it does unveil the available surface area of an electrocatalytic material, it bases its measurements on N_2 adsorption and desorption. Consequently, not all sites accommodating N_2 molecules are inherently electrochemically active. Thus, normalizing with the BET surface area can yield misleading results.

Furthermore, the BET method gauges surface area in an unaltered state. In contrast, electrochemical investigations occur after electrode fabrication, encompassing various universal procedures, such as catalyst ink preparation, modifications, and dry conditions. These steps collectively often lead to a notable alteration (usually reduction) in the material's initially measured BET surface area in its pristine form. Consequently, this approach is currently not recommended for assessing the water-splitting electrocatalytic capabilities of any nanostructured material [9].

To tackle these challenges, the precise extent of an electrocatalyst's surface needs a comprehensive assessment. This task can be accomplished through several methodologies, including the electrochemically active surface area (ECSA) approach and the hydrogen adsorption/desorption technique. The ECSA method, a well-established procedure, discerns an electrode's surface area by quantifying the charge essential for oxidizing the active sites on the electrode's surface. Conversely, the hydrogen adsorption/desorption approach quantifies the quantity of hydrogen gas absorbed and released from the electrode's surface, proving to be a dependable technique for ascertaining electrocatalysts' bona fide surface area.

In conclusion, the conventional method of using the BET surface area to normalize the activity of electrocatalytic materials is not advisable for evaluating nanostructured materials in water-splitting electrocatalysis. Instead, the real surface area of the electrocatalyst should be determined using reliable methods such as the ECSA and hydrogen adsorption/desorption methods. By doing so, researchers can obtain more accurate and meaningful data on the catalytic properties of nanostructured electrocatalysts, essential for developing efficient and sustainable water-splitting technologies.

5.7.2 ECSA Normalized Activity

The BET surface area normalized activity has several disadvantages, which have led to the increased importance of the ECSA normalized activity. ECSA refers to the electrochemically active surface area of an electrochemical material, which can be determined in several ways depending on the physicochemical properties of the material under study. While discrepancies can arise when using different methods to measure ECSA, it is believed to reflect the material's surface area exposed to the electrolyte, making activity normalized by ECSA a more accurate measure of catalytic properties [32].

The ECSA normalized activity is designated as j_{ECSA}, and it is a promising activity parameter for evaluating the inherent catalytic properties of different types of electrocatalysts. However, difficulties in the determination and reproducibility of ECSA have led to debates over its use as an essential activity parameter. As a result, only a few researchers use this parameter to evaluate catalysts in water-splitting electrocatalysis. Moreover, variations of considerable magnitude are evident in the ECSA values of identical materials when assessed through diverse methodologies. Despite these limitations, researchers may opt to determine ECSA normalized activity and other pivotal activity parameters in water-splitting electrocatalysis should they anticipate a beneficial contribution. However, it cannot be recommended as a critical activity parameter due to its limitations. In summary, the ECSA normalized activity is a promising measure of catalytic properties that can provide valuable insights into the performance of electrocatalysts. Still, its use should be evaluated carefully based on the specific context and limitations of the studied material [33].

5.8 Faradic Efficiency and Its Significance

Electrocatalysts play a significant role in water-splitting electrocatalysis, and the selectivity of the catalyst is one of the essential qualities. Selectivity reflects how efficiently the supplied energy is utilized for the desired electrochemical reaction, and electrocatalysts must be screened for their selectivity for OER and HER process by determining their Faradaic efficiencies (FE) [34].

Two established techniques are available for ascertaining the Faradaic efficiency (FE) of both OER and HER. The initial approach involves collecting evolved gas within a graduated cylindrical measuring vessel through water–gas displacement. The volume of the collected gas is subsequently converted into moles, relying on Avogadro's principle. In contrast, the number of gas molecules in moles is deduced from the applied charge, adhering to Faraday's second law of electrolysis. Alignment between the estimated and actual collected volumes indicates a 100% FE for the specific gas evolution reaction, implying pronounced selectivity of the catalyst toward the particular electrochemical reaction [17].

$$\text{Faradaic efficiency} = \frac{V_{\text{Experimental}}}{V_{\text{Theoretical}}}$$

While

$$V_{\text{Theoretical}} \text{ for } O_2 = \frac{1}{4} * \frac{Q}{F} * V_m$$

The number "1" signifies one mole of O_2 per mole of H_2O, while "4" represents four moles of electrons per mole of H_2O.

$$V_{\text{Theoretical}} \text{ for } H_2 = \frac{1}{2} * \frac{Q}{F} * V_m$$

The number 1 corresponds to 1 mol of H_2 per mole of H_2O, and 2 signifies 2 mol of electrons per mole of H_2O.

Q It (amount of charged passed through electrode)
F Faraday constant (96,485 C mol^{-1})
V_m Molar volume of gas (24.1 L mol^{-1}, 293 K, 101 kPa).

The second approach is analogous to the first one, involving the analysis of evolved gases using gas chromatography (GC). The GC method has gained wide acceptance due to practical challenges in collecting evolved gases using the water–gas displacement technique. With GC, continuous monitoring of the reaction's progress is achievable, even from the initial stages of gas evolution. This monitoring offers insights into how the reaction rate changes with applied potential, compared against the theoretically expected gas production. Fluorescence spectroscopy has also emerged as an alternative to GC, enabling precise gas measurement during potentiostatic or galvanostatic electrolysis for both OER and HER, facilitating FE determination [17].

For evaluating OER, an alternative dependable electrochemical approach employs the rotating ring-disk electrode (RRDE) configuration to quantify the Faradaic efficiency (FE) of an electrocatalyst functioning in both basic and acidic environments. The basic principle involves capturing and reducing the evolving O_2 molecules from the disk electrode, employing a lower scan rate of 5 mV s^{-1}, while concurrently maintaining a consistent potential at the ring electrode where O_2 can be converted to water wholly and efficiently. In summary, reporting the FE of a water-splitting electrocatalyst is crucial, reflecting its selectivity alongside other vital activity metrics. Diverse methods offer precise FE determination, including water–gas displacement, gas chromatography, and fluorescence spectroscopy. Notably, the rotating ring-disk electrode (RRDE) configuration is a dependable electrochemical technique for assessing the FE of OER catalytic electrocatalysts.

Best practice	Description
Electrode preparation	Ensure consistent electrode preparation methods, such as cleaning and polishing, to eliminate any variability between different samples
Electrocatalyst synthesis	Provide detailed information on the synthesis method used for the electrocatalyst, including precursor materials, reducing agents, and reaction conditions. This information is essential for reproducibility and understanding of the structure–activity relationship
Catalyst characterization	Characterize the catalyst using multiple techniques such as X-ray diffraction (XRD), transmission electron microscopy (TEM), and Fourier-transform infrared (FTIR) spectroscopy to confirm its structure and composition
Standardized testing conditions	Establish standardized testing conditions, like various kinds of electrolyte, scan rate, potential range, and temperature, to allow for accurate comparison between different studies
Control experiments	Conduct control experiments to ensure that observed electrocatalytic activity is due to the catalyst and not other factors such as impurities or side reactions
Reproducibility	Perform multiple trials and report statistical analysis to ensure the reproducibility of the results
Normalization	Normalize the electrocatalytic activity by the mass, surface area, or number of active sites of the catalyst to allow for comparison between different electrocatalysts
Performance metrics	Report appropriate performance metrics such as turnover frequency (TOF), Faradaic efficiency, and overpotential to accurately quantify the electrocatalytic performance
Stability	Evaluate the electrocatalyst stability under electrochemical conditions to determine its long-term durability
Comparison to state-of-the-art	Compare the activity of the electrocatalyst to state-of-the-art materials and provide a detailed discussion on the advantages and disadvantages of the catalyst

References

1. Pal B, Yang S, Ramesh S, Thangadurai V, Jose R (2019) Electrolyte selection for supercapacitive devices: a critical review. Nanoscale Adv. https://doi.org/10.1039/c9na00374f
2. Huang M, Li M, Niu C, Li Q, Mai L (2019) Recent advances in rational electrode designs for high-performance alkaline rechargeable batteries. Adv Funct Mater. https://doi.org/10.1002/adfm.201807847
3. Subbaraman R, Danilovic N, Lopes PP, Tripkovic D, Strmcnik D, Stamenkovic VR, Markovic NM (2012) Origin of anomalous activities for electrocatalysts in alkaline electrolytes. J Phys Chem C. https://doi.org/10.1021/jp3075783
4. Márquez RA, Kawashima K, Son YJ, Castelino G, Miller N, Smith LA, Chukwuneke CE, Mullins CB (2023) Getting the basics right: preparing alkaline electrolytes for electrochemical applications. ACS Energy Lett 1141–1146. https://doi.org/10.1021/acsenergylett.2c02847
5. Vass Á, Endrödi B, Samu GF, Balog Á, Kormányos A, Cherevko S, Janáky C (2021) Local chemical environment governs anode processes in CO_2 electrolyzers. ACS Energy Lett. https://doi.org/10.1021/acsenergylett.1c01937
6. Hausmann JN, Traynor B, Myers RJ, Driess M, Menezes PW (2021) The PH of aqueous NaOH/KOH solutions: a critical and non-trivial parameter for electrocatalysis. ACS Energy Lett. https://doi.org/10.1021/acsenergylett.1c01693
7. Trotochaud L, Young SL, Ranney JK, Boettcher SW (2014) Nickel-iron oxyhydroxide oxygen-evolution electrocatalysts: the role of intentional and incidental iron incorporation. J Am Chem Soc 136(18):6744–6753. https://doi.org/10.1021/ja502379c
8. Kawashima K, Márquez RA, Son YJ, Guo C, Vaidyula RR, Smith LA, Chukwuneke CE, Mullins CB (2023) Accurate potentials of Hg/HgO electrodes: practical parameters for reporting alkaline water electrolysis overpotentials. ACS Catal 1893–1898. https://doi.org/10.1021/acscatal.2c05655
9. Anantharaj S, Ede SR, Karthick K, Sam Sankar S, Sangeetha K, Karthik PE, Kundu S (2018) Precision and correctness in the evaluation of electrocatalytic water splitting: revisiting activity parameters with a critical assessment. Energy Environ Sci 11(4):744–771. https://doi.org/10.1039/c7ee03457a
10. ul Haq T, Mansour SA, Munir A, Haik Y (2020) Gold-supported gadolinium doped CoB amorphous sheet: a new benchmark electrocatalyst for water oxidation with high turnover frequency. Adv Funct Mater 1910309, 1–11. https://doi.org/10.1002/adfm.201910309
11. Mei BA, Munteshari O, Lau J, Dunn B, Pilon L (2018) Physical interpretations of Nyquist plots for EDLC electrodes and devices. J Phys Chem C 122(1):194–206. https://doi.org/10.1021/acs.jpcc.7b10582
12. Anantharaj S, Ede SR, Sakthikumar K, Karthick K, Mishra S, Kundu S (2016) Recent trends and perspectives in electrochemical water splitting with an emphasis on sulfide, selenide, and phosphide catalysts of Fe Co, and Ni: a review. ACS Catal. https://doi.org/10.1021/acscatal.6b02479
13. Suen NT, Hung SF, Quan Q, Zhang N, Xu YJ, Chen HM (2017) Electrocatalysis for the oxygen evolution reaction: recent development and future perspectives. Chem Soc Rev 46(2):337–365. https://doi.org/10.1039/c6cs00328a
14. Shi Y, Zhang B (2016) Recent advances in transition metal phosphide nanomaterials: synthesis and applications in hydrogen evolution reaction. Chem Soc Rev. https://doi.org/10.1039/c5cs00434a
15. Anantharaj S, Karthick K, Kundu S (2017) Evolution of layered double hydroxides (LDH) as high performance water oxidation electrocatalysts: a review with insights on structure, activity and mechanism. Mater Today Energy. https://doi.org/10.1016/j.mtener.2017.07.016
16. Munir A, ul Haq T, Qurashi A, Rehman HU, Ul-Hamid A, Hussain I (2019) Ultrasmall Ni/NiO nanoclusters on thiol-functionalized and -exfoliated graphene oxide nanosheets for durable oxygen evolution reaction. ACS Appl Energy Mater 2(1):363–371. https://doi.org/10.1021/acsaem.8b01375

17. ul Haq T, Pasha M, Tong Y, Mansour SA, Haik Y (2022) Au nanocluster coupling with Gd–Co_2B nanoflakes embedded in reduced TiO_2 nanosheets: seawater electrolysis at low cell voltage with high selectivity and corrosion resistance. Appl Catal B Environ. https://doi.org/10.1016/j.apcatb.2021.120836
18. Anantharaj S, Karthik PE, Noda S (2021) The significance of properly reporting turnover frequency in electrocatalysis research. Angew Chem Int Ed. https://doi.org/10.1002/anie.202110352
19. Anantharaj S, Karthik PE, Kundu S (2017) Petal-like hierarchical array of ultrathin $Ni(OH)_2$ nanosheets decorated with $Ni(OH)_2$ nanoburls: a highly efficient OER electrocatalyst. Catal Sci Technol. https://doi.org/10.1039/c6cy02282k
20. Anantharaj S, Karthik PE, Subramanian B, Kundu S (2016) Pt nanoparticle anchored molecular self-assemblies of DNA: an extremely stable and efficient HER electrocatalyst with ultralow Pt content. ACS Catal. https://doi.org/10.1021/acscatal.6b00965
21. Mayet N, Servat K, Kokoh KB, Napporn TW (2019) Probing the surface of noble metals electrochemically by underpotential deposition of transition metals. Surfaces. https://doi.org/10.3390/surfaces2020020
22. Stoffelsma C, Rodriguez P, Garcia G, Garcia-Araez N, Strmcnik D, Marković NM, Koper MTM (2010) Promotion of the oxidation of carbon monoxide at stepped platinum single-crystal electrodes in alkaline media by lithium and beryllium cations. J Am Chem Soc. https://doi.org/10.1021/ja106389k
23. Kashale AA, Hsu FC, Juang RH, Chen IWP (2021) Coupling of thermal and electrochemical-activated stainless-steel mesh as a highly robust electrocatalyst for oxygen evolution reaction. ACS Appl Energy Mater. https://doi.org/10.1021/acsaem.1c02407
24. ul Haq T, Haik Y, Hussain I, Rehman HU, Al-Ansari TA (2021) Gd-doped Ni-oxychloride nanoclusters: new nanoscale electrocatalysts for high-performance water oxidation through surface and structural modification. ACS Appl Mater Interfaces 13(1):468–479. https://doi.org/10.1021/acsami.0c17216
25. Herrero E, Buller LJ, Abruña HD (2001) Underpotential deposition at single crystal surfaces of Au, Pt, Ag and other materials. Chem Rev. https://doi.org/10.1021/cr9600363
26. ul Haq T, Haik YS (2021) Doped Cu_2O–CuO nanoneedles array: free standing oxygen evolution electrode with high efficiency and corrosion resistance for seawater splitting. Catal Today. https://doi.org/10.1016/j.cattod.2021.09.015
27. ul Haq T, Haik Y (2022) NiO_x–FeO_x nanoclusters anchored on g-C_3N_4 sheets for selective seawater oxidation with high corrosion resistance. ACS Sustain Chem Eng 10(20):6622–6632. https://doi.org/10.1021/acssuschemeng.2c00359
28. McCrory CCL, Jung S, Peters JC, Jaramillo TF (2013) Benchmarking heterogeneous electrocatalysts for the oxygen evolution reaction. J Am Chem Soc 135(45):16977–16987. https://doi.org/10.1021/ja407115p
29. Peugeot A, Creissen CE, Karapinar D, Tran HN, Schreiber M, Fontecave M (2021) Benchmarking of oxygen evolution catalysts on porous nickel supports. Joule. https://doi.org/10.1016/j.joule.2021.03.022
30. Wei C, Rao RR, Peng J, Huang B, Stephens IEL, Risch M, Xu ZJ, Shao-Horn Y (2019) Recommended practices and benchmark activity for hydrogen and oxygen electrocatalysis in water splitting and fuel cells. Adv Mater 31(31):1–24. https://doi.org/10.1002/adma.201806296
31. McCrory CCL, Jung S, Ferrer IM, Chatman SM, Peters JC, Jaramillo TF (2015) Benchmarking hydrogen evolving reaction and oxygen evolving reaction electrocatalysts for solar water splitting devices. J Am Chem Soc. https://doi.org/10.1021/ja510442p
32. Binninger T, Fabbri E, Kötz R, Schmidt TJ (2014) Determination of the electrochemically active surface area of metal-oxide supported platinum catalyst. J Electrochem Soc. https://doi.org/10.1149/2.055403jes
33. Wang Y, Tang J, Kong B, Jia D, Wang Y, An T, Zhang L, Zheng G (2015) Freestanding 3D graphene/cobalt sulfide composites for supercapacitors and hydrogen evolution reaction. RSC Adv. https://doi.org/10.1039/c4ra15912h

34. ul Haq T, Mansour S, Haik Y (2022) Electronic and structural modification of Mn_3O_4 nanosheets for selective and sustained seawater oxidation. ACS Appl Mater Interfaces. https://doi.org/10.1021/acsami.1c24304

Chapter 6
Bottlenecks in Water Electrolysis: A Comprehensive Exploration for Hydrogen Production

6.1 Challenges in Water Electrolysis

Despite its potential as a sustainable and clean method for hydrogen production, there are several bottlenecks that have hindered the commercialization of water electrolysis. Membrane issues, including degradation, fouling, and poor selectivity, and electrocatalyst stability issues are among the main challenges that have limited the efficiency and durability of electrolysis cells. These problems are caused by various factors such as harsh operating conditions, impurities in the feedwater, and membrane material limitations. Therefore, it is essential to understand these issues and develop solutions to overcome them to make water electrolysis viable for large-scale hydrogen production. Addressing these bottlenecks through innovative approaches and advanced materials can help improve the efficiency, durability, and cost-effectiveness of electrolysis cells, thereby accelerating the commercialization of this promising technology.

6.2 Membrane Challenges in Electrolysis

Mostly water electrolysis technologies need an efficient membrane to separate the anode and cathode compartments, which allows for the selective transfer of protons and maintains the electrical neutrality of the cell. The membrane also prevents gas crossover, which can reduce the efficiency of the cell and cause safety hazards [1]. However, several challenges are associated with the use of membranes in water electrolysis, which can affect the performance, durability, and cost-effectiveness of the process.

1. Membrane degradation: Membrane degradation is a major challenge in membrane-based water electrolysis. The membranes used in water electrolysis are typically ion exchange membranes that allow the transport of ions while

T. u. Haq and Y. Haik, *Electrochemical Water Splitting*, Materials Horizons: From Nature to Nanomaterials, https://doi.org/10.1007/978-981-99-9860-9_6

preventing the mixing of reactants and products. The membranes are usually made of polymers such as perfluoro sulfonic acid (PFSA) or hydrocarbon-based materials [2].

During water electrolysis, the operating conditions such as temperature, pH, and current density can cause various degradation mechanisms in the membranes, leading to reduced performance and shorter lifespan. The following are some of the common degradation mechanisms:

- Chemical degradation: The high pH and oxidizing environment of the anode can cause the hydrolysis of the membrane's polymer chains, resulting in a loss of mechanical strength and ion exchange capacity. This chemical degradation can lead to pinholes and cracks in the membrane, allowing the mixing of reactants and products, which reduces the efficiency of the electrolysis process [3].
- Mechanical degradation: The membranes can undergo mechanical degradation due to factors such as high pressure differentials, thermal expansion, and mechanical stresses. Mechanical degradation can result in cracks, pinholes, and reduced thickness, leading to decreased ion selectivity and increased reactant crossover [3].
- Biological degradation: In some cases, bacteria can grow on the membrane's surface and degrade it through enzymatic reactions, leading to fouling and decreased performance.
- Electromigration: The migration of ions across the membrane can lead to an accumulation of charge and cause degradation of the membrane. This degradation can result in the formation of pinholes and reduced selectivity [4].

2. Membrane fouling: Membrane fouling is a common issue that occurs during water electrolysis, particularly in the case of using membrane-based electrolyzers. Fouling refers to the accumulation of unwanted materials on the surface of the membrane, which can lead to a decrease in the efficiency and lifespan of the membrane. This issue arises due to various reasons, including the characteristics of the feedwater, operational conditions, and the membrane's properties. Membrane fouling is mainly categorized into three types: scaling, biofouling, and particulate fouling [5].

    - Scaling occurs when the solubility of the dissolved salts in the feedwater is exceeded, causing precipitation and accumulation on the membrane's surface. The formation of scales can cause membrane pore blocking, reducing the membrane's transport properties, and ultimately leading to a drop in the system's efficiency. Scaling can be prevented by maintaining the feedwater's pH and temperature within the recommended range and using anticipants [6].
    - Biofouling is the accumulation of biological material, such as bacteria and algae, on the membrane surface, causing pore blocking and surface damage. Biofouling can be prevented by maintaining the feedwater's biocide levels within the recommended range and using pre-treatment techniques such as ultraviolet irradiation, chlorination, or ozonation [7].
    - Particulate fouling occurs when suspended solid particles in the feedwater accumulate on the membrane surface, leading to pore blocking and membrane

damage. Particulate fouling can be prevented by using pre-treatment techniques such as sedimentation or filtration [8].

3. Cost: The cost of membrane-based water electrolysis is higher than other hydrogen production technologies such as steam methane reforming or coal gasification. The high cost is due to the cost of the membrane, electrodes, and other components. The high cost of the technology makes it less attractive for commercial applications [9].
4. Low efficiency: The efficiency of membrane-based water electrolysis is lower than other hydrogen production technologies. The low efficiency is due to the high overpotential required for water electrolysis in the presence of the membrane. The overpotential results in increased energy consumption and decreased efficiency [10].
5. Membrane stability: The stability of the membrane is a key factor in the efficiency of membrane-based water electrolysis. The membrane must be stable in the presence of the electrolyte, which can be acidic or basic. Acidic electrolytes can lead to the degradation of the membrane, while basic electrolytes can lead to the formation of unwanted compounds on the membrane surface. Membrane stability is crucial for the long-term efficiency of membrane-based water electrolysis [11].
6. Gas crossover: Gas crossover is a critical issue associated with membrane-based water electrolysis. The gaseous products of the electrolysis process, hydrogen, and oxygen have the tendency to pass through the membrane and mix with each other. This results in an inefficient production of the desired end product and can also be dangerous due to the risk of explosion. The gas crossover issue is mostly associated with the degradation or failure of the membrane. When the membrane degrades or is damaged, its selectivity is compromised, allowing gases to pass through it. One of the main causes of membrane degradation is exposure to highly oxidative environments, which occurs during the water oxidation reaction. This can lead to the formation of reactive oxygen species (ROS) that attack the membrane's functional groups and degrade the polymer structure. Another cause of membrane degradation is mechanical damage caused by the pressure generated during gas evolution, which can lead to cracking, tearing, and delamination of the membrane [12]. The mechanical damage can also occur due to the presence of foreign particles in the feedwater, which can cause abrasion to the membrane surface. In addition to membrane degradation, the membrane's properties, such as pore size and thickness, can also affect gas crossover. Membranes with larger pore sizes or thinner thickness are more prone to gas crossover, as they offer less resistance to gas diffusion [13].
7. Scale-up: The scale-up of membrane-based water electrolysis is challenging due to the complexity of the technology. The efficiency and performance of the technology can be affected by the size of the cell, the composition of the electrolyte, and the type of membrane used. Scaling up the technology can also increase the cost of the system.

6.3 Metal Corrosion

Oxygen evolution reaction (OER) involves a series of electron transfer reactions between the electrocatalyst and reaction intermediates. Throughout the OER process, interactions between the metal and these reaction intermediates give rise to the creation of M–O bonds, significantly influencing the electrocatalyst's catalytic performance. Specifically, under certain environmental conditions, such as in electrolytes with a basic pH or subjected to high potentials, the integrity of M-M bonds might be compromised, consequently causing the metal to undergo dissolution into the electrolyte solution.

This phenomenon can be attributed to the oxidative dissolution mechanism of the electrocatalyst. This mechanism entails the movement of electrons from the surface of the metal to the reaction intermediates, ultimately resulting in the generation of metal ions that become dispersed within the electrolyte solution. The presence of these metal ions within the electrolyte subsequently facilitates reactions with the electrolyte components, leading to the development of soluble complexes. Consequently, this complex formation triggers the gradual dissolution of the metal into the surrounding electrolyte solution [14].

One way in which the interaction of electrons between the metal and the intermediates during the oxygen evolution reaction (OER) can have an impact is by causing a weakening of the M-M bonds. This weakening occurs due to the creation of active forms of oxygen, such as hydroxyl radicals and superoxide anions. These forms of oxygen are able to react with the metal's surface, leading to the formation of metal–oxygen compounds. Unfortunately, these compounds are not stable and contribute to the degradation of the M-M bonds. The reason behind this bond weakening lies in the partial transfer of electrons from the metal's surface to the oxygen-containing forms. This transfer leads to the development of metal–oxygen bonds as electrons are redistributed. This process is facilitated by the interaction between the forms of oxygen and the existing M-M bonds, which causes these bonds to break. As the new metal–oxygen bonds take shape, the stability of the metal's surface diminishes. Consequently, this reduced stability makes the catalyst more prone to dissolving when exposed to the surrounding electrolyte [15].

Electrocatalytic dissolution within the OER pertains to the creation and subsequent breakdown of metal oxide or hydroxide entities on the electrocatalyst's surface. This phenomenon transpires as an intrinsic part of the OER process. The catalyst's susceptibility to electrocatalytic dissolution is attributed to the elevated potentials necessitated by OER, which can surpass the inherent thermodynamic steadiness of the metal oxide or hydroxide species that emerge during the OER process. The dissolution of these metal oxide or hydroxide species follows several conceivable pathways. Among them is the development of intermediary species, like metal oxyhydroxides, during the OER course. These intermediaries exhibit the potential to undergo successive oxidative processes and subsequent dissolution [16].

A range of factors influences the process of transition metal dissolution in the context of OER. These factors encompass characteristics like the specific type of metal, the surrounding conditions of the reaction environment, and the conditions under which the process occurs. An illustrative instance is provided by investigations indicating that iron dissolution within the OER mechanism hinges on the creation of a notably soluble intermediate known as FeOOH. In contrast, the dissolution of nickel in the same process is attributed to the generation of a comparatively less soluble compound identified as NiO [17]. The mechanism of metal corrosion during water electrolysis can be described as an electrochemical process that occurs at the interface between the metal electrode and the electrolyte. The corrosion process can be divided into two main reactions: an oxidation reaction and a reduction reaction. The oxidation reaction occurs at the anode, which is typically made of a metal such as nickel, iron, or titanium. The metal atoms lose electrons and are oxidized to form metal ions in the electrolyte. This process can be represented by the following Eq. 6.1 [18]:

$$M \rightarrow M^{+} + e^{-} \tag{6.1}$$

where M represents the metal electrode and e^{-} represents an electron.

The reduction reaction occurs at the cathode, which is typically made of a metal such as platinum or gold. The hydrogen ions in the electrolyte gain electrons and are reduced to form hydrogen gas. This process can be represented by the following equation (Eq. 6.2):

$$2H^{+} + 2e^{-} \rightarrow H_2 \tag{6.2}$$

where H^{+} represents a hydrogen ion and H_2 represents hydrogen gas. During the corrosion process, the metal ions that are formed at the anode can react with other ions in the electrolyte, such as hydroxide ions or oxygen ions, to form metal oxide or metal hydroxide compounds. For example, if the anode is made of iron, the following reaction can occur (Eq. 6.3) [19]:

$$4Fe^{2+} + O_2 + 2H_2O \rightarrow 4Fe(OH)_2 \tag{6.3}$$

where Fe^{2+} represents an iron ion and O_2 represents an oxygen ion. The metal oxide or hydroxide that is formed can then react with more hydrogen ions in the electrolyte, leading to further metal dissolution and the formation of more metal oxide or hydroxide compounds.

Binninger et al. have proposed a mechanism for the metal corrosion during OER, which involves the electrochemical oxidation of the metal cation $M_{LOER}{}^{2n+}$ to a higher oxidation number, resulting in an increased solubility in the electrolyte (Eq. 6.4) [20].

$$M_{LOER}^{2n} \rightarrow M_{(aq)}^{(2n+m)+} + me^{-} \tag{6.4}$$

In specific electrochemical response, the metal cation represented as $M_{LOER}{}^{2n+}$ can experience an oxidation process leading to an elevated oxidation state. This transition consequently contributes to an augmented solubility within the electrolyte solution. This sequence of events is denoted by the formulation presented as Eq. 6.4. The electrochemical process of oxidizing the metal cation induces the shift to a higher oxidation state and prompts the creation of cations bearing higher valences. These cations exhibit an enhanced solubility within the electrolyte medium. This heightened solubility encourages their dispersion from the electrode, as depicted in Eq. 6.5.

$$MO_n \rightarrow M_{(aq)}^{(2n+m)+} + 1/2nO_2 + (2n + m) \tag{6.5}$$

In this cationic cycle, concurrent oxidation is observed in the lattice metal cation and the oxygen anion. This oxidation process is instrumental in creating a three-dimensional boundary layer at the junction where the electrolyte interfaces with the metal oxide electrode. This particular layer is given the term "hydrous amorphous layer." Within this layer, distinct active sites are generated that serve as locations for the adsorption of intermediates. While the cationic cycle triggers dynamic alterations within the hydrous amorphous layer, it is essential to note that it does not result in any substantial mass reduction for the electrode.

The oxygen evolution reaction (OER) is a critical process for renewable energy conversion and storage. However, during the OER process, metal catalysts can undergo a surface reconstruction process due to the high oxidation potentials, leading to metal oxides or (oxy)hydroxides forming. Early studies have shown that transition metal sulfides, selenides, nitrites, and phosphides demonstrate suitable OER activities under alkaline conditions. However, it has been revealed that these catalysts undergo a reconstruction process during the reaction, leading to the formation of corresponding metal (oxy)hydroxides on the surface or even in bulk, which serve as the true catalytically active species and stable phases for OER. For instance, Ni–Fe-based (oxy)hydroxides, derived from transition metal-based alloys, are one of the most efficient OER catalysts under alkaline conditions [21].

On the other hand, surface reconstruction can also occur on OER catalysts under acidic conditions, which poses a challenge for designing stable and active catalysts. Only a few catalysts remain stable during the oxidation process in acidic media; nearly all are noble metal-based catalysts. Among these candidates, Ir- and Ru-based catalysts are two well-known materials that show good catalytic activity and stability for acidic OER. Atom probe tomography (APT) has revealed that the surface of bulk Ir and Ru metal would be oxidized into IrO_2 and RuO_2 during the catalytic process. Under the anodic current during oxygen evolution, the first few atomic layers of the pure Ir film gradually transform into non-stoichiometric Ir–O species and finally into IrO_2 as the stable phase for this electrocatalytic reaction. These findings suggest that the pristine catalysts are not the real active species nor the stable phase under the electrocatalytic environment due to the possible surface reconstruction process. Therefore, designing and developing catalysts that can resist surface reconstruction and maintain their activity and stability is a critical task for efficient OER (Fig. 6.1) [22].

For the electrode to attain a state of dynamism and stability, all metal cations must be engaged in the cyclic progression. Solely, those metal cations departing from this

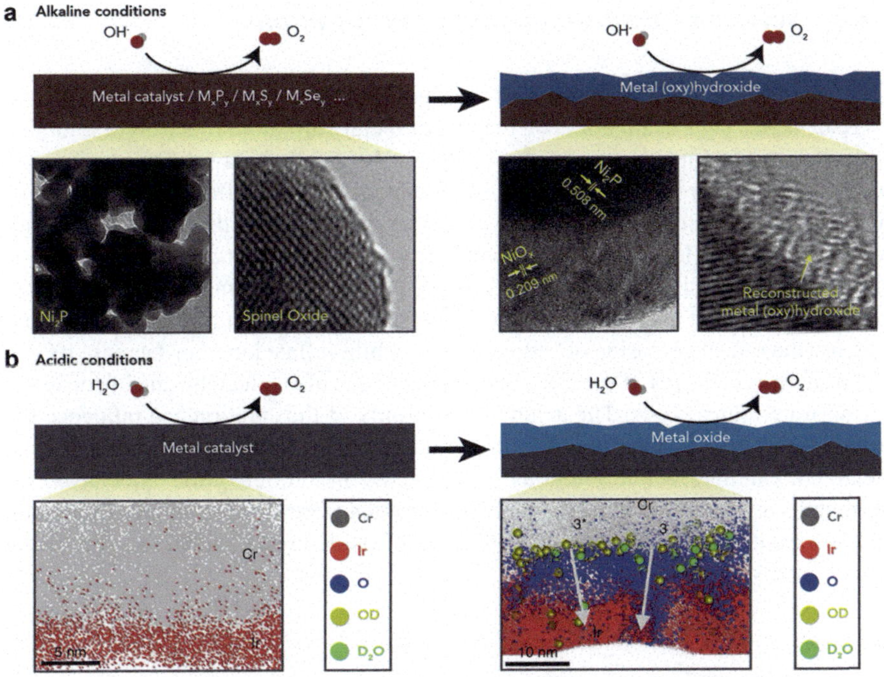

Fig. 6.1 **a** When exposed to alkaline conditions, certain catalysts containing transition metals, metal phosphides, sulfides, and selenides can undergo a reconstruction process that transforms them into metal (oxy)hydroxides. HRTEM images of Ni2P and spinel oxide CoFe0.25Al1.75O4 (shown on the left) illustrate this reconstruction process, with corresponding metal (oxy)hydroxides shown on the right. **b** Alternatively, under acidic conditions, metal-based catalysts (shown on the left) can be reconstructed into metal oxides. APT images of Ir (also shown on the left) being transformed into IrO_2 (shown on the right) are provided as an example. Reproduced with permission from Ref. [22], (c) 2021, cell press

cycle and subsequently dissolving into the electrolyte contribute to the electrode's loss in mass. The metal oxide electrode experiences a chemical progression where oxygen anions undergo oxidation due to the influence of positively charged metal cations known as $M_{LOER}{}^{2n+}$. This oxidation results in the transference of oxygen anions to a region known as the hydrous amorphous layer. This course of action bears semblance to the diffusion of oxygen vacancies from the surface stratum toward the core of the metal oxide. Consequently, this diffusion dynamic fosters the expansion of the boundary layer, endowing it with a three-dimensional character [23].

The following are factors that can affect metal corrosion during water electrolysis.

6.3.1 Solution Composition and Concentrations

The composition and concentration of the electrolyte solution used in water electrolysis can have a significant effect on the rate of metal corrosion. Electrolyte solutions typically contain ions that facilitate the electrochemical reactions that occur on the surface of the electrodes. The concentration of these ions can affect the rate of corrosion by increasing the number of active sites on the surface of the metal. Higher ion concentrations can lead to higher rates of corrosion, as more active sites are available for the electrochemical reactions to occur. Additionally, the composition of the electrolyte solution can also affect the rate of corrosion. For example, chloride ions can increase the rate of corrosion of certain metals, while sulfate ions can decrease it [24].

Furthermore, the pH of the electrolyte solution can also affect the rate of corrosion during water electrolysis. The acidity or alkalinity of the solution can influence the corrosion process by affecting the formation of oxide layers on the surface of the metal. For example, acidic solutions can cause the dissolution of the oxide layers on the metal surface, which can lead to increased rates of corrosion. On the other hand, alkaline solutions can promote the formation of oxide layers, which can reduce the rate of corrosion [25].

6.3.2 Diffusion Rate of Ions

The diffusion rates of ions and oxygen gases in the electrolyte solution can also affect the rate of metal corrosion during water electrolysis. During the electrolysis process, ions and oxygen gases move toward the electrodes and react with the metal surfaces, leading to corrosion. The rate at which these species diffuse toward the electrodes can determine the rate of corrosion. For example, if the diffusion rate of oxygen gas toward the anode is high, it can result in increased corrosion of the anode due to the formation of metal oxides. Similarly, if the diffusion rate of ions toward the cathode is high, it can lead to increased hydrogen evolution, which can cause hydrogen embrittlement of the cathode metal [26].

Additionally, the presence of stagnant regions in the electrolyte solution can also affect the diffusion rates of ions and oxygen gases. If there are stagnant regions in the solution, such as in dead zones or at the bottom of the electrolytic cell, the diffusion rates of ions and oxygen gases toward the electrodes can be reduced. This can result in localized areas of higher corrosion rates, where the metal surface is exposed to a higher concentration of corrosive species. Therefore, it is important to ensure that the electrolyte solution is well-agitated during the electrolysis process to prevent the formation of stagnant regions and promote the diffusion of ions and oxygen gases toward the electrodes [26].

6.3.3 Surrounding Conditions

The surrounding atmosphere can also have an effect on the rate of metal corrosion during water electrolysis. The presence of certain gases, such as oxygen or chlorine, can accelerate the corrosion process by promoting the formation of metal oxides or metal chlorides, respectively. For example, if the electrolysis cell is operated in an atmosphere containing high levels of oxygen, it can lead to the formation of metal oxides on the surface of the electrodes, resulting in increased rates of corrosion. Similarly, if the cell is operated in an atmosphere containing chlorine, it can lead to the formation of metal chlorides, which can also accelerate the corrosion process [24]. In addition to gases, the presence of moisture in the surrounding atmosphere can also affect the rate of metal corrosion during water electrolysis. Moisture can promote the formation of oxide layers on the metal surfaces, which can increase the resistance of the electrodes and reduce the efficiency of the electrolysis process. It can also lead to the formation of localized areas of higher corrosion rates, especially if there are variations in the moisture content of the surrounding atmosphere.

6.3.4 Reaction Conditions

The reaction conditions during water electrolysis can also have a significant effect on the rate of metal corrosion. Factors such as the temperature, current density, and voltage applied to the electrodes can all influence the electrochemical reactions that occur on the metal surfaces, which can lead to changes in the rate of corrosion. For example, higher temperatures can increase the rate of electrochemical reactions, which can lead to increased rates of metal corrosion. Similarly, higher current densities or voltages can increase the rate of ion transfer toward the electrodes, which can also lead to increased rates of metal corrosion [27]. Furthermore, the duration of the electrolysis process can also affect the rate of metal corrosion. Longer electrolysis times can lead to increased exposure of the metal surfaces to the corrosive electrolyte solution, which can result in increased rates of corrosion. Therefore, it is important to carefully control the reaction conditions during water electrolysis to minimize the effects of these factors on the rate of metal corrosion. This can be achieved by optimizing the temperature, current density, voltage, and electrolysis time to minimize the corrosion rate while still achieving the desired level of electrolysis efficiency.

6.3.5 Electrode Configurations

The electrode configuration can also play a role in the rate of metal corrosion during water electrolysis. Different types of electrode configurations can result in different levels of corrosion due to variations in the local electrochemical environment around the electrodes. For example, in a parallel plate electrode configuration, where two electrodes are placed parallel to each other in the electrolyte solution, the electric field lines are straight and uniform, which can lead to a more uniform distribution of the electric field and the flow of ions toward the electrodes. This can result in more uniform corrosion rates across the electrode surfaces [28].

On the other hand, in a concentric electrode configuration, where one electrode is placed inside another electrode, the electric field lines are more complex, which can lead to variations in the local electrochemical environment around the electrodes. This can result in localized areas of higher corrosion rates, where the electric field is more intense, or lower corrosion rates, where the electric field is weaker. Therefore, it is important to carefully consider the electrode configuration when designing an electrolysis cell to minimize the effects of these variations on the rate of metal corrosion.

In the field of heterogeneous catalysis, synthesized materials are commonly characterized through bulk or near-surface characterization methods such as X-ray diffraction (XRD) or X-ray photoelectron spectroscopy (XPS). However, this approach has limitations. Firstly, the conductivity of the catalyst is an important bulk material property that is often overlooked in many fundamental studies. This omission makes it difficult to assess the catalyst's performance in subsequent tests in membrane electrode assemblies. Secondly, heterogeneous catalysis typically occurs only on the surface of the catalyst, making the surface atomic structure and composition crucial characteristics (Fig. 6.2) [29]. Unfortunately, surface analysis of the catalyst is rarely studied in depth experimentally. This issue is particularly significant for catalysts containing low-cost metals, as these elements have a high susceptibility to surface oxidation and decomposition when exposed to air. The resulting oxide layer can change the composition/stoichiometry of the surface and several layers below it. Furthermore, catalyst dissolution is a common cause of performance degradation, particularly under harsh conditions during OER, including strong acidic or alkaline environments and oxidation environments. These challenges highlight the need for comprehensive characterization techniques that incorporate both bulk and surface analyses, particularly for catalysts containing low-cost metals.

6.4 Structural Instability

The structural instability of electrocatalysts refers to the phenomenon where the catalysts undergo structural changes during electrocatalysis, resulting in a decrease in their activity or stability. The structural changes can occur in various forms, such as phase transformation, surface reconstruction, or dissolution. The causes of structural instability can vary, depending on the type of electrocatalyst and the reaction conditions. One of the main causes of structural instability is the high potential and harsh conditions during electrocatalysis. The high potential can induce oxidation or reduction reactions on the surface of the catalyst, leading to the formation of surface oxides or reduction products. These surface products can affect the surface chemistry

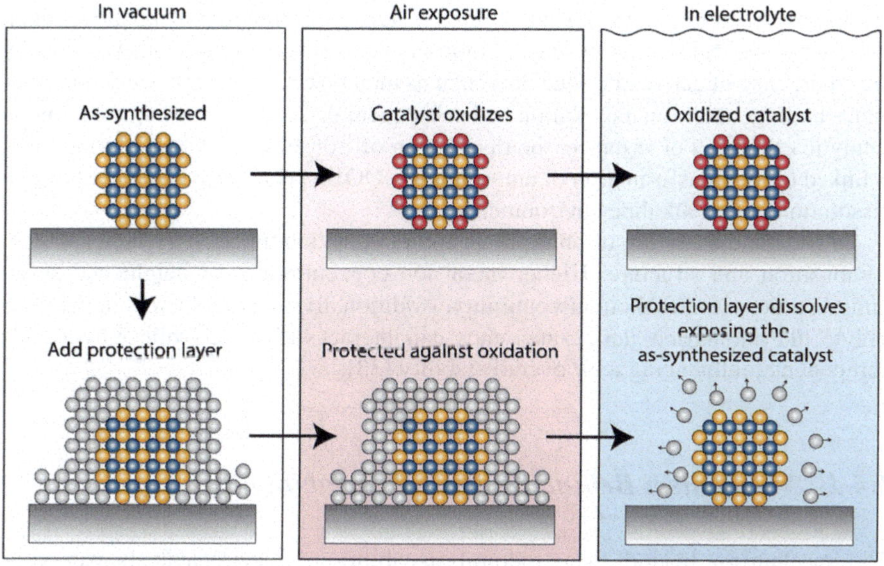

Fig. 6.2 This illustration shows the oxidation of a catalyst during transportation in air (top row), which can cause the oxide to be exposed on the surface during electrochemistry. To prevent this oxidation, a protection layer can be applied (bottom row). This layer can be dissolved when the catalyst is immersed into the electrolyte. Reproduced with permission from Ref. [28]. © 2023, American Chemical Society

of the catalyst, leading to changes in its activity or selectivity. The harsh conditions, such as high temperature, pressure, and pH, can also affect the structural stability of the catalyst by inducing phase transformations or dissolution [30].

In the oxygen evolution reaction (OER) process, the electrocatalyst's surface can restructure, resulting in diminished activity. This change arises from interactions between metal oxide electrocatalysts and protons generated during OER. As per the Lux classification, metal oxides are considered basic, and this interaction with protons can lead to their protonation, causing corrosion, dissolution, or amorphization [31]. Amorphous compounds denote substances characterized by an absence of well-defined atomic arrangement, leading to a disorganized atomic structure. These compounds can emerge through diverse mechanisms, including swift cooling or solution precipitation. When transition metals dissolve during OER in alkaline conditions, the amorphous compounds manifest due to the quick and unregulated expansion of metal oxide or hydroxide species on the catalyst's surface. This rapid expansion of these entities can create irregular and disordered amorphous compounds, diverging from the usual organized crystalline formations [32].

First, the amorphous compounds' disorganized atomic arrangement contributes to a notable structural disorder and defects, amplifying their sensitivity to dissolving within alkaline OER conditions. This phenomenon arises due to the rapid formation and growth of metal oxide or hydroxide components on the catalyst's surface, leading to structural anomalies due to the challenge of arranging metal atoms in an orderly manner. Subsequently, the dissolution of amorphous compounds is driven by the emergence of active intermediates such as metal oxyhydroxides. These intermediates undergo additional oxidation and subsequent dissolution, causing a decline in catalytic efficacy. For instance, the dissolution of iron-based catalysts during OER is linked to the transformation of amorphous FeOOH species into Fe_2O_3, leading to dissolution within alkaline environments.

Lastly, the dissolution of amorphous compounds is influenced by their inherent composition and structure. Higher metal ion concentrations or heightened structural disorder elevates their susceptibility. Additionally, external factors in the electrolyte, like anions or organic compounds, can interact with metal oxide or hydroxide components, influencing their overall stability [33].

6.4.1 Mechanism Behind Structural Instability

The mechanism behind the structural instability of electrocatalysts can vary, depending on the type of catalyst and reaction conditions. However, some general mechanisms can be identified, which are discussed below.

6.4.1.1 Surface Reconstruction

One of the most common mechanisms of structural instability is surface reconstruction. During electrocatalysis, the high potential and harsh conditions can induce surface reconstruction, where the surface atoms rearrange to form a new surface structure. This new surface structure can be different from the initial surface structure, leading to changes in the surface chemistry and activity of the catalyst. The surface reconstruction can occur through various mechanisms, such as adsorption, diffusion, and migration of surface species [34].

Adsorption

Adsorption is the process of a molecule or atom attaching to the surface of a material. During water electrolysis, water molecules adsorb on the surface of the catalyst material. The adsorption of water molecules can lead to surface reconstruction due to the reaction between the water molecules and the surface atoms. For example, on the surface of the Ni–Fe$(OH)_x$ catalyst, the OH^- ions in the water molecules adsorb on the Ni site, and the H^+ ions adsorb on the Fe site, leading to the formation of the Ni–OH and Fe–OH bonds. This reaction can result in the transformation of the catalyst surface from metal-to-metal oxide or hydroxide [35].

Diffusion

Diffusion is the process of the movement of atoms or molecules from high to low concentration regions. During water electrolysis, the diffusion of oxygen atoms and hydroxyl ions on the surface of the catalyst material can lead to surface reconstruction. The diffusion of oxygen atoms on the surface of the catalyst can cause the formation of metal oxides. For example, in the case of the RuO_2 catalyst, the oxygen atoms diffuse on the surface of the catalyst and form Ru–O bonds, leading to the transformation of the catalyst surface from metal-to-metal oxide. Similarly, the diffusion of hydroxyl ions on the surface of the catalyst can also lead to the formation of metal hydroxides [36].

Migration

Migration is the process of the movement of atoms or molecules from one position to another on the surface of a material. During water electrolysis, the migration of surface species such as metal ions, oxygen atoms, and hydroxyl ions on the surface of the catalyst material can lead to surface reconstruction. The migration of metal ions on the surface of the catalyst can cause the formation of metal oxides or hydroxides. For example, on the surface of the Co-based catalyst, the migration of Co ions can lead to the formation of Co–OH bonds, leading to the transformation of the catalyst surface from metal-to-metal hydroxide. Similarly, the migration of oxygen atoms and hydroxyl ions on the surface of the catalyst can also lead to surface reconstruction [37].

During water electrolysis, the surface reconstruction of the catalyst material can occur through various mechanisms, as discussed above. The mechanism of surface reconstruction depends on the nature of the catalyst material and the reaction conditions. For example, on the surface of the Ni–Fe$(OH)_x$ catalyst, the adsorption of water molecules leads to the formation of Ni–OH and Fe–OH bonds, while on the surface of the RuO_2 catalyst, the diffusion of oxygen atoms leads to the formation of Ru–O bonds. Similarly, the migration of surface species such as metal ions, oxygen atoms, and hydroxyl ions on the surface of the catalyst can also lead to surface reconstruction.

6.4.1.2 Phase Transformation

The phase transformation mechanism of electrocatalysts during water splitting involves several steps, including adsorption, diffusion, nucleation, and growth of new phases. The first step involves the adsorption of water and OH^- ions on the catalyst surface, which leads to the formation of surface hydroxides. The surface hydroxides then undergo a diffusion process, which involves the migration of OH^- ions and H_2O molecules across the surface of the catalyst. The migration of these species can cause the catalyst to undergo structural changes, leading to the formation of new phases [38].

Nucleation is the next step in the phase transformation mechanism and involves the formation of new phases on the surface of the catalyst. The nucleation process occurs when the concentration of the surface hydroxides exceeds the solubility limit, leading to the formation of small clusters or islands of the new phase. The size and density of the clusters depend on various factors, such as the concentration of surface hydroxides, the surface energy of the catalyst, and the temperature and pressure conditions [39].

The growth of the new phase occurs through the addition of surface hydroxides to the clusters, leading to the formation of larger and more stable clusters. The growth process is influenced by several factors, including the concentration of surface hydroxides, the surface energy of the catalyst, the temperature and pressure conditions, and the presence of other species, such as CO_2 and O_2.

The stability of the electrocatalyst during water splitting depends on several factors, such as the composition and morphology of the catalyst, the reaction conditions, and the interaction between the catalyst and the electrolyte. The composition of the catalyst can influence its stability by affecting its surface energy and its affinity for surface species, such as water and OH^- ions. For example, the surface energy of the catalyst can affect the adsorption and diffusion of surface species, leading to different phase transformation mechanisms. The morphology of the catalyst can also affect its stability by influencing the diffusion and migration of surface species. For example, a high surface area catalyst can have a higher concentration of surface hydroxides, leading to a higher propensity for phase transformation. Similarly, a catalyst with a rough surface can have a higher surface energy, leading to a higher affinity for surface species and a higher propensity for phase transformation [40]. The reaction conditions can also affect the stability of the catalyst by influencing the concentration of surface species, the temperature and pressure conditions, and the presence of other species in the electrolyte. For example, a high concentration of OH^- ions can lead to a higher concentration of surface hydroxides, increasing the propensity for phase transformation. Similarly, high temperatures and pressures can lead to faster diffusion and growth of the new phase, leading to a higher propensity for phase transformation [41].

6.4.1.3 Dissolution

Dissolution is another mechanism of structural instability, where the electrocatalyst dissolves in the electrolyte during electrocatalysis. The dissolution can occur due to the high potential and harsh conditions, which can induce the oxidation or reduction of the catalyst, leading to the formation of soluble products [42]. The dissolution can affect the activity and stability of the catalyst by reducing its surface area and changing its composition. For example, during OER, the dissolution of nickel-based catalysts can lead to the formation of soluble nickel ions, which can affect the surface chemistry and activity of the catalyst. The dissolution of an electrocatalyst can cause structural instability, which can lead to reduced catalytic activity and efficiency, as well as increased resistance and decreased durability. There are several mechanisms by which dissolution can lead to structural instability of an electrocatalyst. One of these mechanisms is the loss of active sites on the catalyst surface. Active sites are the locations on the catalyst surface where the electrochemical reaction takes place. When the catalyst dissolves, these active sites can be lost, reducing the surface area available for catalysis and lowering the overall catalytic activity [43].

Another mechanism is the formation of surface defects and changes in surface morphology. As the catalyst dissolves, the surface can become rougher, with pores and cracks forming on the surface. These defects can lead to increased resistance and decreased efficiency, as they can interfere with the electrochemical reaction and limit the transport of reactants and products. Dissolution can also lead to the formation of new chemical species that can interfere with the electrochemical reaction. For example, in water electrolysis, the dissolution of the catalyst can lead to the formation of metal ions in the electrolyte, which can then react with other species in the electrolyte to form new compounds. These compounds can clog the catalyst pores and inhibit the reaction [44].

6.4.2 Agglomerations

Agglomeration of particles can lead to a loss of structural stability of the electrocatalyst and negatively affect the performance of the water electrolysis system. In this report, we will discuss the basic mechanisms of particle agglomeration and its effects on the structural stability of electrocatalysts during water electrolysis. Particle agglomeration can occur through various mechanisms, including van der Waals forces, electrostatic interactions, and chemical bonding. Van der Waals forces are weak intermolecular forces that attract particles together when they are in close proximity. These forces are proportional to the distance between particles, and they increase as the distance between particles decreases. When particles are in close proximity, the van der Waals forces can cause them to stick together, leading to particle agglomeration [27].

Electrostatic interactions are another mechanism of particle agglomeration. Electrostatic interactions can occur between particles with opposite charges or between particles and charged surfaces. When particles have opposite charges, they can be attracted to each other and stick together. Similarly, when particles are in contact with charged surfaces, they can be attracted or repelled, leading to particle agglomeration or dispersion, respectively.

Chemical bonding is a third mechanism of particle agglomeration. Chemical bonding can occur when the surfaces of particles are reactive and can form chemical bonds with each other. This can happen through covalent bonding or through the formation of intermolecular forces, such as hydrogen bonding [45].

Effects of Particle Agglomeration on Electrocatalyst Structural Stability:

Particle agglomeration can have several negative effects on the structural stability of electrocatalysts during water electrolysis. One effect is the formation of larger particles, which can lead to a reduction in the surface area available for catalysis. Catalytic activity is directly proportional to the surface area of the catalyst, and as the surface area decreases due to particle agglomeration, the catalytic activity of the electrocatalyst is reduced.

Particle agglomeration can also lead to the formation of large agglomerates that are difficult to disperse in the electrolyte. Large agglomerates can settle at the bottom of the cell, leading to a loss of contact between the electrocatalyst and the electrolyte. This can result in a reduction in the catalytic activity and efficiency of the electrocatalyst. In addition, particle agglomeration can lead to the formation of cracks and defects in the electrocatalyst structure. These cracks and defects can increase the resistance of the electrocatalyst, reducing the efficiency of the water electrolysis system. Furthermore, agglomerated particles can block the pores of the electrocatalyst, inhibiting the diffusion of reactants and products and reducing the overall efficiency of the system.

Particle agglomeration can lead to the loss of structural stability of the electrocatalyst, which in turn can affect its performance and efficiency. Therefore, understanding the thermodynamics of particle agglomeration is important to prevent this problem from occurring [46].

6.4.2.1 Thermodynamics of Particle Agglomeration

The thermodynamics of particle agglomeration is governed by the free energy change associated with the process, which can be determined by the following equation (Eq. 6.6):

$$\Delta G = \Delta H - T\Delta S \tag{6.6}$$

where ΔG is the free energy change, ΔH is the enthalpy change, T is the temperature, and ΔS is the entropy change. The enthalpy change represents the energy associated with the interactions between the particles, which can be attractive or repulsive in nature. When the particles are attracted to each other, the enthalpy change is negative, indicating that the interaction is exothermic. On the other hand, if the particles repel each other, the enthalpy change is positive, indicating that the interaction is endothermic. The entropy change represents the degree of disorder in the system, which can also be positive or negative. When the particles are dispersed in the electrolyte, the entropy change is positive, indicating that the system is becoming more disordered. In contrast, when the particles agglomerate, the entropy change is negative, indicating that the system is becoming more ordered. The temperature also plays an important role in the thermodynamics of particle agglomeration. At higher temperatures, the thermal energy of the particles increases, making it easier for them to overcome the barriers to agglomeration. At lower temperatures, the thermal energy is reduced, making it more difficult for the particles to agglomerate [27].

Driving Forces Behind Particle Agglomeration: Particle agglomeration can occur due to various driving forces that depend on the electrocatalyst and electrolysis conditions. One of the main driving forces is the attractive forces between the particles, which can be due to various reasons, such as van der Waals forces, hydrogen bonding, or electrostatic interactions. Van der Waals forces are the weak attractive forces that exist between all atoms and molecules due to fluctuations in electron density. These forces can cause the particles to stick together when they come close enough to each other. Hydrogen bonding is a type of attractive force that occurs between the hydrogen atoms of one molecule and the electronegative atoms of another molecule, such as oxygen or nitrogen. These forces can cause the particles to adhere to each other. Electrostatic interactions occur when the particles have a charge or a dipole moment, which can attract or repel other charged or polar particles in the electrolyte. These forces can cause the particles to aggregate or repel each other, depending on their polarity [47].

6.4.2.2 Kinetics of Particle Agglomeration

The kinetics of particle agglomeration refers to the rate at which the particles aggregate to form larger clusters. The rate of particle agglomeration is influenced by various factors such as the concentration of the particles, the size and shape of the particles, the properties of the electrolyte, and the applied potential.

The rate of particle agglomeration can be described by the following equation (Eq. 6.7) [48]:

$$\frac{\mathrm{d}N}{\mathrm{d}t} = KN^2 \tag{6.7}$$

where $\mathrm{d}N/\mathrm{d}t$ is the rate of agglomeration, K is the rate constant, and N is the number of particles. The rate constant, K, is influenced by several factors such as the strength of the attractive forces between the particles, the diffusion rate of the particles, and the presence of inhibitors or promoters that can affect the rate of particle agglomeration.

Factors influencing the kinetics of particle agglomeration

The kinetics of particle agglomeration is influenced by several factors that can be broadly categorized into two types, namely intrinsic factors and extrinsic factors.

Intrinsic factors: The intrinsic factors that influence the kinetics of particle agglomeration include the size and shape of the particles, the concentration of the particles, and the surface properties of the particles [49].

- The size and shape of the particles can affect the rate of particle agglomeration. Smaller particles have a higher surface area-to-volume ratio, which increases the probability of their interactions and promotes particle agglomeration. Similarly, the shape of the particles can influence the rate of agglomeration. Particles with more irregular shapes have more contact points, which promote particle agglomeration.
- The concentration of the particles in the electrolyte can also affect the rate of particle agglomeration. Higher particle concentrations increase the probability of their interactions, which can promote particle agglomeration.
- The surface properties of the particles can also influence the kinetics of particle agglomeration. The surface chemistry of the particles can affect the strength of

the attractive forces between them, which can promote or inhibit particle agglomeration. Similarly, the surface charge of the particles can also influence their interactions and promote or inhibit particle agglomeration.

Extrinsic factors: The extrinsic factors that influence the kinetics of particle agglomeration include the properties of the electrolyte and the applied potential [50].

- The properties of the electrolyte can affect the rate of particle agglomeration. Electrolytes with higher viscosities can inhibit particle agglomeration by reducing the diffusion rate of the particles. Similarly, the presence of inhibitors or promoters in the electrolyte can also affect the rate of particle agglomeration.
- The applied potential can also influence the kinetics of particle agglomeration. Higher potentials can increase the strength of the attractive forces between the particles, which can promote particle agglomeration. Similarly, lower potentials can reduce the rate of particle agglomeration by reducing the probability of particle interactions.

The Markovic group's highlighted the inverse activity-stability relationship in transition metal oxides, specifically with respect to $SrRuO_3$ catalysts (Fig. 6.3b) [51]. The study demonstrated that the most active crystallographic facets exhibited the most rapid degradation, indicating a correlation between OER activity and stability. The degradation of the catalyst was measured in terms of the number of dissolved cations (i.e., Sr and Ru cations) from the catalyst into the electrolyte, with (111)-oriented facets being the most severely affected. The lower overpotential required to drive the water-splitting reaction (@100 $\mu A/cm^2$) observed for the (111)-oriented sample indicates enhanced activity of the rapidly degrading catalysts. The behavior of $SrRuO_3$ is therefore representative of the inverse activity-stability relationship observed in OER catalysts, wherein a chemical dissolution process distorts and destabilizes the lattice, leading to a vanishing perovskite structure and eventual amorphization. Studies by Danilovic et al. (2014) and Lei et al. (2020) also demonstrate a similar inverse correlation between activity and stability for pure metal catalysts [52]. This trend poses a dilemma, as materials can typically either exhibit "low activity" with high chemical stability or "high activity" with low stability, while both characteristics are necessary for an ideal catalyst (Fig. 6.3c) [52]. This behavior is widespread, as also evidenced by advanced $SrRuO_3$ bilayer structures that have managed to stabilize $SrRuO_3$-based catalysts, albeit with reduced activity. These structures involve the encapsulation of $SrRuO_3$ within a unit cell thick $SrTiO_3$ capping layer, which effectively prevents excessive leaching of cations from the catalyst. Bin Liu and colleagues have developed a solution-based technique to synthesize an amorphous nickel–iron (Ni–Fe) alloy catalyst for water oxidation. The amorphous catalyst can be transformed into a crystal structure through thermal annealing without altering its composition. The amorphous Ni–Fe alloy catalyst demonstrated superior catalytic activity for water oxidation, requiring a lower overpotential to reach a current density of 10 mA/cm^2 compared to the crystalline counterpart. The electrochemical activation of the amorphous catalyst enhances the number of active sites, improving its water oxidation activity. The amorphous catalyst also exhibited excellent stability during prolonged electrochemical reactions, making it a practical choice for water electrolysis. Furthermore, by loading the catalyst onto a 3D porous nickel foam support, the electrode could reach high current densities, making it suitable for industrial applications. The transformation of the amorphous catalyst to a crystal structure enhances the active centers in the bulk form, contributing to its superior performance [53].

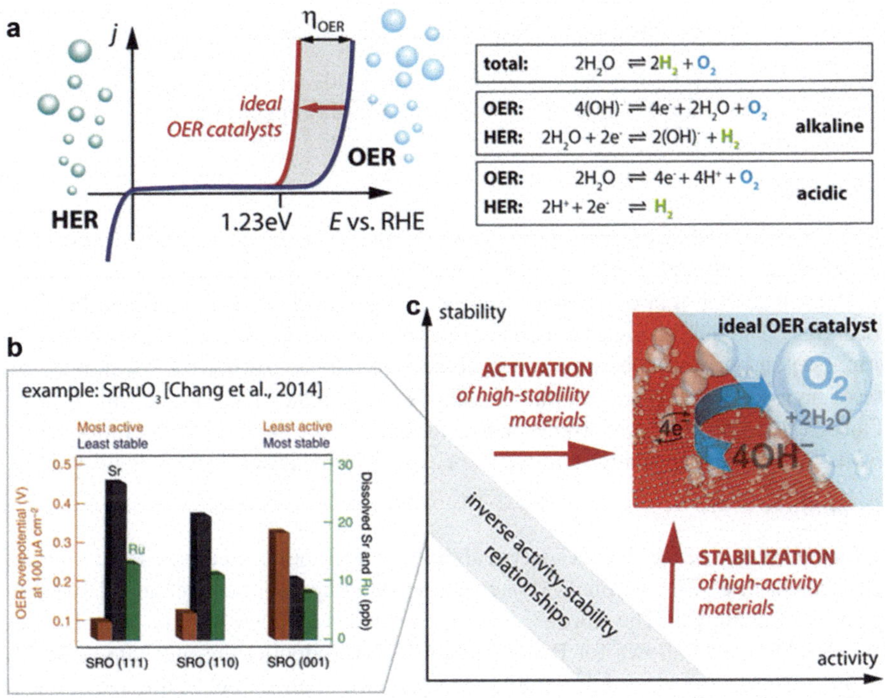

Fig. 6.3 Figure **a** illustrates the water electrolysis process in liquid electrolyte, depicting HER as well as OER processes for both acidic and alkaline media. The schematic shows the behavior of ideal (red) and real (blue) OER catalysts. In Fig. **b**, the enhanced cation dissolution of strontium and ruthenium ions from SrRuO3 thin films is shown for the most active crystal facets (11.1 The inverse relationship between chemical dissolution and required overpotential demonstrates the inverse stability-activity relationship of standard OER catalysts). Figure **c** emphasizes the need for material design strategies to activate and stabilize oxygen evolution reaction catalysts, which is necessary to overcome stability-activity dilemma

6.5 Mechanical Strength of Electrocatalysts

Mechanical strength refers to the ability of a material to resist deformation or failure under mechanical stress. During water electrolysis, an electrocatalyst is subjected to mechanical stress due to the formation and release of gas bubbles on its surface. Gas bubbles can cause mechanical deformation, cracking, or detachment of the electrocatalyst from the electrode surface. Therefore, the mechanical strength of an electrocatalyst is critical to its stability and performance during water electrolysis [54].

Factors Affecting the Mechanical Strength of Electrocatalysts:

The mechanical strength of an electrocatalyst depends on several factors, including its composition, morphology, microstructure, and the operating conditions of the electrolysis system.

Composition

The composition of an electrocatalyst plays a crucial role in determining its mechanical strength. The addition of a ductile metal, such as nickel or copper, to a brittle metal, such as platinum or iridium, can improve the mechanical strength of the electrocatalyst. The ductile metal can absorb the mechanical stress caused by gas bubble formation and prevent cracking or detachment of the electrocatalyst. Moreover, the use of nanoparticles instead of bulk materials can enhance the mechanical strength of an electrocatalyst. Nanoparticles have a higher surface area to volume ratio, which results in stronger interparticle interactions and improved mechanical stability [55].

Morphology

The morphology of an electrocatalyst also affects its mechanical strength. The use of porous or three-dimensional (3D) structures can improve the mechanical stability of an electrocatalyst. The porous structure provides a larger surface area for gas bubble release and allows for better gas diffusion, reducing the mechanical stress on the electrocatalyst. Additionally, 3D structures can enhance the mechanical stability of an electrocatalyst by providing mechanical support and reducing stress concentration [56].

Microstructure

The microstructure of an electrocatalyst, including its crystal structure, grain size, and defects, can also influence its mechanical strength. The crystal structure of an electrocatalyst can affect its mechanical properties, as different crystal structures have different mechanical properties. Moreover, the grain size of an electrocatalyst can affect its mechanical strength, with smaller grain sizes generally associated with higher mechanical strength. The presence of defects, such as dislocations or grain boundaries, can also affect the mechanical properties of an electrocatalyst [56].

Operating Conditions

The operating conditions of an electrolysis system, including the temperature, pressure, and electrolyte composition, can also affect the mechanical strength of an electrocatalyst. Higher temperatures and pressures can increase the mechanical stress on an electrocatalyst, leading to decreased mechanical strength. The electrolyte composition can also affect the mechanical stability of an electrocatalyst, with high ionic strength electrolytes generally associated with decreased mechanical strength. There are several techniques that can be used to measure the mechanical strength of an electrocatalyst. The choice of technique depends on the specific properties of the catalyst and the conditions under which it will be used. Some commonly used techniques include [57]:

Tensile Testing: This technique involves stretching a sample of the catalyst until it breaks, while measuring the force applied and the resulting deformation. This allows the determination of parameters such as yield strength, ultimate tensile strength, and elongation at break. Tensile testing is particularly useful for studying the mechanical properties of thin films or coatings.

Compression Testing: This technique involves compressing a sample of the catalyst until it deforms or fractures, while measuring the force applied and the resulting deformation. This allows the determination of parameters such as compressive strength and modulus of elasticity. Compression testing is particularly useful for studying the mechanical properties of bulk materials.

Bend Testing: This technique involves bending a sample of the catalyst until it fractures, while measuring the force applied and the resulting deformation. This allows the determination of parameters such as bending strength and modulus of elasticity. Bend testing is particularly useful for studying the mechanical properties of thin films or coatings.

Scratch Testing: This technique involves dragging a sharp or blunt object across the surface of the catalyst at a controlled speed and force, while measuring the resulting depth and width of the scratch. This allows the determination of parameters such as scratch resistance and adhesion strength. Scratch testing is particularly useful for studying the wear resistance and durability of coatings or thin films.

Nanoindentation: This technique involves pressing a small probe into the surface of the catalyst at a controlled force and measuring the resulting indentation depth and width. This allows the determination of parameters such as hardness, elastic modulus, and indentation creep. Nanoindentation is particularly useful for studying the mechanical properties of thin films or coatings at small scales.

Atomic Force Microscopy: This technique involves scanning the surface of the catalyst with a small probe that can detect forces at the nanometer scale. This allows the determination of parameters such as surface roughness, adhesion, and stiffness. Atomic force microscopy is particularly useful for studying the surface properties and mechanical behavior of thin films or coatings.

Overall, the choice of measurement technique depends on the specific mechanical properties of the electrocatalyst that are of interest and the scale at which they need to be studied. Combining multiple techniques can provide a more comprehensive understanding of the mechanical behavior of an electrocatalyst under different conditions.

6.6 Support Degradation

During water electrolysis, the support material used in the electrocatalyst can undergo degradation due to various factors such as the harsh electrochemical environment, high temperature, and mechanical stress. The support degradation can significantly affect the performance and stability of the electrocatalyst, which in turn affects the overall efficiency of the water electrolysis process [58].

Mechanisms of Support Degradation: There are several mechanisms by which the support material can undergo degradation during water electrolysis [2]:

1. Corrosion: Corrosion of the support material can occur due to the high oxidative potential at the anode and the high reduction potential at the cathode. Corrosion can lead to the formation of metal oxides, which can reduce the conductivity of the support material and affect its mechanical strength.
2. Ion Migration: During water electrolysis, ion migration can cause support degradation, leading to decreased mechanical strength and catalytic activity. Ion migration refers to the movement of ions in a solution under an applied electric field. In an electrolytic cell, ions can migrate toward the electrode surface, leading to electrochemical reactions that can cause support degradation.

One mechanism for ion migration and support degradation is through the formation of metal oxides at the interface between the catalyst and support. The formation of metal oxides can occur through the migration of metal ions from the catalyst to the support, leading to a reduction in the mechanical strength of the support. Additionally, the formation of metal oxides can cause a decrease in the effective surface area of the catalyst, leading to decreased catalytic activity. Another mechanism for ion migration and support degradation is through the formation of metal hydroxides. Metal hydroxides can form at the interface between the catalyst and support due to the migration of metal ions and the reaction of water with the metal ions. The formation of metal hydroxides can cause a decrease in the mechanical strength of the support, as well as a decrease in the effective surface area of the catalyst.

Factors that can affect ion migration and support degradation during water electrolysis include the pH of the solution, the temperature of the electrolyte, the current density, and the composition of the catalyst and support. For example, at high pH values, metal ions are more likely to migrate toward the support, leading to increased support degradation. Similarly, at high temperatures, the rate of ion migration and support degradation can increase.

3. **Hydrogen Embrittlement**
   Hydrogen embrittlement is another factor that can contribute to the degradation of the support material during water electrolysis. Hydrogen embrittlement is a phenomenon in which hydrogen atoms diffuse into the metal lattice and cause the material to become brittle and more susceptible to cracking and fracture. During water electrolysis, the hydrogen gas that is produced at the cathode can diffuse into the support material, causing hydrogen embrittlement. The degree of susceptibility to hydrogen embrittlement varies depending on the type of material used as the support. Materials with high hydrogen solubility, such as titanium, are particularly susceptible to hydrogen embrittlement. Other materials, such as nickel and stainless steel, have lower hydrogen solubility and are less susceptible to embrittlement [59].
   The mechanism of hydrogen embrittlement involves the diffusion of hydrogen atoms into the metal lattice, where they can combine with other atoms to form hydrogen gas bubbles. These bubbles can create voids and microcracks within the material, which can eventually lead to macroscopic cracking and failure. The hydrogen gas can also interact with other materials in the system, such as oxides or impurities, which can exacerbate the embrittlement process.

Factors Affecting Support Degradation: Several factors can affect the degradation of support materials during water electrolysis, including:

1. Electrolyte Composition: The composition of the electrolyte can affect the rate of corrosion of the support material. High acidity, high salinity, and high temperature can all increase the rate of corrosion.
2. Temperature: High temperatures can accelerate the degradation of support materials. This is because high temperatures can cause thermal expansion and contraction, which can result in the formation of cracks and defects.
3. Mechanical Stress: Mechanical stress can occur due to the expansion and contraction of the support material during water electrolysis. This can cause cracking and deformation of the support material, leading to reduced mechanical strength.
4. Electrocatalyst Loading: The loading of the electrocatalyst on the support material can also affect its mechanical strength. Higher loading can result in increased stress on the support material, leading to faster degradation.

6.7 Electrode Aerophilic Nature

Gas evolution reactions involve the vigorous and continuous formation of gas bubbles, which have a significant impact on these processes. Over the past few decades, extensive research has been conducted to gain a better understanding of the evolution of gas bubbles, elucidate the underlying mechanisms by which they influence gas evolution reactions, and develop new strategies that utilize bubbles to enhance reaction efficiency. The typical progression of a gas bubble's life cycle at a gas-evolving electrode can be categorized into four phases: nucleation, growth, coalescence, and detachment. Nucleation denotes the random formation of a cluster of gas molecules within a solution holding excess dissolved gas. After nucleation, the bubble expands by absorbing additional dissolved gas molecules. During this growth, the bubble experiences an augmented buoyant force. As this buoyant force exceeds the adhesion force, keeping the bubble anchored to the electrode's surface, the bubble detaches and rises. Coalescence transpires when two bubbles encounter, either on the electrode or within the solution. This union aims to diminish the overall surface energy. Recently, the Ren group provided a brief overview of these bubble dynamics in the context of electrochemical gas evolution reactions [60].

6.7.1 Nucleation

The process known as nucleation serves as the initial phase in bubble formation and is conventionally explained through classical nucleation theory (CNT). The energy necessary to initiate a bubble's presence in a solution can be understood as the summation of energies needed to establish both a gas–liquid interface and a bulk gas phase. The energy imperative for constructing a gas–liquid interface escalates with the increment in interface surface area, and this energy is approximately proportional to the square of the bubble's radius, as depicted in Fig. 6.4. Conversely, the energy requisite for establishing a bulk gas phase from a solution exhibiting supersaturation is energetically favorable and correlates with the cube of the bubble's radius. However, an energy barrier emerges for bubble formation due to the opposing directional changes in volume and surface energy concerning the bubble's radius. The highest point of this energy barrier corresponds to the activation energy essential for bubble nucleation. This point is reached at a critical bubble radius (r_{crit}). The rate equation can be derived by substituting this activation energy into the Arrhenius equation, as exemplified in Eq. 6.8 [60].

$$J = J_0 \exp[-\frac{16\pi\gamma^3\phi\theta}{3k_B T(P_{gas} - P_0)^2}] \tag{6.8}$$

The provided equation includes various terms: J_0 represents the pre-exponential factor, γ stands for surface tension, and $\Phi(\theta)$ is the geometric factor contingent on the contact angle (θ) of the critical nucleus on the surface. The equation incorporates other constants like Boltzmann's constant (k_B), temperature (T), partial pressure of gas in the bubble (P_{gas}), and ambient pressure (P_0). The nucleation rate, J, is notably influenced by P_{gas}, which aligns with the concentration of dissolved gas by Henry's law. This implies that a certain degree of dissolved gas supersaturation is necessary to yield a meaningful nucleation rate.

Multiple research endeavors have sought to quantify the essential level of dissolved gas supersaturation for bubble nucleation at gas-evolving electrodes. Notably, these reported values exhibit substantial variation, potentially attributed

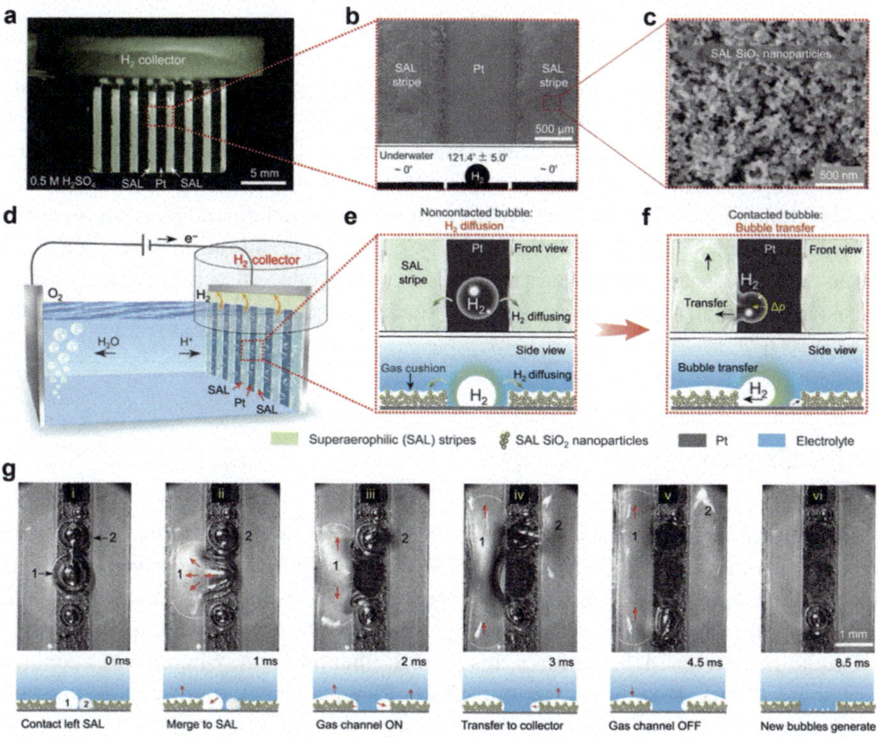

Fig. 6.4 **a** This image shows the SAL/flat Pt electrode, which consists of SAL stripes and Pt, with a glass vessel placed above it as an H_2 collector. **b** The SEM image and wettability characterization reveal that the SAL stripe has superaerophilicity with a bubble contact angle of approximately 0°, while the Pt exhibits aerophobicity with a bubble contact angle of 121.4° ± 5.0°. **c** The SAL stripe comprises numerous SAL SiO_2 nanoparticles, as shown in the high-magnification SEM image. **d** The SAL/flat Pt electrode has enhanced mass transfer, enabling H_2 bubbles generated on the Pt region to be transported to the H_2 collector through the SAL stripes, as indicated by yellow arrows. **e** Initially, the H_2 bubbles do not contact the SAL stripe, and the dissolved H_2 can diffuse out of the reaction system through the gas cushion on the SAL stripe, as indicated by green arrows. **f** As the H_2 bubbles grow, they contact the SAL stripes and are timely transferred through them. **g** In-situ optical observation confirms the successful transfer of H_2 bubbles into the gas collector through the SAL stripes, as indicated by red arrows. The left SAL stripe (i–ii–iii–iv–v–vi) and right SAL stripe (iii–vi) show a similar transfer process. The SAL/flat Pt electrode then revives the Pt region to continue HER

to variations in nucleation site properties and numbers on the electrodes. A nanoelectrode-based approach has been developed to address the intricacies of multiple bubble nucleation events on the gas-evolving electrode. This method has successfully generated individual H_2, N_2, and O_2 nanobubbles through electrochemical processes like proton reduction, hydrazine oxidation, and hydrogen peroxide oxidation. The peak current (ip) is harnessed for determining the necessary dissolved gas concentration for nucleation, as outlined in Eq. 6.9 [60].

$$i_p = 4nFD(C - C_{\mathrm{bulck}})a \tag{6.9}$$

The concentrations required to initiate the formation of gas bubbles on various Pt electrodes demonstrate uniformity: 0.23 ± 0.02 M for H_2, 0.11 ± 0.01 M for N_2, and 0.24 ± 0.04 M for O_2. These concentrations correspond to the necessary levels of supersaturation crucial for the nucleation of bubbles. The nanoelectrode-based technique presents a distinctive approach to electrochemical measurement for individual bubbles. This method effectively gages supersaturation levels, enabled by the rapid mass transfer to and from the nanoelectrode. This, in turn, establishes consistent concentration profiles for both reactants and products in a steady-state manner [60].

6.7.2 Growth

After nucleation, gas bubbles progress in size due to their elevated internal pressure compared to the Laplace pressure, coupled with the inflow of dissolved gas. This enlargement encompasses three distinct growth modes: inertia-controlled growth, diffusion-controlled growth, and direct injection growth. A widely applicable empirical equation, $R(t) = At\beta$, captures this growth phenomenon, wherein $R(t)$ denotes the bubble radius, β signifies the growth coefficient, and A represents a constant factor.

The inertia-controlled growth, though brief (around 0.01 s), swiftly transitions to diffusion-controlled growth. Inertia-driven expansion initially hinges on momentum interaction between the bubbles and the encompassing fluid, characterized by a β of 1. Subsequent growth is directed by gas diffusion into the bubbles, indicating an envisaged β of 0.5. The third growth mode, direct injection, denotes the immediate infusion of gas species into a bubble, culminating in $R \approx t^{1/3}$ and a corresponding β value of 1/3. This model has been particularly noted in H_2 and O_2 bubbles atop electrodes operating under high current densities.

Exploration of bubble growth extends to scenarios encompassing microgravity, magnetic fields, and fluid dynamics. The bulk of previous research into bubble growth relied predominantly on conventional analytical methods, thus providing limited insights into the developing growth stages due to the constraints of short time scales and minute dimensions. It is important to underscore that effective bubble growth necessitates a solution rich in gas supersaturation. Failure to fulfill this condition might lead to bubble dissolution, particularly in cases where no current is applied to the gas-evolving electrode [60].

6.7.3 Coalescence Top of Form

When two gas bubbles come close to each other, they can join together through coalescence, forming a single bubble. This happens because the combined bubble has less surface energy compared to the total surface area of the two separate bubbles. The coalescence process is affected by different things, such as the amount of salts and the types of positive and negative ions in the liquid. Recent improvements in high-speed optical microscopy have given us new information about how coalescence works. The process begins by pushing aside the liquid between the two bubbles. This creates a thin layer of liquid that collapses into a narrow connection called a "neck." The thin liquid layer gets thinner until it breaks at a certain thickness (<100 nm). After that, the neck forms and moves like a small wave on the surface of the liquid. At the start, viscosity plays a significant role in forming the neck in the first 10 billionths of a second after contact. Then, the process switches to being controlled by the momentum of the liquid [61].

6.7.4 Bubble Detachment

Bubble detachment is vital in gas evolution reactions as it unveils previously covered active areas. The process involves a bubble's contact line getting stuck on the surface, followed by a thinning of the bubble's connection with the substrate. This narrowing ultimately results in detachment. Methods like high-speed imaging and electrochemical noise analysis of current, resistance, and potential have been employed to understand bubble detachment from electrodes during gas evolution. In theory, bubble detachment happens when the upward buoyant force becomes stronger than the adhesive force on the bubble. When a bubble detaches from a flat surface, the balance between these forces provides the detachment radius, RF. This is determined by the bubble's shape and the interplay of buoyant and adhesive forces. However, the analysis becomes more intricate when a bubble detaches from a vertical surface due to the bubble's asymmetrical shape and changing contact angles. Furthermore, the electrostatic force between a charged bubble and the electrode is significant. This force depends on solution pH and electrode potential. It influences bubble detachment differently at various pH levels, leading to distinct detachment radii. Other factors like Marangoni stress, surfactants, and magnetic fields have been explored for their roles in bubble detachment [62].

Gas bubbles can have a significant negative impact on the performance of an electrocatalyst for water electrolysis. When gas bubbles accumulate on the surface of the electrocatalyst, they can block active sites, reducing the catalyst coverage and its activity. This can decrease the efficiency of the electrochemical reaction by hindering the transport of reactants and products to and from the electrocatalyst surface. Gas bubbles can also cause mechanical stress on the electrocatalyst, leading to its degradation and reduced performance over time. Additionally, gas bubbles can displace the electrolyte around the electrocatalyst, leading to a depletion of the electrolyte and a decrease in the electrochemical reaction rate. In order to maintain efficient electrocatalysis for water electrolysis, it is crucial to manage gas bubbles. This can be achieved by optimizing the operating conditions and designing the electrode and electrolyte system to minimize gas bubble accumulation [63].

The Cao group has recently developed a highly sustainable water-splitting catalyst comprised of porous Ni/Ni–Fe–MoO_x nanoplates grown over a Ni foam substrate [64]. The catalyst's remarkable efficiency, as the team notes, can be attributed to multiple findings from their experiments. Firstly, the amorphous Ni/Ni–Fe–MoO_x exhibited oxygen vacancies that adjusted the d-band center position relative to the Fermi level. This tuning optimized the chemisorption energy of the active component, thus stabilizing the OER intermediates at lower input potentials. The conductive substrate further bolstered the catalyst's effectiveness by establishing a proficient pathway for charge transfer, effectively limiting non-conductive layer formation and elevating performance.

Moreover, the catalyst's hydrophilic and aerophobic porous surface structure created concise diffusion channels, expediting mass transfer kinetics. Notably, the porous free-standing electrode with aerophobic properties contributed to mechanical robustness and prevented active site detachment from the substrate during high current density scenarios. Ultimately, the catalyst's combination of hydrophilic and aerophobic attributes played a pivotal role in enhancing water adsorption and gas desorption, consequently amplifying reaction kinetics and overall durability. Recently, the utilization of FeCoNiP@C catalyst supported on metallic Ni has also been explored for water splitting, showcasing notable intrinsic activity and endurance for water oxidation [65]. The catalyst's exceptional performance stems from various insights garnered through experimentation. Firstly, the Ni foam current collector, being conductive, alongside metal phosphide-based active components, alleviates mass transport constraints via strong conductive adhesion. Furthermore, the catalyst's nanoarray morphology expedites bubble nucleation, coalescence, and subsequent desorption from active sites, facilitating cyclic reaction progress. Secondly, the catalyst's microscale amorphous structure boasts a high specific surface area with an optimal adsorption energy, consequently enhancing active site reactivity. Thirdly, the N-doped carbon matrix stabilizes active sites via robust σ and π bonding, enhancing mechanical resilience, and safeguarding against leaching in challenging electrolytic conditions. Fourthly, a thin oxyhydroxide layer on the catalyst's surface prompts an OH- attack, initiating water dissociation with minimal input energy. Lastly, the catalyst's hydrophilic-aerophobic surface drives water adsorption and gas desorption, counteracting support passivation and augmenting structural and chemical robustness throughout water oxidation. It is important to acknowledge that the wetting behavior of the electrolyte and the speed at which bubbles form on the electrode surface during water splitting notably impact the catalysts' activity [66]. Creating a "dead zone" on the electrode surface is possible when a strong bond forms between the electrode and bubbles. In such cases, electrolyte diffusion can be hindered and active site exposure reduced, consequently impairing the efficiency and stability of water splitting. However, studies indicate that modifying electrode wettability can enhance water splitting's effectiveness and strength.

This can be accomplished by engineering the electrode to become superhydrophilic, enhancing wettability and superaerophobic, facilitating quicker bubble release. Recent investigations have showcased that constructing micro–nanostructures on flat substrates or applying hydrophobic coatings can influence electrode wettability. These modifications affect various bubble dynamics, encompassing processes like heterogeneous nucleation, bubble growth, bubble layer thickness, and the size at which bubbles detach [67].

Carbon-based materials with porous frameworks, layered with active components, showcase distinctive attributes of hydrophilicity and aerophobicity. These features aid in electron migration, diminishing ohmic resistance, and accelerating gas desorption rates. Consequently, they effectively reduce the inactive region on the electrode surface. Gosh et al. introduced a bimetallic heterostructure, comprising Ni_2P-Cu_2P active sites grown on NF graphene carbon nanotubes (Ni-Gr-CNTs) exhibiting superaerophobicity. This specialized design demonstrated notable structural, chemical, and mechanical resilience, along with impressive corrosion resistance [68]. The sustainability of the electrode's performance can be attributed to its superhydrophilic and aerophobic designs. These features enhance the adhesion of the electrocatalyst to the electrolyte while reducing bonding with the produced gases. Measurements of contact angles revealed that the bimetallic heterojunction decoration on Ni-Gr-CNTs significantly heightened the hydrophilicity of active sites and concurrently reduced bubble size. Integrating a porous carbon-based framework created shorter pathways for mass movement, crucial for bolstering support durability. Carbon nanotubes (CNTs) and graphene layered over Ni foam provided an added surface area for anchoring the nanostructure, curbing the risk of accumulation—a common catalyst activity decline factor. The electrode's 3D structure, showcasing a flower-like morphology, generated densely concentrated edges with optimal d-band positioning. This configuration lowered the activation energy for OER and heightened the catalyst's resistance to corrosion.

References

1. Park EJ, Arges CG, Xu H, Kim YS (2022) Membrane strategies for water electrolysis. ACS Energy Lett. https://doi.org/10.1021/acsenergylett.2c01609
2. Feng Q, Yuan XZ, Liu G, Wei B, Zhang Z, Li H, Wang H (2017) A review of proton exchange membrane water electrolysis on degradation mechanisms and mitigation strategies. J Power Sources. https://doi.org/10.1016/j.jpowsour.2017.09.006
3. Rakousky C, Reimer U, Wippermann K, Carmo M, Lueke W, Stolten D (2016) An analysis of degradation phenomena in polymer electrolyte membrane water electrolysis. J Power Sources. https://doi.org/10.1016/j.jpowsour.2016.06.082
4. Sun S, Shao Z, Yu H, Li G, Yi B (2014) Investigations on degradation of the long-term proton exchange membrane water electrolysis stack. J Power Sources. https://doi.org/10.1016/j.jpowsour.2014.05.117

5. Wang J, Tang X, Cheng X, Xing J, Wang H, Li G, Liang H (2021) In-Situ Crystallization generated by CEM electrolysis for NF concentrate softening along with the alleviation of ceramic membrane fouling. Desalination. https://doi.org/10.1016/j.desal.2021.115243
6. Paidar M, Fateev V, Bouzek K (2016) Membrane electrolysis—history, current status and perspective. Electrochim Acta. https://doi.org/10.1016/j.electacta.2016.05.209
7. Abid HS, Lalia BS, Bertoncello P, Hashaikeh R, Clifford B, Gethin DT, Hilal N (2017) Electrically conductive spacers for self-cleaning membrane surfaces via periodic electrolysis. Desalination. https://doi.org/10.1016/j.desal.2017.04.018
8. Ding A, Fan Q, Cheng R, Sun G, Zhang M, Wu D (2018) Impacts of applied voltage on microbial electrolysis cell-anaerobic membrane bioreactor (MEC-AnMBR) and its membrane fouling mitigation mechanism. Chem Eng J. https://doi.org/10.1016/j.cej.2017.09.190
9. Wu D, Zhou C, Lu G, Zhou Y, Shen Y (2019) Simultaneous membrane fouling mitigation and emerging pollutant benzophenone-3 removal by electro-peroxone process. Sep Purif Technol. https://doi.org/10.1016/j.seppur.2019.115715
10. Miller HA, Bouzek K, Hnat J, Loos S, Bernäcker CI, Weißgärber T, Röntzsch L, Meier-Haack J (2020) Green hydrogen from anion exchange membrane water electrolysis: a review of recent developments in critical materials and operating conditions. Sustainable Energy Fuels. https://doi.org/10.1039/c9se01240k
11. Lee SA, Kim J, Kwon KC, Park SH, Jang HW (2022) Anion exchange membrane water electrolysis for sustainable large-scale hydrogen production. Carbon Neutralization. https://doi.org/10.1002/cnl2.9
12. Schalenbach M, Carmo M, Fritz DL, Mergel J, Stolten D (2013) Pressurized PEM water electrolysis: efficiency and gas crossover. Int J Hydrogen Energy. https://doi.org/10.1016/j.ijhydene.2013.09.013
13. Grigoriev SA, Porembskiy VI, Korobtsev SV, Fateev VN, Auprêtre F, Millet P (2011) High-pressure PEM water electrolysis and corresponding safety issues. Int J Hydrogen Energy. https://doi.org/10.1016/j.ijhydene.2010.03.058
14. Knöppel J, Möckl M, Escalera-López D, Stojanovski K, Bierling M, Böhm T, Thiele S, Rzepka M, Cherevko S (2021) On the limitations in assessing stability of oxygen evolution catalysts using aqueous model electrochemical cells. Nat Commun. https://doi.org/10.1038/s41467-021-22296-9
15. Li A, Sun Y, Yao T, Han H (2018) Earth-abundant transition-metal-based electrocatalysts for water electrolysis to produce renewable hydrogen. Chem Eur J. https://doi.org/10.1002/chem.201803749
16. Lei Z, Wang T, Zhao B, Cai W, Liu Y, Jiao S, Li Q, Cao R, Liu M (2020) Recent progress in electrocatalysts for acidic water oxidation. Adv Energy Mater. https://doi.org/10.1002/aenm.202000478
17. Cherevko S (2018) Stability and dissolution of electrocatalysts: building the bridge between model and "real world" systems. Curr Opin Electrochem. https://doi.org/10.1016/j.coelec.2018.03.034
18. Lee WH, Han MH, Ko YJ, Min BK, Chae KH, Oh HS (2022) Electrode reconstruction strategy for oxygen evolution reaction: maintaining Fe-CoOOH phase with intermediate-spin state during electrolysis. Nat Commun. https://doi.org/10.1038/s41467-022-28260-5
19. Van Genuchten CM, Peña J, Amrose SE, Gadgil AJ (2014) Structure of Fe(III) precipitates generated by the electrolytic dissolution of Fe(0) in the presence of groundwater ions. Geochim Cosmochim Acta. https://doi.org/10.1016/j.gca.2013.11.044
20. Binninger T, Mohamed R, Waltar K, Fabbri E, Levecque P, Kötz R, Schmidt TJ (2015) Thermodynamic explanation of the universal correlation between oxygen evolution activity and corrosion of oxide catalysts. Sci Rep. https://doi.org/10.1038/srep12167
21. Zeng Y, Zhao M, Huang Z, Zhu W, Zheng J, Jiang Q, Wang Z, Liang H (2022) Surface reconstruction of water splitting electrocatalysts. Adv Energy Mater. https://doi.org/10.1002/aenm.202201713
22. Chen FY, Wu ZY, Adler Z, Wang H (2021) Stability challenges of electrocatalytic oxygen evolution reaction: from mechanistic understanding to reactor design. Joule 5(7):1704–1731. https://doi.org/10.1016/j.joule.2021.05.005

23. Wang J, Hu J, Niu S, Li S, Du Y, Xu P (2022) Crystalline-amorphous $Ni_2P4O_{12}/NiMoO_x$ nanoarrays for alkaline water electrolysis: enhanced catalytic activity via in situ surface reconstruction. Small. https://doi.org/10.1002/smll.202105972
24. Haq T, Haik Y (2022) Strategies of anode design for seawater electrolysis: recent development and future perspective. Small Sci 2(9):2200030. https://doi.org/10.1002/smsc.202200030
25. Haq T, Pasha M, Tong Y, Mansour SA, Haik Y (2022) Au nanocluster coupling with Gd-Co_2B nanoflakes embedded in reduced TiO_2 nanosheets: seawater electrolysis at low cell voltage with high selectivity and corrosion resistance. Appl Catal B Environ. https://doi.org/10.1016/j.apcatb.2021.120836
26. Carpinteri A, Borla O, Manuello A, Veneziano D, Goi A (2013) Hydrogen embrittlement and piezonuclear reactions in electrolysis experiments. In: 18th International conference on cold fusion 2013, ICCF 2013
27. Jin H, Ruqia B, Park Y, Kim HJ, Oh HS, Choi SI, Lee K (2021) Nanocatalyst design for long-term operation of proton/anion exchange membrane water electrolysis. Adv Energy Mater. https://doi.org/10.1002/aenm.202003188
28. Wei C, Rao RR, Peng J, Huang B, Stephens IEL, Risch M, Xu ZJ, Shao-Horn Y (2019) Recommended practices and benchmark activity for hydrogen and oxygen electrocatalysis in water splitting and fuel cells. Adv Mater 31(31):1–24. https://doi.org/10.1002/adma.201806296
29. Hochfilzer D, Chorkendorff I, Kibsgaard J (2023) Catalyst stability considerations for electrochemical energy conversion with non-noble metals: do we measure on what we synthesized? ACS Energy Lett. https://doi.org/10.1021/acsenergylett.3c00021
30. Chen Z, Yang H, Kang Z, Driess M, Menezes PW (2022) The pivotal role of S-, p-, and f-block metals in water electrolysis: status quo and perspectives. Adv Mater. https://doi.org/10.1002/adma.202108432
31. Thorarinsdottir AE, Veroneau SS, Nocera DG (2022) Self-healing oxygen evolution catalysts. Nature Commun. https://doi.org/10.1038/s41467-022-28723-9
32. Feng C, Wang F, Liu Z, Nakabayashi M, Xiao Y, Zeng Q, Fu J, Wu Q, Cui C, Han Y et al (2021) A self-healing catalyst for electrocatalytic and photoelectrochemical oxygen evolution in highly alkaline conditions. Nat Commun. https://doi.org/10.1038/s41467-021-26281-0
33. Surendranath Y, Nocera DG (2011) Oxygen evolution reaction chemistry of oxide-based electrodes. Prog Inorg Chem. https://doi.org/10.1002/9781118148235.ch9
34. Liang L, Cheng H, Lei F, Han J, Gao S, Wang C, Sun Y, Qamar S, Wei S, Xie Y (2015) Metallic single-unit-cell orthorhombic cobalt diselenide atomic layers: robust water-electrolysis catalysts. Angew Chemie Int Ed. https://doi.org/10.1002/anie.201505245
35. Candelaria SL, Bedford NM, Woehl TJ, Rentz NS, Showalter AR, Pylypenko S, Bunker BA, Lee S, Reinhart B, Ren Y et al (2017) Multi-Component Fe-Ni hydroxide nanocatalyst for oxygen evolution and methanol oxidation reactions under alkaline conditions. ACS Catal 7(1):365–379. https://doi.org/10.1021/acscatal.6b02552
36. Stoerzinger KA, Diaz-Morales O, Kolb M, Rao RR, Frydendal R, Qiao L, Wang XR, Halck NB, Rossmeisl J, Hansen HA et al (2017) Orientation-dependent oxygen evolution on RuO_2 without lattice exchange. ACS Energy Lett 2(4):876–881. https://doi.org/10.1021/acsenergylett.7b00135
37. ul Haq T, Mansour SA, Munir A, Haik Y (2020) Gold-supported gadolinium doped CoB amorphous sheet: a new benchmark electrocatalyst for water oxidation with high turnover frequency. Adv Funct Mater 1910309:1–11. https://doi.org/10.1002/adfm.201910309
38. Dehnavi V, Liu XY, Luan BL, Shoesmith DW, Rohani S (2014) Phase transformation in plasma electrolytic oxidation coatings on 6061 aluminum alloy. Surf Coatings Technol. https://doi.org/10.1016/j.surfcoat.2014.04.010
39. Tambunan OT, Lee MY, Kim DH, Parwanta KJ, Jung CU (2014) Phase transformation of the brownmillerite $SrCoO_{2.5}$ thin film through alkaline water electrolysis. J Korean Phys Soc. https://doi.org/10.3938/jkps.64.1845
40. Yu M, Budiyanto E, Tüysüz H (2022) Principles of water electrolysis and recent progress in cobalt-, nickel-, and iron-based oxides for the oxygen evolution reaction. Angew Chemie Int Ed. https://doi.org/10.1002/anie.202103824

41. Chen P, Xu K, Tao S, Zhou T, Tong Y, Ding H, Zhang L, Chu W, Wu C, Xie Y (2016) Phase-transformation engineering in cobalt diselenide realizing enhanced catalytic activity for hydrogen evolution in an alkaline medium. Adv Mater. https://doi.org/10.1002/adma.201601663
42. Kasian O, Grote JP, Geiger S, Cherevko S, Mayrhofer KJJ (2018) The common intermediates of oxygen evolution and dissolution reactions during water electrolysis on iridium. Angew Chemie Int Ed. https://doi.org/10.1002/anie.201709652
43. Suffredini HB, Cerne JL, Crnkovic FC, MacHado SAS, Avaca LA (2000) Recent developments in electrode materials for water electrolysis. Int J Hydrogen Energy. https://doi.org/10.1016/S0360-3199(99)00049-X
44. Shiva Kumar S, Lim H (2022) An overview of water electrolysis technologies for green hydrogen production. Energy Rep. https://doi.org/10.1016/j.egyr.2022.10.127
45. Gilson MK (1995) Theory of electrostatic interactions in macromolecules. Curr Opin Struct Biol. https://doi.org/10.1016/0959-440X(95)80079-4
46. Liu X, Meng J, Ni K, Guo R, Xia F, Xie J, Li X, Wen B, Wu P, Li M et al (2020) Complete reconstruction of hydrate pre-catalysts for ultrastable water electrolysis in industrial-concentration alkali media. Cell Reports Phys Sci. https://doi.org/10.1016/j.xcrp.2020.100241
47. Park DH, Kim MH, Lee HJ, Lee WJ, Byeon JH, Kim JH, Jang JS, Park KW (2022) Development of Ni-Ir oxide composites as oxygen catalysts for an anion-exchange membrane water electrolyzer. Adv Mater Interfaces. https://doi.org/10.1002/admi.202102063
48. Hubert M, Laurencin J, Cloetens P, Morel B, Montinaro D, Lefebvre-Joud F (2018) Impact of nickel agglomeration on solid oxide cell operated in fuel cell and electrolysis modes. J Power Sources. https://doi.org/10.1016/j.jpowsour.2018.06.097
49. Zeng K, Zhang D (2014) Evaluating the effect of surface modifications on Ni based electrodes for alkaline water electrolysis. Fuel. https://doi.org/10.1016/j.fuel.2013.08.070
50. Márquez RA, Kawashima K, Son YJ, Rose R, Smith LA, Miller N, Carrasco Jaim OA, Celio H, Mullins CB (2022) Tailoring 3D-printed electrodes for enhanced water splitting. ACS Appl Mater Interfaces. https://doi.org/10.1021/acsami.2c12579
51. Markovic NM (2015) (Invited) electrochemistry at metal-oxide interfaces: at the crossroads of activity and stability. ECS Meet Abstr. https://doi.org/10.1149/ma2015-01/36/1945
52. Akbashev AR, Zhang L, Mefford JT, Park J, Butz B, Luftman H, Chueh WC, Vojvodic A (2018) Activation of ultrathin $SrTiO_3$ with subsurface $SrRuO_3$ for the oxygen evolution reaction. Energy Environ Sci. https://doi.org/10.1039/c8ee00210j
53. Xie Y, Cao J, Wang X, Li W, Deng L, Ma S, Zhang H, Guan C, Huang W (2021) MOF-derived bifunctional Co0.85Se nanoparticles embedded in N-doped carbon nanosheet arrays as efficient sulfur hosts for lithium-sulfur batteries. Nano Lett. https://doi.org/10.1021/acs.nanolett.1c02037
54. Mayerhöfer B, Speck FD, Hegelheimer M, Bierling M, Abbas D, McLaughlin D, Cherevko S, Thiele S, Peach R (2022) Electrochemical- and mechanical stability of catalyst layers in anion exchange membrane water electrolysis. Int J Hydrogen Energy. https://doi.org/10.1016/j.ijhydene.2021.11.083
55. Wang M, Wang Z, Gong X, Guo Z (2014) The intensification technologies to water electrolysis for hydrogen production—a review. Renew Sustain Energy Rev. https://doi.org/10.1016/j.rser.2013.08.090
56. Margalef R, Bosque M, Minaya-Muñoz F, Valera-Garrido F, Santafe MM (2021) Safety analysis of percutaneous needle electrolysis: a study of needle composition, morphology, and electrical resistance. Acupunct Med. https://doi.org/10.1177/0964528420988007
57. Zhang QB, Zhao J (2014) A review of dynamic experimental techniques and mechanical behaviour of rock materials. Rock Mech Rock Eng. https://doi.org/10.1007/s00603-013-0463-y
58. Paciok P, Schalenbach M, Carmo M, Stolten D (2017) On the mobility of carbon-supported platinum nanoparticles towards unveiling cathode degradation in water electrolysis. J Power Sources. https://doi.org/10.1016/j.jpowsour.2017.07.033
59. Laadel NE, El Mansori M, Kang N, Marlin S, Boussant-Roux Y (2022) Permeation barriers for hydrogen embrittlement prevention in metals—a review on mechanisms, materials suitability and efficiency. Int J Hydrogen Energy. https://doi.org/10.1016/j.ijhydene.2022.07.164

60. Zhao X, Ren H, Luo L (2019) Gas bubbles in electrochemical gas evolution reactions. Langmuir. https://doi.org/10.1021/acs.langmuir.9b00119
61. Angulo A, van der Linde P, Gardeniers H, Modestino M, Fernández Rivas D (2020) Influence of bubbles on the energy conversion efficiency of electrochemical reactors. Joule. https://doi.org/10.1016/j.joule.2020.01.005
62. Wang Y, Gordon E, Ren H (2019) Mapping the nucleation of H_2 bubbles on polycrystalline Pt via scanning electrochemical cell microscopy. J Phys Chem Lett. https://doi.org/10.1021/acs.jpclett.9b01414
63. Soto ÁM, German SR, Ren H, Van Der Meer D, Lohse D, Edwards MA, White HS (2018) The Nucleation rate of single O_2 nanobubbles at Pt nanoelectrodes. Langmuir. https://doi.org/10.1021/acs.langmuir.8b01372
64. Li YK, Zhang G, Lu WT, Cao FF (2020) Amorphous Ni–Fe–Mo suboxides coupled with ni network as porous nanoplate array on nickel foam: a highly efficient and durable bifunctional electrode for overall water splitting. Adv Sci. https://doi.org/10.1002/advs.201902034
65. Zhang X, Hu J, Cheng X, Nartey KA, Zhang L (2021) Double metal-organic frameworks derived Fe–Co–Ni phosphides nanosheets as high-performance electrocatalyst for alkaline electrochemical water splitting. Electrochim Acta. https://doi.org/10.1016/j.electacta.2020.137536
66. Shan X, Liu J, Mu H, Xiao Y, Mei B, Liu W, Lin G, Jiang Z, Wen L, Jiang L (2020) An engineered superhydrophilic/superaerophobic electrocatalyst composed of the supported $CoMoS_x$ chalcogel for overall water splitting. Angew Chemie Int Ed 59(4):1659–1665. https://doi.org/10.1002/anie.201911617
67. Xu W, Lu Z, Sun X, Jiang L, Duan X (2018) Superwetting electrodes for gas-involving electrocatalysis. Acc Chem Res. https://doi.org/10.1021/acs.accounts.8b00070
68. Riyajuddin S, Azmi K, Pahuja M, Kumar S, Maruyama T, Bera C, Ghosh K (2021) Super-hydrophilic hierarchical Ni-foam-graphene-carbon nanotubes-Ni_2P-CuP_2 nano-architecture as efficient electrocatalyst for overall water splitting. ACS Nano 15(3):5586–5599. https://doi.org/10.1021/acsnano.1c00647

Chapter 7
Electronic Modulation of Electrocatalysts for Enhanced Water Electrolysis

7.1 Electronic Modification

Heterogeneous catalysis is a promising technique for electrochemical water splitting because of its high efficiency, cost-free, and extensive material's availability. However, the electrocatalytic performance of the catalyst is limited by the kinetics of the electrochemical reactions, particularly the oxygen evolution reaction (OER) and the hydrogen evolution reaction (HER). To improve catalytic activity of the catalyst, researchers have developed various strategies, including electronic modulation, to alter the electronic characteristics of the catalyst and, consequently, amplify its electrocatalytic capabilities.

7.2 Cation Doping

The configuration of electrons within a catalyst's active sites significantly influences their positioning concerning the Fermi energy level. This positioning has a direct impact on the energy barrier for O–O coupling, ultimately shaping the catalyst's stability as well as its activity. Altering their electronic structure stands out as an effective approach for optimizing chemisorption energy and improving active sites recyclability. This is achieved through adjustments in the electronic cloud enveloping the active components.

Norskov's theorem underscores that the speed of adsorption as well as desorption of intermediate hinges on the metal d band location concerning the Fermi level. The electronic attributes of active components can be regulated by altering the characteristics, position, and dopants density within the original infrastructure. Markovic and his research team delved into the utilization of clusters composed of conductive metals (M) combined with hydroxide/oxyhydroxide (O_xH_y) and M-containing O_xH_y (M=Cu, Fe, Co, and Ni) to study trends in activity and stability for water oxidation

T. u. Haq and Y. Haik, *Electrochemical Water Splitting*, Materials Horizons: From Nature to Nanomaterials, https://doi.org/10.1007/978-981-99-9860-9_7

(OER) [1]. The research demonstrated that the dissolution and subsequent repositioning of Fe at the interface of the Fe-containing electrolyte led to the development of resilient Fe active sites. This innovation overcame the constraints arising from instability of oxide mcomposites under anodic conditions. It established a fundamental principle encompassing both electrode's activity as well as its durability. The team conducted in-situ ICP-MS analysis and observed that, within the OER potential range, the pure Cu active centers in Cu–O_xH_y exhibited relatively limited resilience compared to the Cu active centers in Ni-Cu O_xH_y. Moreover, the incorporation of Fe into the system led to an enhancement in Faraday efficiency and stability of catalyst by reducing the rate of dissolution. The increase in current density (ASF) was ascribed to synergistic intermetallic interactions, modifications in the electronic structure, an increase in stable active centers, and a reduction in the potential energy barrier for O–O coupling. This highlighted how achieving high intrinsic activity and stability simultaneously efficiently elevated the ASF value for anode-based composites. The same group conducted a separate investigation to investigate the origins of the remarkably active and stable components within the composite $La_{1\text{-}x}Sr_xCoO_3$ (LSCO) [2]. Their research uncovered that Co sites undergo a chemical transformation, forming Co-OOH. Subsequently, Co-OOH interacts with small quantities of Fe present in the electrolyte, resulting in the formation of persistently stable active constituents. The chemical oxidation process, transforming Co into Co–OOH, serves as the initial trigger for the Fe interaction, which brings about alterations in electronic distribution. This interaction effectively restrains Co dissolution and, notably, enhances stability by a factor of ten when compared to the nanoscale cobalt hydroxide structure. Moreover, calculations of current dissolution density (S) illustrated that the interaction between Co-active sites and La, Sr reduces the dissolution rate of metal significantly from 23 pg $cm^{-2}s^{-1}$ to 1.8 pg $cm^{-2}s^{-1}$. Density functional theory (DFT) computations unveiled that the interactions in LSCO raised the free energy of Co-OOH active sites by 0.3 eV when compared to Co–OOH stabilized over Pt.

The active surface area (ASF) estimation indicated a three-fold development after doping of Sr and La. Furthermore, the scientists put forward a framework for electronic conduction occurring at the junction between the perovskite oxide and the alkaline water interface. This framework facilitates the chemical oxidation of the transition metal (TM)-hydroxide/oxy(hydroxide) shell encircling the perovskite core, thereby enabling their interaction with iron. This interaction generates dynamically stable active centers, promoting the oxygen evolution reaction (OER). They highlighted that manipulating electronic density through dopant composition and nature while stabilizing active sites could further elevate their relative stability. Lastly, exploring epitaxial strain was suggested for developing a sustainable anode. In a recent investigation, researchers explored the augmentation of electrochemical performance and the long-term stability of active sites involved in water oxidation. This improvement was realized by establishing electronic interactions between Cu nanoclusters (NCs) and CoS, which were constructed on a Cu foam substrate [3]. Through spectroscopic investigation, it was observed that Cu and Co metals underwent chemical oxidation to Cu–O and Co–O, respectively, during high anodic potential water oxidation. This oxidation process not only preserved the inner structure but also heightened the material's resistance to corrosion. Electrochemically activated catalysts displayed notably enhanced efficiency of catalyst in comparison to Co-based oxide and pristine Cu counterparts. In light of X-ray Absorption Fine Structure (XAFS) analysis, the researchers noted that the exposed Co–O component had the lowest coordination number of 2. Despite its notably disordered structure, this arrangement consistently maintained a high current density throughout continuous alkaline water electrolysis. The cooperative synergy between the Co and Cu elements prevented the development of partially or nonconductive layers. This synergy preserved the maximum contact area at the electrode–electrolyte interface, minimized metal dissolution, and consequently, amplified the active surface area (ASF) (Fig. 7.1a–h).

Recent research by Menezes and colleagues has investigated an intermetallic approach to enhance the intrinsic catalytic performance and sustainability of Fe catalysts by developing a $FeSn_2$ anode (Fig. 7.1i) [4]. The $FeSn_2$ catalyst displays structural irregularities that lead to an abundance of active sites and an augmented active surface area (ASF). The sustainability of $FeSn_2$ stems from its numerous active sites, each exhibiting high intrinsic activity. These active centers enable substantial current output even at low input voltage, protect against passivation, and notably, prevent metal dissolution in contrast to the original Fe catalyst. Studies suggest that crystalline αFe–O(OH) exhibits increased resistance to solvation under high overpotential conditions when compared to its amorphous counterpart. Furthermore, this crystalline phase requires a lower input voltage to overcome the energy barrier for O–O coupling. The significant increase in the active surface area (ASF) can be attributed to the restructuring or adjustment of active site surfaces. Under anodic bias, Sn may leach, oxidize, and stabilize the Fe into a more active state, namely αFe–O(OH). Throughout the OER, the oxidation of Fe $^{\delta+}$ leads to the formation of $Fe^{2+}(OH)_2$, which subsequently transforms into αFe–O(OH). This newly formed component becomes actively involved in the process of water oxidation.

$$FeSn_2 + 15OH^- \rightarrow Fe^{III}O(OH) + 2[Sn^{IV}(OH)_6]^{2-} + 11e^- + H_2O$$

The experimental findings showcased a clear link between the well-organized pattern of αFe–O(OH) and the structural alignment of the $FeSn_2$ core. This association plays a role in converting the surface into a crystalline phase. Electronic fine-tuning through intermetallic interactions enhances the activity factor by adjusting the position of the metal d band center relative to the O 2p orbitals. This optimization yields tailored chemisorption energy and effectively hinders metal dissolution by the robust interaction within the outer orbitals. Including catalytically less active metals into the catalyst's active site has demonstrated noteworthy influence on its activity and durability. Huang et al. explored the effects of incorporating Zn^{2+} (which lacks an empty orbital for catalysis) into a Co–OOH catalyst (Fig. 7.1j–p) [5]. The introduction of Zn introduces oxygen nonbonding states with a variety of local configurations. Both empirical and theoretical investigations have revealed that specific local arrangements and the presence of oxygen vacancies significantly influence

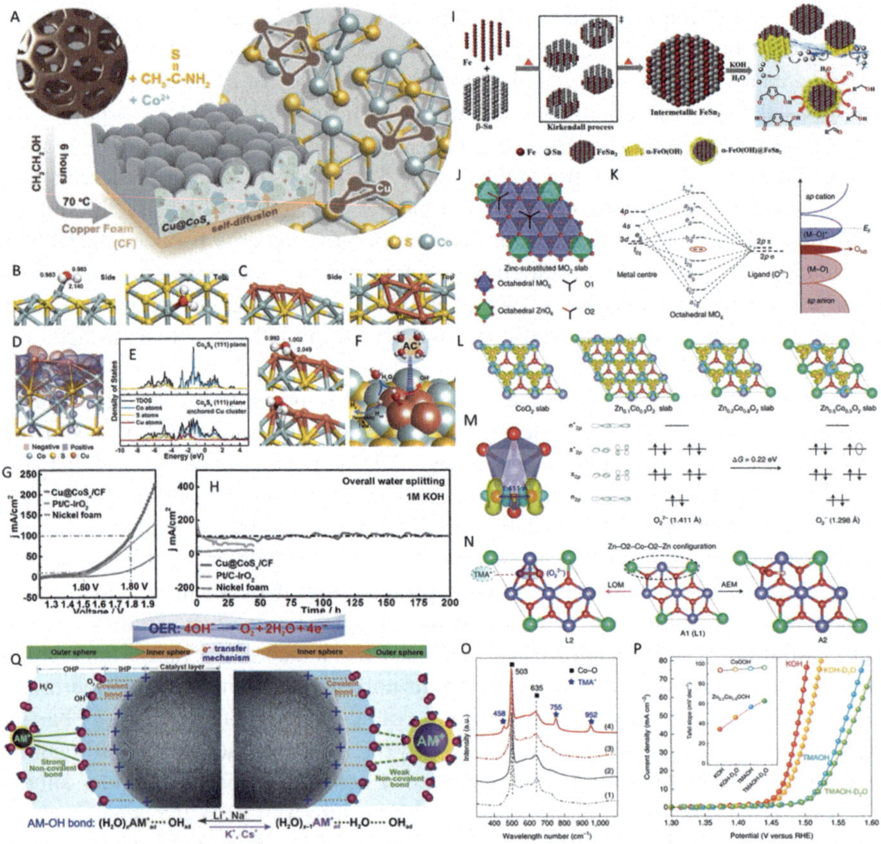

Fig. 7.1 **a** Cu@CoS_x/CF synthesis: Formation of films of CoS_x at Cu species and CF interdiffusion. **b** Adsorption of H_2O molecules on the Co_9S_8 plane surface (111). **c** Cu (4clusters of 4 atoms) stabilization on the plane surface (111) of Co_9S_8. **d** Charge density difference within Cu@Co_9S_8 clusters: blue region indicating depletion of electrons and red part depict accumulation of electrons. **e** Calculated density of states (DOS) of Co_9S_8. **f** Adsorption, reorientation, and dissociative adsorption of H_2O molecules on Cu@ Co_9S_8 clusters at (111) plane. **g** Comparison of cell voltage of comparative electrocatalysts, Pt/C–IrO_2, Cu@CoS_x/CF and bare NF couples in 1 M KOH electrolyte. **h** Current density verses time curve at voltage of 1.8 V for water splitting. Reproduced from Ref. [3] with permission. Copyright 2016, Wiley–VCH. **i** Intermetallic $FeSn_2$ structural description with close atomic Fe and Sn packing. Reproduced from Ref. [4] with permission. Copyright 2020, Wiley–VCH. **j** MO_2 model in substitution with Zn: Denoted by different balls Zinc (green), oxygen (red), and metal atoms (purple). **k** Formation of O non-bonding states by extrapolation of MOT diagram for MO_6 with octahedral geometry. **l** Projected density of states (DOS). **m** The species of O_2^- and O_2^{2-} showing thermodynamic stabilities: blue and yellow colors represent charge depletion and accumulation. **n** Chemical recognition of O_2^{2-} species using probe (TMA^+). **o** Raman spectral analysis (1 and 2) for CoOOH and $Zn_{0.2}Co_{0.8}OOH$ (4 and 3) electrodes under different conditions. Reproduced from Ref. [5] with permission. Copyright 2019, Nature Publishing Group. **p** Polarization curves and Tafel slopes of $Zn_{0.2}Co_{0.8}OOH$ in TMAOH and 1 M KOH mixed in H_2O (99%). Reproduced from Ref. [6] with permission. Copyright 2019, Nature Publishing Group. **q** Adsorption of alkali metal cations for greater OER process. Reproduced from Ref. with permission. Copyright 2021, Elsevier

the reaction pathway for O–O coupling. The authors suggest that the feasibility of the Late Oxygen Reduction (LOM) reaction mechanism depends on the capacity of two oxidized oxygen atoms (located in proximity, closer than the van der Waals radii) to merge their respective oxygen vacancies while maintaining active site-O 2p hybridization. These unique local configurations dictate the reversible oxygen refilling process and determine the durability of the catalyst.

Intermetallic interaction-mediated electronic modulation elevates the activity factor by adjusting the metal d band center's alignment compared to the O 2p orbitals. This fine-tuning yields an optimized chemisorption energy and effectively curbs metal dissolution through robust interactions in the outer orbitals. Menezes and colleagues have documented a strategic approach involving intermetallic interactions to enhance intrinsic catalytic activity and stability, exemplified by $FeSn_2$ anode development [4]. The catalyst displays structural irregularities that yield an abundance of active sites, thereby elevating the ASF. In contrast to the unmodified Fe catalyst, no evidence of metal dissolution was observed in the case of $FeSn_2$. This durability can be ascribed to the ample active centers, each possessing enhanced intrinsic performance. As a result, these active sites generate a significant current output even when subjected to low input voltages and prevent passivation from occurring.

The involvement of atomic orbitals, specifically O 2p orbitals, contributes to M–O bonds formation in MOOH, creating bonding and antibonding states. The difference between these states, often referred to as charge transfer energy, depends on the energy levels of oxygen ions and the metal's orbitals. Moreover, late transition metal oxides (such as Fe, Cu, Co, and Ni), which are electrochemically active, split the antibonding states due to d-d Coulomb interactions (U), leading to the formation of a filled lower Hubbard band (LHB) and an empty upper Hubbard band (UHB). The intensity of these interactions exhibits an inverse relationship with orbital size and is, therefore, highly contingent on the metal's oxidation state, which reflects the effective nuclear charge. In the context of MOOH, the charge transfer energy exceeds that of the d-d Coulomb interaction. As a result, the lower Hubbard band (LHB) undergoes oxidation, revealing a significant increase in the metal's valence charge after the initial OH attack.

Researchers have unveiled two essential prerequisites for catalyzing the LOM process while ensuring mechanical and structural robustness: (1) the introduction of a buffer band derived from the oxygen nonbonding state to maintain the initial bond order of M–O throughout oxidation and (2) the establishment of a covalent nature within the M–O bond to diminish the likelihood of removal of electron from the LHB. The research conducted by the authors illustrates that the introduction of a catalytically inactive metal (Zn) into CoOOH increases Co–O covalent bonding, creates readily available oxygen nonbonding states, and maintains a consistently high current density over extended periods.

Chuang group has found that the catalyst materials IrO_x and $NiCo_2O_3$ exhibit significant differences in OER activity, ranging from 4- to eightfold, after introducing different alkali metal cation (AM^+) [6]. Their investigations showed that noncovalent interactions between OH_{ad} and AM^+ ad are the critical factor governing the OER process in terms of accessible electrochemical surface area (ECSA), reaction kinetics, mechanistic pathway, and intermediate formation, regardless of the catalyst materials and electrolyte concentration. Furthermore, the findings demonstrated that the strength of the AM-OH bond led to different complexes, such as $(H_2O)_xAM^+{}_{ad}$-OH_{ad} (for AM = Li or Na) and $(H_2O)_{x-1}AM^+{}_{ad}\cdots H_2O\cdots OH_{ad}$ (for AM = K or Cs) (Fig. 7.1q). With a strong bonding, the former hinders OH_{ad} dissociation, which promotes an inner-sphere electron transfer to form the OOH_{ad} intermediates via AEM. However, the intermediate linear scaling relation significantly improves the OER performance. Apart from it, the latter facilitates hydrogen bonding between adsorbed OH_{ad} and hydration water. The formation of hydrogen bonds weakens the influence of nonspecifically adsorbed $AM^+{}_{ad}$ on accessible OHad, improves OER kinetics by bypassing the OOH_{ad} formation, and overcomes the limitation of linear scaling relation via LOM. However, the larger size of Cs^+ leads to a lower OER activity than K^+ due to the blocking effect of the nonspecific $AM^+{}_{ad}$.

The Ce doping effect on the catalytic performance and durability of Co oxides was investigated by Yu et al. [7] The structural analysis of the $Co_{3-x}Ce_xO_4$ composites revealed that the incorporation of Ce caused a partial disruption of the Co oxide lattice, leading to improved performance and stability of the Co-active centers. Spectroscopy investigations indicated that the interaction between Ce- and Co-induced modifications in electronic infrastructure of the surface, facilitating improved kinetics of charge transportation. The Co_3O_4 and $Co_{3-x}Ce_xO_4$ samples exhibited the presence of Co in both its divalent (Co^{2+}) and trivalent (Co^{3+}) states. However, the Ce interaction promoted an increase in the Co^{2+}/Co^{3+} ratio, converting active centers of Co in the +2-valence state into Co-OOH under anodic bias. This transformation led to these sites functioning as active centers during O–O coupling, followed by a return to their initial state, thus promoting metal redox recyclability.

The in-depth spectroscopic analysis revealed that samples containing Ce maintained a consistent Co^{2+}/Co^{3+} ratio throughout the oxygen evolution reaction (OER), resulting in a sustained high current density during gas evolution. The study also observed the redox transformation of $Ce^{3+/4+}$ due to its interaction with Co and the existence of oxygen vacancies within the material. Moreover, a significant fraction of Ce^{3+} contributed to generating oxygen vacancies, which introduced a band gap state because of appearance of unpaired d electrons. This modification improved the interaction among the O 2p orbital and antibonding state of the active element. The results also indicated that the pairing of Ce^{3+} and Co^{2+} strengthened the Co–O π bonding by promoting electronic delocalization. This, consequently, accelerated the charges transportation from Ce to Co, facilitated adsorption–desorption processes of intermediates, and curtailed the dissolution of the metal.

7.2.1 Heteroatom Doping

Heteroatom doping has been recognized as a highly promising approach to effectively manipulate the electronic structure of a catalyst. This technique is known to enhance the intrinsic catalytic activity of active centers and boost the electrochemical performance of the catalyst. By introducing heteroatoms into the electrocatalyst, charge redistribution occurs, which leads to electronic structure as well as surface modification. This process also enhances the electroconductivity of the catalyst and assists chemisorption energy optimization. Consequently, this leads to the formation of dynamically stable and active sites, effectively catalyzing the reaction. Moreover, the introduction of foreign elements has notably influenced the physicochemical characteristics of the catalyst. In particular, this procedure augments the catalyst's specific surface area and roughness factor, subsequently revealing more reactive sites for the concentration of electrolyte. This also hinders the development of nonconductive layers at elevated anodic potentials. Researchers have, therefore, extensively explored the impact of different heteroatoms on the catalytic stability and activity. The results of these studies have demonstrated that catalysts doped with heteroatoms have shown higher activity and stability than their pure counterparts.

The addition of heteroatoms into the catalyst alters its electronic properties through introduction of new energy levels and modifying the band structure of the electrode. Heteroatom doping can change the oxidation state of the catalyst, creating new active sites, enhancing electrons number available for reactions, and enhancing the electron transfer properties of the catalyst. The dopants can either donate or accept electrons from the catalyst, leading to changes in the electron density and distribution within the catalyst. The type and concentration of heteroatoms introduced into the catalyst influences its electronic structure and catalytic properties. Some common heteroatoms used for doping include nitrogen (N), phosphorus (P), boron (B), fluorine (F), and sulfur (S). These elements possess unique electronic configurations and chemical properties that can influence the catalyst's electronic structure. For instance, N-doped catalysts tend to have higher catalytic activity because of robust electron-donating properties of the N atom, creating additional active sites for reactions. On the other hand, S and P dopants tend to modify the electronic density around the metal atoms and enhance electron transfer properties of the catalyst.

The mechanism behind the enhancement of catalytic activity by heteroatom doping can be explained by the changes in the catalyst's electronic structure. One of noteworthy changes is the modification of the catalyst's band structure, which affects its electronic features. The band structure of a catalyst describes the arrangement of the energy levels of its electrons. These energy levels determine the electronic and catalytic features of electrocatalyst, such as its conductivity and reactivity. The introduction of heteroatoms creates new energy levels or modifies the existing ones, leading to a change in the band structure. This change can either increase or decrease the energy barrier for reactions, making the catalyst more efficient. Heterogeneous catalysts are materials with a complex electronic structure that is depending upon the morphology, composition, and surface properties. In general, the band structure of a heterogeneous catalyst determines its electronic properties and its ability to participate in catalytic reactions. The band structure of a solid is typically described by the density of states (DOS), which is the number of allowed electronic states per unit energy per unit volume. The DOS of a heterogeneous catalyst is usually composed of two main regions: the conduction band (C_B) and the valence band (V_B). The V_B

contains the occupied states, while the C_B contains unoccupied states. The energy difference between the V_B and C_B is called the bandgap (E_g), which determines the ability of the material to absorb or emit light.

The doping process can modify the electronic structure of the catalyst by new energy levels introduction or by shifting the existing ones. The changes in the band structure induced by the doping process can affect the selectivity and performance of the catalyst. One of the most important effects of heteroatom doping is the modification of edges of the C_B and V_B. The introduction of heteroatoms can create new electronic states within the bandgap, which can act as either donor or acceptor states. These new states can modify the electronic catalytic properties, such as the Fermi level, the work function, and the charge density. In addition, heteroatom doping can shift the position of C_B and V_B edges, which can affect the ability of the catalyst to participate in redox reactions. Another important effect of heteroatom doping is the creation of new active sites for catalysis. The doping process can create new surface sites or modify the existing ones, which can improve the adsorption properties of the catalyst. For example, N-doping of carbon materials can create new basic sites that can enhance the adsorption of acidic molecules, such as CO_2. Similarly, P-doping of metal oxides can create new Lewis's acid sites that can improve the adsorption of basic molecules, such as NH_3. Another factor that influences the catalytic properties of the doped catalyst is the changes in the surface properties of the catalyst. Heteroatom doping can modify the surface charge density and enhance number of active centers on surface of electrocatalyst. The dopants can also modify the surface morphology and structure, leading to variations in the electrocatalyst adsorption properties. In addition to altering the electronic and surface properties, heteroatom doping can also influence the stability of the catalyst. The introduction of heteroatoms can improve the catalyst's resistance to corrosion and deactivation by preventing the formation of metal oxides or other unwanted species.

Heteroatom doping such as N, S, or O into metal phosphide structures has been shown to increase their OER and HER activity and stability. This is due to the coordination of the heteroatoms with the least stable P^{3-} ions, which enhances electrical conductivity and generates more active sites for the reactions. Recently, Lee and coworkers investigated the S doping influence on the electrocatalytic response of CoP for water electrolysis [8]. The researchers calculated the total density of states (TDOS) values for both the bulk crystal and stable low Miller index planes (i.e., (011) and (111) planes) of CoP and CoP-S using density functional theory (DFT). The DOS at the Fermi level (E^0) was found to be nonzero, indicating the formation of a metallic band by the overlapping Co 3d and P 3p orbitals of bare CoP. The S 3p orbital contributes to enhancing its metallic properties by increasing the states within both the valence band and the conduction band, somewhat closer to the Fermi level. Additionally, sulfur doping was observed to localize electron densities within the Co 3d orbital of CoP near the Fermi level, leading to lower oxidation state, as indicated by the shift in Co XPS peaks. Furthermore, the increase in the charge-carrier density near the Fermi level facilitates the electrocatalytic electrolysis of water into O_2 and H_2. The TDOS of S: CoP at the Fermi level was higher (15.10 vs. 14.14) for the (011) and (111) planes compared to those for bare CoP. Additionally, the integrated DOS up to 0.5 eV above the Fermi level was also higher (7.78 vs. 6.93) for the (011) and (111) planes of S: CoP, confirming that a more substantial electrons number, present in the conduction band, making them readily available for water adsorption. In general, the outcomes of this research demonstrate that sulfur doping into CoP enhances its electrocatalytic activity for the water electrolysis process by increasing the charge-carrier density nearby the Fermi level, facilitating water adsorption, and improving electrical conductivity. These findings could potentially be applied to developing more durable and efficient composites for water-splitting applications. In a recent investigation, the combination of Mn–S/NiS with selenium (Se) doping has demonstrated a substantial improvement in electrical conductivity. This enhancement leads to quicker charge transfer and elevated catalytic activity. The study attributes this to two main factors. Firstly, the in-situ growth of Mn–S on NF promotes the formation of a heterojunction structure, where the nickel foam acts as a direct Ni source aiding in building the Mn–S/NiS heterojunctions. The combination of Mn–S with NiS in this structure efficiently benefits charge transfer, resulting in higher activity achieved with respect to the Se–NiS, NiS, and Ni–Se electrodes. Secondly, Se introduction assists to modify the surface structure, resulting in ultrathin nanosheets of Se-Ni-Mn oxide, which further sulfurizes to form an amorphous overlayer. This overlayer enhances active sites number as well as its surface area, accessible for catalysis by facilitating the adsorption of H_2O for HER and OH for OER on the surface. The removal of the amorphous layer resulted in reduced OER and HER electrocatalytic activities, demonstrating the positive effect of this layer on the adsorption of H_2O for better OER and HER catalysis. Furthermore, density functional theory (DFT) outcomes have shown that Se doping on MnS/NiS led to ideal H* adsorption kinetics for the HER reaction [9]. A recent study has shown that the addition of the nonmetal element selenium (Se) can enhance the OER performance of iron oxyhydroxide (FeOOH) and nickel-doped FeSe pre-catalyst. The researchers developed a facile on-site electrochemical activation method to achieve Se-doped FeOOH electrode for efficient OER. They pre-prepared FeSe nanosheet arrays on an iron foam substrate and then converted FeSe into amorphous Se-doping FeOOH during the OER process while maintaining the morphology of 3D vertical nanosheet array for prompt gas release, resulting in a highly active low-cost full-Fe OER electrode. This strategy enhanced the activity of FeOOH and Ni-doped FeSe pre-catalyst, with the electrode capable of delivering a commercial level 500 mA cm^{-2} current density for OER process at a lower 348 mV overpotential at a cost of only 3.3 at% Ni doping. The electrode was further used in a water-alkali electrolyzer, presenting stable hydrogen production with a small cell voltage of 1.62 V at 50 mA cm^{-2}. Finally, a practical solar-driven water splitting system was demonstrated to steadily convert solar energy to hydrogen with a high efficiency of 18.55% [10].

Wang et al. reported a P-doped $CoMoO_4$ vertical microarrays on Ni foam using hydrothermal and phosphorization methods [11]. The spectroscopic analysis showed the substitution of partial $MoO_4{}^{2-}$ groups with $PO_4{}^{3-}$ groups and a change in the valence state of Mo-containing species from single-valence $Mo_6{}^+$ to valence-rich Mo ($+2 < \delta < +4, +4, +5, +6$). The resulting catalyst exhibited ultralow overpotentials of 44 mV for HER (attaining 10 mA cm^{-2} current density) and 260 mV for OER process (attaining 20 mA cm^{-2} current density), correspondingly, in basic media. In addition, in-situ surface reconstruction of P-CMO was carefully studied, which revealed that the P-doped catalyst undergoes fast self-reconstruction, resulting in the generation of ultrathin CoOOH nanosheets under oxidation potential that act as real active centers for O–O coupling (Fig. 7.2).

Chen et al. explored N doping influence on electrocatalytic water splitting. They have synthesized a catalyst composed of N-anion decorated metallic Ni_3S_2 and exhibits excellent catalytic activity for both OER as well as HER processes in basic media. The electrode's unique 3D configuration enhanced electrical behavior, larger surface area, and optimal Gibbs free-energy changes ($\Delta G\mathrm{H}^*$ and $\Delta G\mathrm{H_2O}^*$) contributing its enormous bifunctional performance, stability, and favorable reaction kinetics. The authors used density functional theory (DFT) outcomes to investigate N atoms influence on composite's activity. The analyses showed that the N–Ni_3S_2/NF electrode had a lower $\Delta G\mathrm{H}^*$ and $\Delta G\mathrm{H_2O}^*$, indicating that the catalyst-H^* state and catalyst-H_2O state might occur more efficiently on its surface. This finding suggests that the N-anion decorated Ni_3S_2 material has a lower energy barrier for the

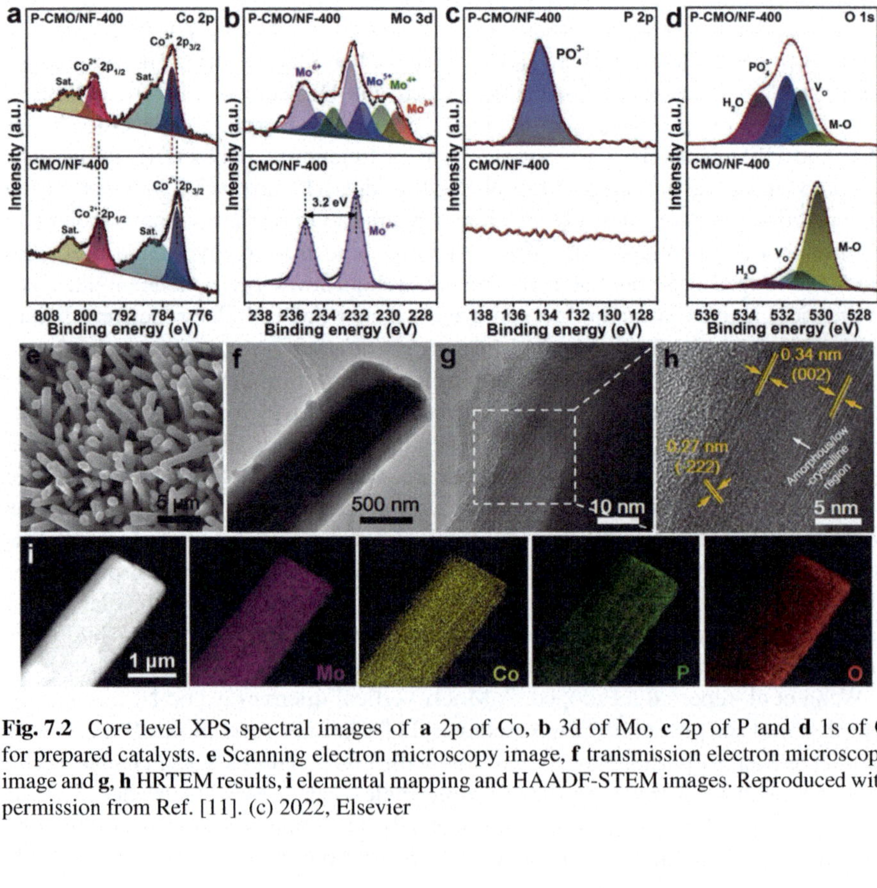

Fig. 7.2 Core level XPS spectral images of **a** 2p of Co, **b** 3d of Mo, **c** 2p of P and **d** 1s of O for prepared catalysts. **e** Scanning electron microscopy image, **f** transmission electron microscopy image and **g**, **h** HRTEM results, **i** elemental mapping and HAADF-STEM images. Reproduced with permission from Ref. [11]. (c) 2022, Elsevier

HER and OER processes, which is beneficial for achieving higher electrocatalytic performance.

The higher density of states (DOS) near the Fermi level of the N-anion decorated Ni_3S_2 material enhances its electrical conductivity, facilitating electron transfer and improving catalytic activity. The authors also found that the N- Ni_3S_2/NF electrode possessed larger active surface areas that favored achieving higher electrocatalytic performance. In addition, the electrode's 3D conductive network further enhances the penetration of the electrolyte and electron transfer process, contributing to improved reaction kinetics [12].

7.2.2 Oxygen Vacancies

Oxygen vacancies are sites in the crystal lattice of materials where oxygen atoms are missing. These vacancies can occur due to various reasons, such as thermal reduction, exposure to reactive gases, or mechanical damage. The oxygen vacancies presence can affect electronic and catalytic properties of materials, making them highly relevant in various technological applications, including catalysis, electrochemistry, and solid-state electronics.

The existence of oxygen vacancies can alter the electronic configuration of a catalyst through various mechanisms. First, oxygen atoms in a lattice site act as electron donors, and their absence in a lattice site creates a hole that can accept electrons. Consequently, oxygen vacancies can give rise to the creation of localized states within the catalyst's bandgap, thereby exerting a substantial influence on the material's electronic and catalytic characteristics. Oxygen vacancies can create localized states within the bandgap of the catalyst, which can substantially affect the electronic and catalytic properties of the material. These localized states arise due to defect levels, which are energy levels that occur due to defects in the crystal lattice. The position and width of these defect levels are dependent on the type and concentration of defects present, and they have the potential to influence the material's electronic and catalytic features. The formation of localized states may influence the composite's electronic conductivity by creating additional energy levels that can facilitate electron transfer between the reactants and the catalyst. Moreover, the formation of localized states can affect the reactivity of the material by facilitating the activation as well as adsorption of reactant molecules. Finally, the presence of localized states can enhance the catalytic activity of electrode, which leads to enhanced selectivity and rate of reaction.

The catalyst's electronic density of states (DOS) is also affected by oxygen vacancies. Electronic DOS describes the distribution of electronic states in the energy spectrum of the material, and it determines the electronic conductivity and reactivity of the catalyst. The oxygen vacancies may alter the DOS of the catalyst by creating additional energy levels in the bandgap. The creation of additional energy levels may influence electronic conductivity of the electrocatalyst by increasing the density of states near the Fermi level. The Fermi level denotes the energy level at which there is a 50% likelihood of locating an electron, and it dictates the electronic conductivity of the material. Therefore, the presence of additional energy levels can enhance the electron transfer rate within the catalyst, leading to improved electronic conductivity.

Moreover, the creation of additional energy levels can affect the catalyst's reactivity by altering the material's electronic properties. The presence of additional energy levels can facilitate the adsorption and activation of reactant molecules by creating additional binding sites. Furthermore, the creation of additional energy levels can affect the redox features of electrocatalyst by altering the position of the energy levels to the redox potentials of the reactants. Oxygen vacancies in heterogeneous catalysts can create additional sites for redox reactions by providing a location for the adsorption and activation of reactants. When an oxygen atom is removed from the crystal structure of a metal oxide or perovskite catalyst, it creates a vacancy in the lattice structure. This vacancy site can then serve as an active site for the adsorption of reactant molecules, such as oxygen, water, or hydrocarbons, and initiate a redox reaction. Oxygen vacancies can additionally modify the catalyst surface's localized electronic structure, resulting in shifts in the catalyst's redox behavior. The formation of oxygen vacancies can give rise to the emergence of fresh electronic states in proximity to the Fermi level, thereby aiding transfer of electrons between the reactants and the catalyst surface. This can result in improved catalytic activity, selectivity, and changes in the reaction mechanism.

Additionally, oxygen vacancies can also change the oxidation state of the catalyst. When an oxygen atom is removed from the crystal structure, it creates a positively charged vacancy site, which can be compensated for by incorporating additional cations or electrons. Incorporating cations or electrons can lead to electrocatalyst oxidation state variations, which can influence its redox properties. The degree of oxygen vacancy formation can also have an impact on the redox behavior of the catalyst. In certain instances, having a moderate concentration of oxygen vacancies can boost the catalyst's redox activity, whereas an excessive formation of oxygen vacancies can result in a decline in both catalytic activity and selectivity. The degree of oxygen vacancy formation can be controlled through various synthesis methods, such as annealing and reduction processes.

The activity of OER on transition metal oxides (TMOs) is primarily determined by intermediates having specified adsorption energy such as OH^*, O^*, and OOH^* on the catalyst surface. However, the factors that govern the adsorption property or surface reactivity of TMOs still need to be clarified, and the critical working condition of OER makes it extremely difficult to measure the intermediates' adsorption properties directly. Therefore, effectively tailoring the adsorption energy of intermediates to achieve desirable OER activity is highly challenging.

In contrast to OER, the reverse reaction of OER, i.e., oxygen reduction reaction (ORR), can be efficiently catalyzed by transition metals (TMs), whose reactivity may be described by the d-band model. This has led to the successful design of catalysts for ORR. However, no general model exists to represent the surface reactivity of TMOs, which makes it challenging to tailor the OER composite performance. One of the main challenges in developing a surface reactivity descriptor for TMOs is the extreme divergence in features and structure of these composites, which contrasts with the relatively constant differences in the properties of TMs. Nonetheless, several attempts have been made to correlate adsorption properties with several features in the electronic structure of the oxides. For example, the e.g., filling has been proposed as a descriptor for perovskites (AMO_3), and the states' near-Fermi level has been correlated with the adsorption property for TiO_2 and 4d and 5d late TMOs. However, while these correlations describe the trends for the investigated materials, they do not provide predictability for well-studied OER catalysts such as Co_3O_4 and MnO_2. One way to overcome the complexity of TMOs and tune the surface electronic structure continuously is to create model systems that can mimic the surface of TMOs. For example, model systems such as thin films, single crystals, and nanoparticles can be used to examine the effect of electronic structure of surface on the adsorption property of intermediates. In addition, computational approaches such as density functional theory (DFT) outcomes may assist to predict the adsorption energy of intermediaries on catalyst surface.

Bin Liu and coworkers present a model aimed at elucidating the connection among the surface configuration of transition metal oxides (TMOs) and their reactivity in OER process [13]. They suggest that the surface structure of TMOs, particularly the presence of undercoordinated metal cation sites (MCUS), is an essential factor in determining their reactivity in OER. To understand the electronic origin of this descriptor, the authors propose a model based on molecular orbital and band structure theory. They illustrate the bonding of atomic oxygen, a key intermediate in OER, to the surface of a TMO in Fig. 7.3. Atomic oxygen, the second-highest electronegative element, behaves as an electron acceptor in bonding with metal cations. The adsorption sites for atomic oxygen are typically MCUS, which only have d valence electrons for 3d and 4d TMOs. Therefore, the authors assume that the interaction between oxygen and TMO (restricted to 3d and 4d TMOs) is mainly contributed by coupling oxygen's 2p orbitals to the highest occupied d-states of the metal cations (denoted as Ed). This coupling results in bonding and antibonding states, as shown in Fig. 7.3. The bonding states are generally far below the Fermi level (EF) and are filled. In contrast, the filling of antibonding states depends on the relative energy of Ed to EF [denoted as (Ed–EF)]. The higher the energy of Ed close to EF, the less filling of the antibonding states will occur which leads to maximum adsorption. The proposed model for understanding the surface reactivity of transition metal oxides (TMOs) is dependent on molecular orbital and band structure theory. According to this model, the surface structure of TMOs, specifically under-coordinated metal cation sites (MCUS), affects the atomic oxygen adsorption on the surface. The key intermediate is atomic oxygen in oxygen evolution reaction (OER). Its bonding with TMOs is mainly contributed by coupling O 2p to the highest occupied d-states (Ed) of 3d and 4d TMOs. The coupling results in bonding and antibonding states, with the filling of antibonding states depending on the relative energy of Ed to the Fermi level (EF).

The interpretation of the model is divided into three categories: n-type, p-type, and metallic TMOs. In the stoichiometric n-type TMOs, Ed in the valence band is far below EF, resulting in filled antibonding states and weak adsorption. However, a surface with higher MCUS than the stoichiometric surface can introduce band

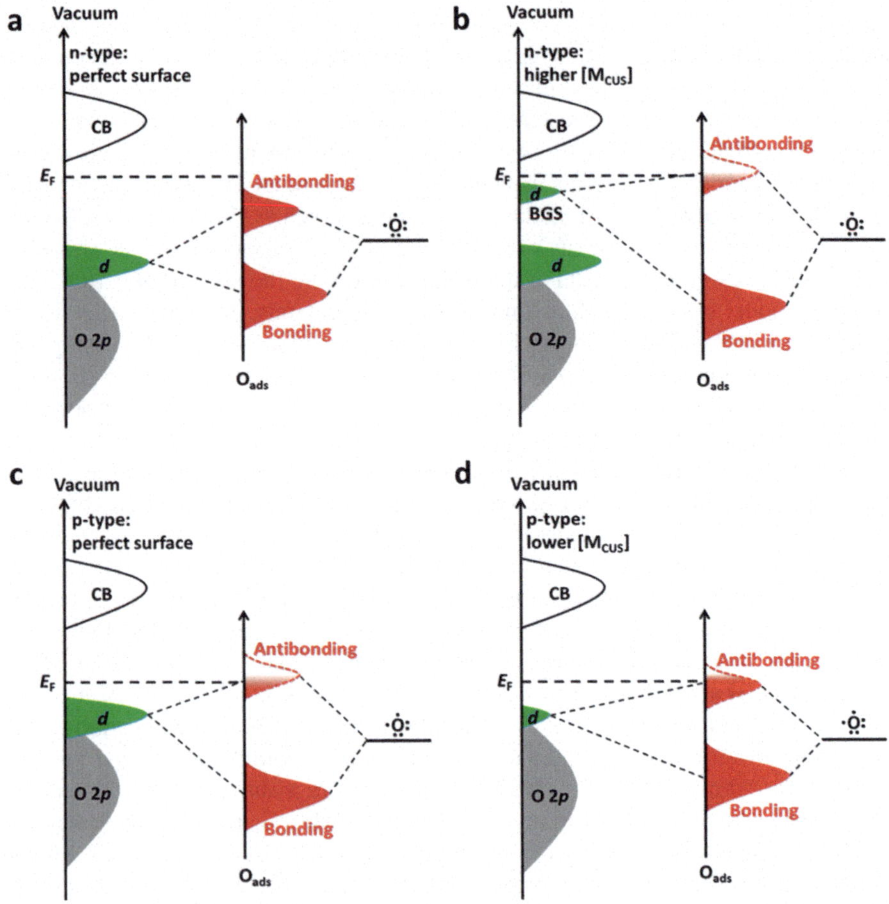

Fig. 7.3 Formation of atomic oxygen bonds on various types of TMOs occurs through the coupling of O 2p to the highest occupied d-states. Most stable TMOs are semiconductors or insulators, and their valence band (VB) is composed of O 2p states and the d-states of the transition metal cation. The bonding of O to an n-type TMO is shown in (**a**, **b**), where the relatively low energy of highest occupied d-states (Ed) to the Fermi level (EF) fills the antibonding states on the stoichiometric surface of an n-type TMO. However, a surface oxygen deficiency (higher [MCUS]) introduces band gap states (BGS) that generate d-states and lead to stronger O adsorption due to an upshift of the antibonding states relative to EF. The bonding of O to a p-type TMO is shown in (**c**, **d**), where Ed is much closer to EF, leading to less filling of the antibonding states as compared to stoichiometric n-type TMOs. This results in weaker bonding due to the downshift of Ed relative to EF. This model suggests that only the highest occupied d-states contribute to the bonding strength of adsorbates and that (Ed–EF) should be considered as the electronic origin of surface reactivity for TMOs. Reproduced with permission from Ref. [13]. ©2016, American Chemical Society

gap states (BGS) due to unpaired d-electrons, which can increase Ed and push the antibonding states above EF, leading to less filling of antibonding states. Therefore, higher MCUS gives rise to higher surface reactivity, evidenced by diminished BGS upon oxygen adsorption. Examples of stoichiometric n-type TMOs include TiO_2, α-Fe_2O_3, V_2O_5, and MoO_3.

In the case of stoichiometric p-type TMOs, Ed is close to EF, and the antibonding states are partially filled, resulting in higher surface reactivity than stoichiometric n-type TMOs. However, some p-type TMOs, such as CuO, have a different band structure where the d band in the valence band is lower than the O 2p band. Therefore, coupling between Cu 3d and O 2p results in filled antibonding states, making the bonding of O to CuO the weakest among all 3d TM monoxides. Therefore, higher MCUS on the surface of p-type TMOs leads to higher surface reactivity, while lower MCUS resulting from extra surface oxygen weakens the adsorption. Examples of stoichiometric p-type TMOs include MnO_2, Co_3O_4, and NiO. For metallic TMOs such as TiO, V_2O_3, and MoO_2, partially occupied d-states extend to EF, resulting in very high surface reactivity, making them easily oxidized in the presence of O_2.

Wang et al. investigated sulfur (S) doping-induced oxygen vacancies effect on the OER performance of CuO-based catalysts [14]. Their findings demonstrated that introducing S-doping led to notable improvements in CuO's electrical conductivity and proton–electron transfer rate, resulting in the augmentation of OER efficiency. Notably, the incorporation of sulfur induced the oxygen vacancies creation, which heightened the concentration of defective active sites. These sites preferred binding with OER intermediates and exhibited favorable thermodynamics for dissociating O–H bonds at lower cell voltages. A more detailed examination unveiled that the oxygen vacancies near Ni/Co atoms increased the polarizability of H–OH bonds, thus enhancing OER kinetics. Employing density functional theory (DFT) calculations, Wang et al. uncovered a synergistic effect arising from oxygen vacancies and selective metals with appropriately occupied states, amplifying the efficiency of the OER. Their investigation encompassed nine distinct models, each considering various electrolyte routes initiating from M–OH creation, then M–O moieties, and finally M–OOH formation for formation of O–O bond and intermediate adsorption processes. Geometries were optimized for models without defects and models with defects, and the adsorption energy of O intermediates on the surface of an ideal catalyst was employed for comparative evaluation. The results indicated that the presence of sulfur vacancies and the insertion of Ni within the standard structural configuration enhanced the performance of the oxygen evolution reaction.

However, the robust adsorption of the OH* intermediate on the Cu–S site hindered the process of electron removal (OH to O). Examination of the free energy profile showed that the formation of vacancies improved the OH* bonding strength with the active sites. This, in turn, lowered the binding strength, while nearby Ni sites aided in the formation of O–O bonds due to the impact of the solid hydrogen binding energy. The defects also had an impact on neighboring Cu atoms, reducing the energy gap between the Fermi levels and 3d orbital, thus enhancing the kinetics of the reaction. Moreover, the existence of oxygen vacancies not only narrowed this disparity but also heightened the rate of reaction, primarily due to the electronegative nature of oxygen

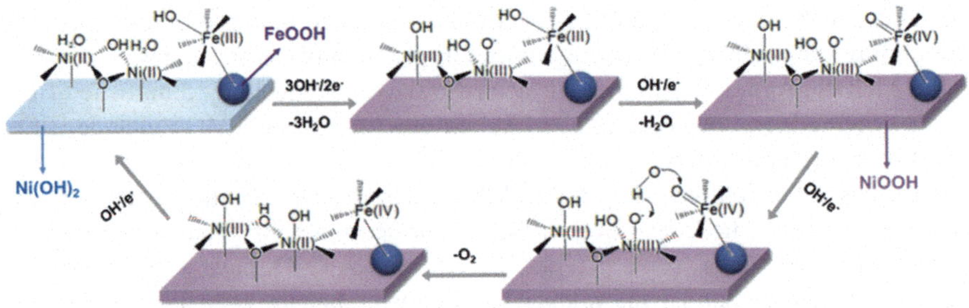

Fig. 7.4 Mechanism of OER on Mo-Ni foam surface, supported by Ni–Fe LDH dopes by S. Reproduced with permission from Ref. [15]. Copyright 2021, Elsevier

vacancies. The doping of anions, mainly through the generation of oxygen vacancies via sulfur doping, emerged as a positive factor influencing the OER process. This effect stemmed from boosted bond polarizability, OER intermediates stabilization, and desorption rate acceleration. In this context, Chai and colleagues conducted a study exploring the electronic structure modifications brought about by sulfur doping in the context of the OER catalyst [15]. Their investigation revealed a significant reduction in free energy, enhancing the bonding between OER intermediates and active centers, as clearly depicted in Fig. 7.4. Their study's outcome highlighted that iron (Fe) undergoes oxidation during the anodic potential to form Fe-OOH, establishing a covalent binding with $Ni(OH)_2$. This oxidation process induces the formation of a Ni^{3+} dimer by release of protons from H_2O molecules or OH. The preequilibrium stage involves a proton-coupled electron transfer process at Fe^{3+} sites, ultimately leading to oxidation and forming Fe^{4+}=O. In the subsequent step, Fe^{4+}=O sites interact with Ni^{3+}-O and OH, generating Fe^{+2}-Ni^{+2}-OH sites. This sequential progression eventually regenerates and exposes the reaction sites for a cyclic response.

The investigation of the mechanism uncovered that sulfur, which possesses electronegative properties, forms coordination bonds with hydroxyl anions, thereby initiating O–H…S interactions. This specific interaction prompts a transfer of electronic density from sulfur to hydrogen atoms, thereby creating an abundance of electrophilic active sites that facilitate the presence of OH^- on MO^*. The O–H…S hydrogen bonding demonstrated superior strength compared to O–H…O interactions, establishing its thermodynamic feasibility. Recent discoveries have unveiled the potential of integrating electronegative atoms with densely charged positive defects, as generated through the anodization process, within the lattice structure. In a recent study, scientists designed amorphous nanoneedles of S-Cu_2O-CuO composite for natural electrolysis of seawater [16]. The researchers employed a direct growth technique to synthesize Cu_2O-CuO nanoneedles on a copper substrate using a $NaBH_4$-based electrolyte. This strategy generated an abundant oxygen vacancies and a robust interaction between support and metal, with an expanded energy gap, thus promoting effective charge and mass transfer at the interface. The introduction of sulfur (S) through an anodization procedure established a sturdy connection between S and the copper framework, further amplifying the number of oxygen vacancies and introducing structural disorder. Examination of the electrolyte following electrolysis revealed that the S content remained consistent, even after stability tests in harsh seawater conditions. This observation underscored the persistent positive effect of sulfur in entire procedure. The findings suggested that extended exposure to the oxygen evolution reaction (OER) resulted in surface oxidation or the partial oxygen-related group's adsorption. This observation was consistent with the behavior of transition metals doped with heteroatoms in the form of oxyhydroxide or metal hydroxide, which exhibit metal redox activity. The S–$Cu^{2+/3+}$ redox couple was postulated as the active phase governing OER and ensuring the regularity of the metal redox cycle. Measurements using Inductively Coupled Plasma Optical Emission Spectroscopy (ICP-OES) of the KOH electrolyte following constant 20-h electrolysis, provided additional support for this theory. These measurements confirmed the lack of sulfur in the electrolyte, suggesting that species of S remained embedded within the hybrid material, functioning as an active phase instead of dissolving into the electrolyte, which is a common occurrence in nonoxide-based catalysts [17, 18].

The latest research indicates that oxygen-deficient surface oxides (Cu–O_x) are spontaneously generated by electrocatalytic oxidation and are efficient catalysts. Meanwhile, the persisting presence of sulfur (S) within the oxidized catalyst exerts an equally beneficial promotional effect, enhancing the performance of metal oxo-/hydroxide OER active centers on S-doped CuO surface. Furthermore, the introduction of oxygen vacancies enhances OER kinetics and selectivity by facilitating the formation of O–O bonds via two MO^* intermediates, deviating from the conventional peroxide pathway.

7.2.3 Multimetallic Electrocatalyst

Multimetallic catalysts, which consist of multiple metal species, are more efficient than single-metal catalysts in water electrolysis. The catalyst surface's electronic structure governs intermediates' adsorption energies, with the valence states of metal (d states) and oxygen (p states) near the Fermi level playing a crucial role in bond formation and determining the adsorption energies of metal oxides. Theoretical investigations indicate that the incorporation of foreign metals causes two significant changes in the electronic infrastructure including: the ligand effect and the strain effect.

7.2.3.1 Ligand Effect

The ligand effect is a crucial factor that impacts the selectivity and activity of multimetallic catalysts. In the context of water electrolysis, the ligand effect refers to the perturbations of the electronic structure induced by charge transfer between neighboring metals. Charge transfer occurs through shared oxygen ions, which change the electronic system of the catalyst surface. The different electron-donating abilities of multimetal can significantly impact the ligand effect. For example, in a multimetallic catalyst consisting of cobalt and nickel, the cobalt atoms have a higher electron-donating ability than the nickel atoms. Consequently, a transfer of charge takes place from the cobalt atoms to the nickel atoms via the oxygen ions they share. This results in an alteration of the electronic configuration of the catalyst's surface, potentially influencing the catalyst's catalytic selectivity as well as activity.

The ligand effect can also impact the HER and the OER processes, the two half-reactions of water electrolysis. For example, in a multimetallic catalyst consisting of iron and nickel, the HER activity is enhanced when the iron is incorporated into the catalyst. This is because iron has a higher electron-donating ability than nickel, which leads to charge transfer from iron to nickel. This charge transfer enhances active sites number for the HER, which enhances the catalyst's performance. The ligand effect can also impact the OER activity of multimetallic catalysts. For example, in a multimetallic catalyst consisting of ruthenium and iridium, the OER activity is enhanced when iridium is incorporated into the catalyst. This is because iridium has a higher electron-donating ability than ruthenium, which leads to charge transfer from iridium to ruthenium. This charge transfer enhances activation and adsorption of oxygen on the catalyst surface, which enhances the OER activity.

Strategies to Control the Ligand Effect:

The ligand effect is a crucial factor that impacts the selectivity and activity of multimetallic catalysts. The electronic and geometric characteristics of the catalyst surface can be altered by charge transfer between neighboring metals, which leads to variations in the adsorption energies of intermediates, therefore influencing the overall catalytic activity. Therefore, controlling the ligand effect in multimetallic catalysts is crucial for designing efficient and selective catalysts for various catalytic reactions, including water electrolysis. In this essay, we will explore different strategies that can be employed to control the ligand effect in multimetallic catalysts.

Rational Design of Catalysts:

The rational design of catalysts is a promising strategy for controlling the ligand effect in multimetallic catalysts. The rational design approach involves the specific control of the structure, shape, size, and chemical composition of the catalysts to achieve specific catalytic properties. The chemical composition of the catalysts can be controlled by incorporating metals with similar electron-donating abilities, which can minimize charge transfer between neighboring metals and reduce the ligand effect. For example, the iron and cobalt atoms have similar electron-donating abilities in a multimetallic catalyst consisting of iron and cobalt. Therefore, charge transfer between iron and cobalt is minimal, resulting in a reduced ligand effect. The rational design approach can also be used to control the catalysts' size, shape, and structure. For instance, the shape as well as size of the catalyst nanoparticles can be tuned to optimize the adsorption properties and number of active centers, which may control the ligand effect.

Surface Modification:

Surface modification is another effective strategy for controlling the ligand effect in multimetallic catalysts. Surface modification involves catalyst surface modulation by introducing functional groups or ligands, which can control the catalysts' electronic properties and surface chemistry. For example, introducing a specific ligand can reduce the ligand effect by stabilizing the adsorbed intermediates on the surface of catalyst and reducing the charge transfer between neighboring metals. For instance, introducing phosphoric acid as a surface modifier in a multimetallic catalyst consisting of platinum and nickel can reduce the ligand effect. Phosphoric acid acts as a ligand and coordinates with the platinum atoms, reducing the charge transfer between platinum and nickel and stabilizing the adsorbed intermediates. This surface modification can enhance the catalyst's activity and selectivity in the oxygen reduction reaction.

Alloying:

Alloying is a well-known strategy for controlling the ligand effect in multimetallic catalysts. Alloying involves mixing two or more than two metals to develop new composites with specific catalytic features. Alloying can reduce the ligand effect by altering the electronic and geometric characteristics of the catalyst surface, which can enhance the catalytic activity and selectivity. For example, alloying can reduce the ligand effect by forming a solid solution of platinum and palladium in a multimetallic catalyst consisting of platinum palladium. This solid solution can alter the electronic and geometric features of catalyst surface, reducing the transportation of charge between neighboring metals and enhancing the catalytic selectivity as well as its activity.

Control of Surface Composition:

The control of the surface composition of multimetallic catalysts is another effective strategy for controlling the ligand effect. The surface composition of multimetallic composite may be controlled by varying the order of deposition of the metals or by selectively removing one metal from the surface. This can alter the spatial arrangement of the metals and reduce the charge transfer between neighboring metals, thereby reducing the ligand effect. For example, in a multimetallic catalyst consisting of copper and platinum, the deposition order can be controlled to minimize the ligand effect. If copper is deposited first and then platinum, the platinum atoms will be dispersed on the copper surface, leading to a reduced ligand effect. Similarly, a selective removal technique, such as electrochemical leaching, removes one metal from the surface. In that case, the remaining metal can have a different spatial arrangement, reducing the ligand effect.

Support Effects:

The support material on which the multimetallic catalyst is deposited can also influence the ligand effect. The support material can alter the electronic and geometric properties of the catalyst surface, leading to changes in the charge transfer between neighboring metals and the ligand effect. For example, in a multimetallic catalyst consisting of platinum and ruthenium, the support material can influence the ligand effect. When the catalyst is supported on carbon, the charge transfer between platinum and ruthenium is minimal, leading to a reduced ligand effect; however, when the catalyst is supported on a metal oxide such as ceria, the charge transfer between platinum and ruthenium increases, leading to a higher ligand effect.

Thermal Treatment:

Thermal treatment is another effective strategy for controlling the ligand effect in multimetallic catalysts. Thermal treatment involves heating the catalysts to high temperatures to induce structural changes that can alter the geometric and electronic features of the catalyst surface, leading to a reduced ligand effect. For example, thermal treatment can reduce the ligand effect in a multimetallic catalyst consisting of iron and nickel. Heating the catalyst to high temperatures causes the formation of an iron-nickel alloy with a specific crystal structure, which can alter the electronic and geometric properties of the catalyst surface and reduce the ligand effect.

7.2.3.2 Strain Effect

The strain effect originates from variations in the lattice parameters on the surface of a catalyst. Incorporating foreign metals into a metal oxide can alter the lattice parameters, distorting the lattice structure. This distortion can cause a shift in the extent of overlap between atomic orbitals, ultimately altering the electronic catalytic infrastructure. The adjustment in the electronic structure may lead to shifts in the electrochemical activity and selectivity of the catalyst. The Strain effect is generally inseparable from the ligand effect because they occur concurrently in multimetal oxides. The modified electronic structure of multimetal oxides is considered a complete result of both effects. The impact of the strain effect and the ligand effect on a catalyst's catalytic activity can vary depending on the reaction conditions and the catalyst's composition.

Strategies to Control the Ligand Effect:

The strain effect in multimetal catalysts can significantly impact their catalytic activity for various chemical reactions, including water electrolysis. Therefore, it is essential to understand and control the strain effect to design stable and efficient composites for electrolysis of water. One of the primary strategies to control the strain effect is to choose suitable foreign metals that can induce the desired level of strain without compromising the stability of the catalyst. The choice of foreign metals should be based on their size, electronic properties, and compatibility with the host metal lattice. For example, smaller foreign metals can induce tensile strain, whereas larger ones can cause compressive strain. Moreover, the electronic properties of the foreign metal should be compatible with those of the host metal to ensure a smooth integration of the two metals. The selection of foreign metals can be guided by computational modeling, which can predict the lattice parameters, strain, and electronic structure of the resulting multimetal catalyst. The other scheme to control the strain effect is to optimize the synthesis method of the multimetal catalyst. The conditions for synthesis, like reaction time, temperature, and pressure, can significantly influence the lattice parameters and strain of the resulting composite. For instance, high-temperature annealing can induce compressive strain, while low-temperature annealing can induce tensile strain. The addition of surfactants or templates during synthesis can also affect the lattice parameters and strain by regulating growth and nucleation of the catalyst particles.

Surface modification is another approach to control the strain effect in multimetal catalysts. The surface modification involves depositing a thin layer of another composite on the catalyst's surface to alter its properties, including the lattice parameters and strain. For example, the deposition of a metal oxide layer on the metal catalyst surface can induce compressive strain and improve its catalytic activity for oxygen evolution. Similarly, the deposition of a metal sulfide layer can cause tensile strain and enhance the catalytic activity for hydrogen evolution. Various techniques can achieve surface modification, such as atomic layer deposition, electrodeposition, and chemical vapor deposition. Finally, the strain effect can be controlled by tuning the morphology of the multimetal catalyst. The morphology of the catalyst, including its size, shape, and surface structure, can influence the strain and catalytic electronic structure. For example, nanocrystals or nanoparticles with different shapes can induce strain due to the differences in their surface energies. Moreover, using porous or nanoscale architectures can increase the surface area and enhance the interaction among foreign metal and the host metal, leading to a more significant strain effect.

Tanveer et al. have innovatively designed a new approach to construct a self-supported bifunctional composites for seawater splitting. This catalyst involves the integration of gold nanoclusters (Au NCs) onto nanoflakes of Gd-Co_2B, which are intercalated within nanosheets of TiO_2 grown on a titanium (Ti) foil substrate. (Au-Gd-Co_2B@TiO_2) (Fig. 7.5) [19]. The electrochemical behavior and subsequent analysis demonstrate the promising potential of the electrocatalyst Au-Gd-Co_2B@TiO_2 for efficient seawater splitting. The outstanding performance of this catalyst can be attributed to several key factors. Firstly, developing well-developed and branched TiO_2 nanosheets directly onto a titanium foil substrate enhances structural stability, offers a substantial surface area for active site formation, and boosts resistance to corrosive seawater conditions. Secondly, Gd's effective electronic conduction, which can be ascribed to its unoccupied 4f orbitals within the 5d orbital, enhances the coordination characteristics of the catalyst's active sites. Thirdly, the distinctive electronic attributes of Au, which affect oxidation of Co as well as its affinity for OH intermediates, give rise to a selective and exceedingly efficient electrochemical activity in seawater splitting. This effect is attributed to the relativistic influence. Lastly, microwave treatment ensures interaction between hydrophilic surfaces of Gd-Co_2B and Au nanoclusters (Au NCs). This interaction facilitates efficient catalysis by enabling appropriate space for electrolyte diffusion, thus enhancing gas evolution

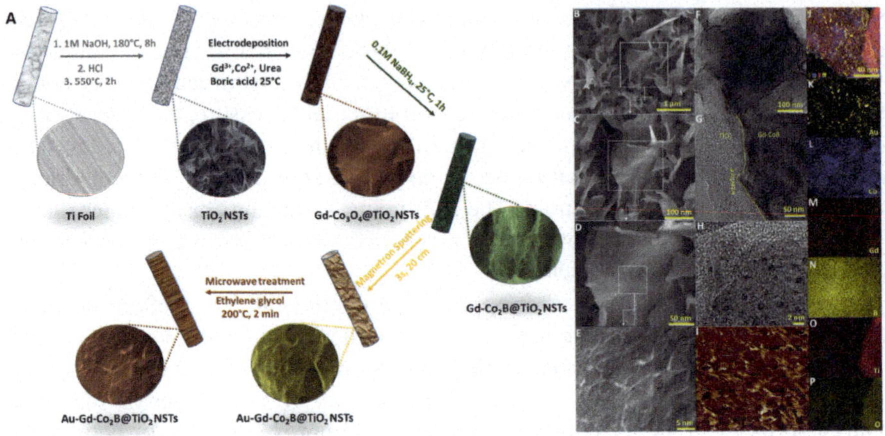

Fig. 7.5 **a** Systematic diagram for the development of Au-Gd-Co_2B@TiO_2 nanosheets. **b–e** SEM, **f–g** TEM, **h** HRTEM, **i** AFM images Au-Gd-Co_2B@TiO_2 nanosheets, **j** STEM and **k–p** corresponding elemental mapping. Reproduced with permission from [19] Copyright 2021, Elsevier

kinetics from the active sites. Consequently, due to B modification, the electrocatalyst Au-Gd-Co_2B@TiO_2 presents many active sites, rapid kinetics, and an efficient transfer coefficient. This optimized structure and surface properties combination provides distinctive active sites, offering favorable OH intermediate adsorption–desorption and Au–Co interface interactions characterized by high conductivity and transfer coefficient. These attributes position the catalyst as a promising candidate for practical seawater electrolysis applications.

References

1. Chung DY, Lopes PP, Farinazzo Bergamo Dias Martins P, He H, Kawaguchi T, Zapol P, You H, Tripkovic D, Strmcnik D, Zhu Y, et al (2020) Dynamic stability of active sites in Hydr(Oxy)oxides for the oxygen evolution reaction. Nat Energy 5(3):222–230. https://doi.org/10.1038/s41560-020-0576-y
2. Lopes PP, Chung DY, Rui X, Zheng H, He H, Farinazzo Bergamo Dias Martins P, Strmcnik D, Stamenkovic VR, Zapol P, Mitchell JF et al (2021) Dynamically stable active sites from surface evolution of perovskite materials during the oxygen evolution reaction. J Am Chem Soc. https://doi.org/10.1021/jacs.0c08959
3. Liu Y, Li Q, Si R, Li GD, Li W, Liu DP, Wang D, Sun L, Zhang Y, Zou X (2017) Coupling subnanometric copper clusters with quasi-amorphous cobalt sulfide yields efficient and robust electrocatalysts for water splitting reaction. Adv Mater. https://doi.org/10.1002/adma.201606200
4. Chakraborty B, Beltrán-Suito R, Hausmann JN, Garai S, Driess M, Menezes PW (2020) Enabling iron-based highly effective electrochemical water-splitting and selective oxygenation of organic substrates through in situ surface modification of intermetallic iron stannide precatalyst. Adv Energy Mater. https://doi.org/10.1002/aenm.202001377

5. Huang ZF, Song J, Du Y, Xi S, Dou S, Nsanzimana JMV, Wang C, Xu ZJ, Wang X (2019) Chemical and structural origin of lattice oxygen oxidation in Co–Zn oxyhydroxide oxygen evolution electrocatalysts. Nat Energy 4(4):329–338. https://doi.org/10.1038/s41560-019-0355-9
6. del Rosario JAD, Li G, Labata MFM, Ocon JD, Chuang PYA (2020) Unravelling the roles of alkali-metal cations for the enhanced oxygen evolution reaction in alkaline media. Appl Catal B Environ 2021(288):119981. https://doi.org/10.1016/j.apcatb.2021.119981
7. Li L, Hu Z, Tao L, Xu J, Yu JC (2020) Efficient electronic transport in partially disordered Co_3O_4 nanosheets for electrocatalytic oxygen evolution reaction. ACS Appl Energy Mater. https://doi.org/10.1021/acsaem.0c00190
8. Anjum MAR, Okyay MS, Kim M, Lee MH, Park N, Lee JS (2018) Bifunctional sulfur-doped cobalt phosphide electrocatalyst outperforms all-noble-metal electrocatalysts in alkaline electrolyzer for overall water splitting. Nano Energy. https://doi.org/10.1016/j.nanoen.2018.08.064
9. Zhu J, Sun M, Liu S, Liu X, Hu K, Wang L (2019) Study of active sites on Se-MnS/NiS heterojunctions as highly efficient bifunctional electrocatalysts for overall water splitting. J Mater Chem A. https://doi.org/10.1039/c9ta10860b
10. Niu S, Jiang WJ, Wei Z, Tang T, Ma J, Hu JS, Wan LJ (2019) Se-doping activates FeOOH for cost-effective and efficient electrochemical water oxidation. J Am Chem Soc 141(17):7005–7013. https://doi.org/10.1021/jacs.9b01214
11. Wang J, Hu J, Liang C, Chang L, Du Y, Han X, Sun J, Xu P (2022) Surface reconstruction of phosphorus-doped cobalt molybdate microarrays in electrochemical water splitting. Chem Eng J 446(P3):137094. https://doi.org/10.1016/j.cej.2022.137094
12. Chen P, Zhou T, Zhang M, Tong Y, Zhong C, Zhang N, Zhang L, Wu C, Xie Y (2017) 3D nitrogen-anion-decorated nickel sulfides for highly efficient overall water splitting. Adv Mater. https://doi.org/10.1002/adma.201701584
13. Tao HB, Fang L, Chen J, Yang HB, Gao J, Miao J, Chen S, Liu B (2016) Identification of surface reactivity descriptor for transition metal oxides in oxygen evolution reaction. J Am Chem Soc 138(31):9978–9985. https://doi.org/10.1021/jacs.6b05398
14. Wang XH, Ling Y, Wu B, Li BL, Li XL, Lei JL, Li NB, Luo HQ (2021) Doping modification, defects construction, and surface engineering: design of cost-effective high-performance electrocatalysts and their application in alkaline seawater splitting. Nano Energy. https://doi.org/10.1016/j.nanoen.2021.106160
15. Zhou YN, Yu WL, Cao YN, Zhao J, Dong B, Ma Y, Wang FL, Fan RY, Zhou YL, Chai YM (2021) S-doped nickel-iron hydroxides synthesized by room-temperature electrochemical activation for efficient oxygen evolution. Appl Catal B Environ. https://doi.org/10.1016/j.apcatb.2021.120150
16. ul Haq T, Haik Y (2021) S doped Cu_2O-CuO nanoneedles array: free standing oxygen evolution electrode with high efficiency and corrosion resistance for sea water splitting. Catal Today. https://doi.org/10.1016/j.cattod.2021.09.015
17. Wang X, Li W, Xiong D, Petrovykh DY, Liu L (2016) Bifunctional nickel phosphide nanocatalysts supported on carbon fiber paper for highly efficient and stable overall water splitting. Adv Funct Mater 26(23):4067–4077. https://doi.org/10.1002/adfm.201505509
18. Tang C, Cheng N, Pu Z, Xing W, Sun X (2015) NiSe nanowire film supported on nickel foam: an efficient and stable 3D bifunctional electrode for full water splitting. Angew Chemie Int Ed 54(32):9351–9355. https://doi.org/10.1002/anie.201503407
19. ul Haq T, Pasha M, Tong Y, Mansour SA, Haik Y (2022) Au nanocluster coupling with Gd-Co_2B nanoflakes embedded in reduced TiO_2 nanosheets: seawater electrolysis at low cell voltage with high selectivity and corrosion resistance. Appl Catal B Environ. https://doi.org/10.1016/j.apcatb.2021.120836

Chapter 8
Structural Modification of Electrocatalysts for Enhanced Water Electrolysis

8.1 Catalyst Surface Structure

The surface structure chemistry of a heterogeneous catalyst refers to the chemical features of the composite's active sites responsible for catalyzing reactions. These properties include surface acidity, basicity, redox properties, and the availability and accessibility of active sites. Surface acidity is a critical property of many heterogeneous catalysts, as it can influence the reaction mechanism and selectivity. Acidic centers in surface of electrocatalyst can facilitate the protonation of reactant molecules, which leads to more reactive intermediates formation. The strength and number of acidic sites can be tuned through surface modification or support materials. For example, introducing acidic groups onto the surface of a metal oxide catalyst can increase its acidity and enhance its activity for acid-catalyzed reactions. Basicity is another surface property that can impact the electrochemical performance of a heterogeneous catalyst. Basic sites on the surface of composite can facilitate the deprotonation of reactant molecules or participate in acid–base-catalyzed reactions. Basic sites can be introduced onto the catalyst's surface through surface modification or using support materials with basic properties [1]. Redox properties are critical for catalysts that participate in redox process, such as reduction or oxidation reactions. The presence of redox active sites on the electrocatalyst's surface can facilitate electron transfer reactions and enable the catalyst to participate in redox reactions. Metal catalysts, such as platinum or palladium, are often used in redox reactions due to their ability to contribute in oxidation as well as reduction process. The availability and accessibility of active sites on surface of catalyst are also important considerations in catalyst design. Dynamic sites that are not accessible or are blocked by surface species or contaminants may be inactive or have reduced activity. Strategies to enhance the availability as well as accessibility of active centers include using support materials with high surface area, introducing functional groups onto the catalyst's surface, and using porous materials such as zeolites [2].

T. u. Haq and Y. Haik, *Electrochemical Water Splitting*, Materials Horizons: From Nature to Nanomaterials, https://doi.org/10.1007/978-981-99-9860-9_8

The surface structure of heterogeneous catalysts can be influenced by various factors, including the catalyst's chemical composition, crystal structure, morphology, and synthesis method. These factors can affect the active site distribution, surface area, surface charge, and electronic properties of the catalyst, which can influence its catalytic activity, selectivity, and stability [3].

8.1.1 Chemical Composition

The chemical composition of a catalyst plays a crucial role in determining its surface structure, which in turn can impact its catalytic activity, selectivity, and stability. The chemical composition of a catalyst refers to the types and amounts of elements or compounds present in the catalyst's design. The chemical composition can influence the surface structure of a catalyst in several ways, including the surface charge, active site distribution, and electronic properties. One of the keyways in which the chemical composition of a catalyst affects its surface structure is through the formation of surface defects. Surface defects can arise due to impurities or vacancies in the catalyst's crystal structure. These defects can create sites for chemical reactions to occur and alter the electronic properties of the composite's surface. Like, defects in the crystal structure of metal oxides can lead to variations in the surface charge as well as electronic features, affecting their catalytic activity. A catalyst's chemical composition can also affect its surface charge, which plays an essential role in maintaining the interaction between the electrocatalyst and molecules of reactant. The surface charge of a catalyst depends on the presence of charged species such as metal ions or oxygen vacancies. For instance, doping a metal oxide catalyst with transition metal ions can modify the surface charge and electronic features, leading to enhanced electrocatalytic performance [4].

Similarly, the presence of oxygen vacancies in composites based on metal oxides can also alter the surface charge, leading to improved catalytic activity for reactions such as water oxidation. A catalyst's chemical composition can also influence the active site distribution on its surface. The active site refers to the region on the catalyst's surface where a chemical process occurs. The distribution and availability of active sites may suggestively impact the electrocatalyst performance and selectivity of the catalyst. For example, incorporating cobalt into a metal oxide catalyst can enhance the distribution and accessibility of active sites, which leads to maximum electrochemical performance for water oxidation [5].

8.1.2 Crystal Structure

The crystal catalytic structure is an essential aspect that significantly impacts the surface structure and, consequently, the electrochemical performance, selectivity, and catalytic durability. It refers to the arrangement of atoms or molecules in a repeating pattern, which influences the chemical as well as physical features of the catalyst. The crystal structure affects the catalytic structure in several ways, including the availability and accessibility of active centers, the surface charge, and the electronic properties. Furthermore, the crystal structure influences the density of active centers, which are the regions on the catalyst's surface where chemical reactions occur. Catalysts with a magnificent surface area and a well-defined morphology can have a higher density of active sites, leading to enhanced catalytic activity [6].

Additionally, the crystal structure of a catalyst can impact its surface charge, which has essential role in the interaction between the catalyst and reactant molecules. Changes in the crystal structure can affect the distribution of charged species, such as metal ions or oxygen vacancies, which alter the catalyst's surface charge and electronic properties. The electronic properties refer to the surface's energy levels and electron configuration, which can influence its interaction with reactant molecules. The crystal structure can alter the electronic properties, which leads to variations in the reaction kinetics and selectivity of the catalyst. Furthermore, the crystal structure can impact the stability of the catalyst under different conditions. Changes in the crystal structure can affect thermal stability, chemical stability, and resistance to catalyst poisoning [7].

8.1.3 Morphology

The morphology of a catalyst is another critical factor that can affect its surface structure, which in turn influences its selectivity, electrochemical activity, and durability. The morphology refers to the size, morphology, and surface characteristics of the catalyst particles, which can vary according to synthetic protocol and processing conditions. The morphology can affect the surface structure of the catalyst by altering the surface area, surface charge, and surface composition. A catalyst's morphology affects its surface structure through changes in the surface area. Catalysts with a maximum surface area can have more active centers available for chemical reactions, leading to enhanced catalytic activity. In addition, the morphology of the composite can impact its surface area by altering the shape and size of the particles. For example, catalyst particles with a higher aspect ratio can have a larger surface area, leading to increased catalytic activity. The morphology of the catalyst can also affect the surface charge and surface composition, which are crucial factors that influence the interaction between the catalyst and reactant molecules. For example, the distribution of charged species, such as metal ions or oxygen vacancies, affects the surface charge of the catalyst. Changes in the morphology of the catalyst particles can lead to variations in the distribution of these charged species, thus altering the catalyst's surface charge and electronic properties. The surface composition can also be affected by changes in the morphology of the catalyst particles, which can lead to differences in the types and density of active sites available for chemical reactions [8].

8.2 Catalyst Surface Engineering

Structural optimization of an electrocatalyst is critical in designing and developing efficient catalysts for sustainable energy conversion. The catalytic surface structure has a pivotal role in determining its electrocatalytic performance, selectivity, and durability, making it essential to optimize the structure to achieve desired catalytic properties. This involves the modification of the catalyst's surface at the atomic or molecular level, using innovative strategies such as defects engineering, 2D and 3D nanomaterials, core–shell nanostructures, porous materials, and surface functionalization. Structural optimization may potentially increase the electrocatalyst's catalytic activity, selectivity, and stability, thus enabling its efficient use in various electrochemical reactions, like fuel cells, electrolysis of water, and metal–air batteries. Therefore, a comprehensive understanding of the structural optimization of composite is crucial in advancing the growth of sustainable energy technologies.

8.2.1 Core–Shell Nanostructure

Core–shell nanostructures are a type of nanomaterial consisting of a core material surrounded by a shell of another material, typically with a thickness in the range of a few nanometers. Various methods, like chemical vapor deposition, electrodeposition, or sol–gel synthetic protocol, can create the core–shell structure. The interaction between the shell and core materials drives the chemistry of core–shell nanostructures. The interface between the two materials can lead to unique properties not found in the individual materials alone. For example, the shell material can protect the core from oxidation, corrosion, or chemical attack, while the core material can provide structural stability and desired functionality [9].

The most vital benefits of core–shell nanostructures is their tunable properties. By varying the morphology, size, and structural composition of the core and shell, it's easy to tailor the chemical as well as physical features of the nanostructure to suit a specific application. To illustrate, one can adjust the electronic and optical characteristics of core–shell quantum dots by managing the composition as well as size of the core and shell composites. This adaptability renders them valuable in various applications like solar cells, sensors, and light-emitting diodes. Another advantage of core–shell nanostructures is their enhanced mechanical, electrical, or chemical stability. By coating a core material with a shell material with a various mechanical, electrical, or chemical property, the resulting core–shell structure can exhibit improved properties over individual materials. For example, a core–shell nanowire made of a metal core and oxide shell can have improved mechanical and chemical stability compared to a pure metal nanowire. In addition, core–shell nanostructures can exhibit unique catalytic properties due to the synergistic effect between the core and shell materials. For example, a core–shell nanoparticle made of a metal core and a metal oxide shell can have enhanced catalytic activity due to the interaction between the metal and the oxide [10].

Deng's group has recently reported a novel method for synthesizing core–shell transition metal compounds with tunable lattice strain [11]. The method does not require the introduction of any additional factors, such as heterostructures, vacancies, or substrates. The researchers used this method to prepare a hybrid of nanorods and nanosheets of homogeneous NiS_xSe_{1-x} on Ni foam, referred to NiS_xSe_{1-x} NNH (Fig. 8.1). The strain effect was induced on both the surface and interior of the materials, resulting in improved catalytic performance. The experimental results demonstrated that the $NiS_{0.5}Se_{0.5}$ electrocatalyst with a lattice strain of approximately 2.7% exhibits the best OER and HER performance, with very low overpotentials of only 70 and 257 mV at 10 mA $cm^{-2.}$ Furthermore, the catalyst maintained excellent stability for 300 h at 10 and even 100 mA cm^{-2}, surpassing other benchmarks and state-of-the-art materials like Pt/C + IrO_2. The better activity is attributed to the exceptional architecture of the nanosheet-wrapped nanorod hybrid and the in-situ growth feature on conductive Ni foam, which ensures efficient electron transfer. Furthermore, the developed lattice strain in NiS_xSe_{1-x} decreases the overlap in the d-orbital of metal and leads to a closer d-band center toward the Fermi level. This approach increases the inherent electronic conductivity while simplifying reaction routes and reducing energy obstacles, ultimately resulting in markedly improved electrocatalytic properties. The Strasser team has innovatively proposed a method to create exceptionally efficient and long-lasting composites for OER process [12]. The researchers introduced a novel catalyst-support pair combining an Ir-Ni core achieved through electrochemical dealloying and an IrO_x shell on a support based on corrosion-resistant and mesoporous features. This innovative approach capitalizes on the oxidized Ir shell creation which is hindered kinetically, remarkably enhances OER activity with minimal state-of-the-art content, allowing efficient dispersion of active nanoparticles (NPs). The mesoporous structure with enormous surface area support contributes to the catalyst's impressive corrosion resistance.

Results indicate that the core–shell catalyst-supported water electrolysis composites achieved a 2.5-fold increase in O–O. Throughout a continuous 20-h electrolysis process, the coupling rate at the electrolyzer level showed minimal activity degradation, outperforming several benchmark Ir composites. This positions the core–shell catalyst/support pair as a promising contender for practical proton exchange membrane (PEM) electrolyzer anodes. One possibility is to apply nanostructured core–shell composites and mesoporous support coatings to these anodes, which would enhance their durability and resistance to corrosion. This core–shell composite/oxide-based support perception offers a practical approach to reduce the reliance on rare elements in electrocatalytic nanoparticles. This innovative technique can significantly enhance OER catalyst performance while reducing reliance on scarce parts, promoting sustainability and cost-effectiveness. Recent work by Zhu et al. demonstrated that the core–shell nanostructure can enhance charge transfer, gas desorption rate, and detachment of bubbles, amplifying intrinsic catalytic reactivity and exposed specific surface area thus improving the ASF [13]. The researchers

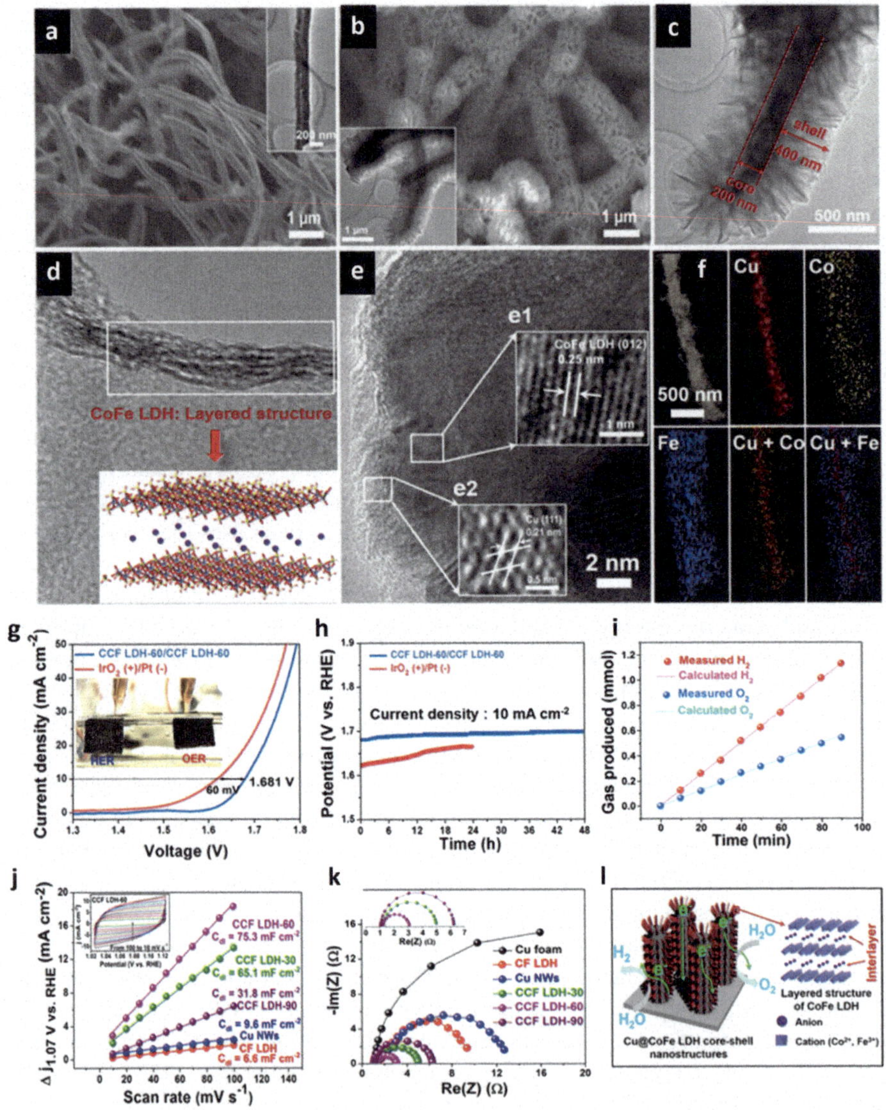

Fig. 8.1 SEM images were taken of copper nanowires (Cu NWs) and cobalt-iron layered double hydroxide (CCF LDH-60) (**a**) and (**b**) respectively, with accompanying TEM insets. Further images show TEM views of CCF LDH-60 (**c**) and HRTEM image (**d**) displaying CoFe LDH layered structure, featuring lattice fringes (**e**) and a CoFe LDH structural model in the inset of (**d**). DF-STEM image of CCF LDH-60 and elemental mapping are presented in (**f**). Electrocatalytic performance in water electrolysis was gauged via polarization and chronopotentiometry curves, as in (**g**) and (**h**), alongside state-of-the-art IrO2 (+)/Pt (*) for comparison. H_2 and O_2 production by CCF LDH-60 electrode was measured experimentally and theoretically in (**i**). Electrical properties were evaluated through capacitance (**j**) and impedance measurements (**k**). Finally, a schematic of Cu@CoFe LDH core–shell structural features is shown in (**l**). Adapted from Ref. [11], Copyright 2017, Elsevier

employed a gas-templated electrodeposition method to create core–shell electrocatalysts, specifically Ni–Fe/Ni–Fe–OH and Ni-Co/Ni-Co–OH. These electrocatalysts demonstrated outstanding efficacy in OER as well as HER process within a 1 M KOH solution, particularly at elevated current densities essential for commercial scale applications. Additionally, the team assembled a water electrolyzer using the Ni-Co/Ni-Co–OH and Ni–Fe/Ni–Fe–OH catalysts, demonstrating prolonged water electrolysis at 1 A cm^2 for more than 300 h with no significant decline in activity. Impressively, the water electrolyzer reached a current density of 500 mA cm^2 with 1.74 V cell voltage in a two-electrode setup. The excellent durability and electrochemical performance of these composites can be attributed to their distinct core–shell heterostructure. This design offers abundant active sites on the surface hydroxides, facilitates efficient electron transport through the inner alloy, and provides robust corrosion resistance. Crafting a core–shell nanostructure in this manner provides a pragmatic approach to tune chemisorption energy and secure reaction intermediates, which, in turn, enhances the intrinsic activity and stability of the catalysts. Core–shell nanostructures have been shown to selectively adsorb the intermediate species of the HER and OER process, primarily due to the different electrochemical potentials required for each reaction. For example, in a Ni–Co/Ni–Co–OH core–shell electrocatalyst system, the Ni–Co core is pathway for electrons conduction and provides a stable substrate for anchoring the active centers. The Ni–Co–OH shell on the other hand serves as the OER active center, where the hydroxide species (OH^-) can easily absorb and desorb due to the low energy barrier. Meanwhile, for the HER, the Ni-Co core provides a low-overpotential site for H-adsorption and the Ni–Co–OH shell serves as a low-overpotential site for H_2O oxidation. This selective adsorption is due to the tunable surface energies of the core and shell, which permits precise control over the adsorption/desorption energies of the intermediates. Through the deliberate design of the core–shell nanostructure, one can modulate chemisorption energy and maintain intermediates with the ideal adsorption energy, ultimately leading to enhanced intrinsic activity and stability for the catalyst [14].

When the core–shell structure is immersed in the electrolyte, the outer shell layer of the catalyst comes into contact with the electrolyte solution. Therefore, the surface chemistry of the outer shell layer can be tailored to absorb either the OER or HER intermediate selectively. For example, suppose the outer shell layer is designed to have a high affinity for oxygen species. In that case, it will selectively adsorb OER intermediates, while if it is intended to have a high affinity for hydrogen species, it will selectively adsorb HER intermediates. Moreover, the composition and electronic structure of the core–shell nanostructure can also influence the adsorption of the intermediates. For instance, a core–shell system with a highly conductive core can facilitate the electrons transportation to and from the catalyst surface, enhancing the adsorption of intermediate species. Additionally, the lattice strain within the core–shell structure can alter the catalyst's electronic structure and impact the intermediates' adsorption energy [15].

Recent research has demonstrated that core–shell architectures can effectively counteract the aggregation of nanostructures caused by van der Waals forces, a phenomenon often detected in 2D nanosheets. To address this issue,

Han and colleagues designed a core–shell structure wherein $MoSe_2$ nanosheets were embedded in conductive metal–organic frameworks (MOFs). This innovative approach prevents sheet aggregation, fostering efficient charge as well as mass transfer at the interface. This is crucial for creating sustainable electrocatalyst designs [16]. In this investigation, researchers utilized Co-MOF nanowalls as a host for 2D $MoSe_2$ nanoflakes at multiple active sites, substantially increasing active centers during redox reaction. During the process of hydrothermal synthesis for $MoSe_2$, Co-MOF underwent a transformation into $CoSe_2$, effectively overcoming the limitations of conductivity associated with Co-MOF. This transformation also led to the emergence of improved electroactive centers as a result of the incorporation of $MoSe_2$ nanoflakes onto the $CoSe_2$ nanowalls. The $MoSe_2$ nanosheets played a dynamic role as a nanosurface composite, enabling electrocatalytic interactions among the electrolytic ions within the nanostructures of MOF-$CoSe_2$@$MoSe_2$. Additionally, the conductive unidirectional pathways formed by the MOF-$CoSe_2$ nanowalls facilitated efficient and low-scattering transportation of electrons between carbon cloth (CC) and $MoSe_2$ nanosheets. Adopting core–shell structures with precise pathways for diffusion of ions proved effective for reactive reaction sites, facilitated by 2D nanosheets of $MoSe_2$'s with porous nature, maintaining nearby electrolytic contact. The developed 2D CC/MOF-$CoSe_2$@$MoSe_2$ nanostructure harnessed the benefits of diverse phases, establishing a robust interfacial link in the outer shell of MOF–$CoSe_2$ and the central core. This configuration proved effective for HER and OER electrochemical activities in basic conditions. Separately, Liu et al. introduced a novel electrocatalyst for the oxygen evolution reaction (OER) by creating a unique core–shell heterostructure consisting of trimetallic Ni–Fe–V disulfide as its core and amorphous Ni–Fe hydroxide as the shell. This innovative electrocatalyst was synthesized through a series of processes including hydrothermal treatment, sulfurization, and in situ electrochemical tuning [17]. The catalyst offers several benefits. To start with, it showcases a strong synergistic connection between the central component (Ni–Fe–V disulfide) and the surrounding layer (amorphous Ni–Fe hydroxide), which promotes a favorable electron arrangement at active sites on the surface of the shell. Secondly, the simultaneous doping of metallic V and Fe enhances the electronic conductive nature of the NiS_2 core, leading to expedited transfer of electrons during the oxygen evolution reaction (OER). The trimetallic $Ni_{0.7}0Fe_{0.10}V_{0.20}S_2$ core–shell nanostructure outperformed mono-/bimetallic core catalysts in OER activity, as demonstrated by electrochemical tests with same mass loading. Developed on a hydrophilic-treated carbon paper (CP), this catalyst forms a unified three-dimensional (3D) electrode with a substantial mass loading and excellent conductivity. This arrangement significantly boosts the overall performance and stability of the material in the oxygen evolution reaction (OER). Consequently, the core–shell structure demonstrates outstanding OER activity and durability.

8.2.2 3D Materials

Three-dimensional (3D) nanostructure materials are widely studied in materials science and engineering because of their novel structural and physical characteristics. The 3D structure of these composites is composed of a network of interconnected nanowires, nanorods, and nanotubes. These materials possess a high surface area-to-volume ratio, which makes them appealing for various applications, including catalysis, sensing, energy storage, and electronics. The structure of 3D nanostructure materials is a crucial aspect that determines their properties and potential applications. These materials have a unique geometry composed of three dimensions, length, width, and height, forming an extended network of interconnected nanoscale objects [18]. This structure offers several advantages over conventional two-dimensional (2D) or one-dimensional (1D) materials, such as enhanced surface area, improved mass transport, and increased active sites for catalysis. The 3D nanostructure materials can be divided into several types on base of their morphologies, such as nanowires, nanotubes, nanocubes, nanorods, nanosheets, and nanoporous materials. The morphology selection is dependent on the desired application and the fabrication method used to produce these materials. 3D nanostructure materials can be synthesized using several techniques, such as chemical vapor deposition, hydrothermal synthesis, template-based synthesis, and solvothermal synthesis [19].

The structure of 3D nanostructure materials can be described based on the edges, facets, and corners that compose them. The edges are the regions where two facets intersect, while the corners are the regions where three or more facets meet. The aspects are the flat surfaces that comprise the bulk of the material. These structural features have vital role in the properties and performance of the materials. For example, the corners and edges of a material can have higher surface energy and be more active for catalytic reactions than the facets. Moreover, the surface area of 3D nanostructure materials can be significantly enhanced compared to their 2D or 1D counterparts because of novel infrastructure. The magnificent surface area offers more sites for chemical reactions, such as adsorption, desorption, and catalysis. For instance, nanocubes with high-index facets exhibit maximum electrocatalytic performance because of greater surface energy and the presence of more active sites [20].

Similarly, nanoporous materials with many interconnected pores offer a high surface area for catalytic reactions, and their pore size can be tuned to fit specific applications. The 3D nanostructure materials can also exhibit improved mass transport properties, which are crucial for many applications, including energy storage and conversion, sensors, and catalysis. The interconnected network of nanoscale objects in 3D nanostructure materials provides a fast and efficient pathway for mass transport, allowing reactants and products to diffuse quickly to and from the active sites. For example, nanotubes with a hollow core can provide a direct pathway for ion transport, enabling fast and efficient charge storage and release in energy storage devices. In addition, the 3D nanostructure materials can exhibit unique features that make them highly attractive for catalysis applications. For instance, the 3D structure can provide an enhanced confinement effect, improving selectivity and efficiency for catalytic reactions. The confined spaces in the 3D materials can also induce a strain effect on the active sites, which can modify their electronic and geometric properties, leading to enhanced catalytic performance. Moreover, the 3D nanostructure materials can offer a tunable pore size and shape, which can tailor the selectivity. 3D nanostructure has recently gained attention as potential material for electrochemical applications because of exceptional structural features and high surface area. A critical aspect of these materials is their redox properties, which can be utilized in various electrochemical reactions, such as water electrolysis [21].

The redox properties of 3D nanostructures arise from their electronic structure, which is heavily influenced by the size and shape of the nanostructure. In 3D nanostructures, quantum confinement effects can be observed due to the confinement of electrons in all three dimensions. This leads to a modification of the electronic band structure, which can result in changes in the material's electronic properties, including its redox properties. One important redox property of 3D nanostructures is their ability to store and transfer charge. Because of magnificent surface area and porous nature, 3D nanostructures can facilitate charge transfer and enhance electrochemical reactions. For example, in a 3D porous carbon structure, numerous edges and corners can create active sites for catalytic reactions. Simultaneously, the greater surface area facilitates more efficient charge exchange between the electrolyte and electrode.

Another important redox property of 3D nanostructures is their capability to undergo reversible redox process. This feature is essential for the OER, where the electrocatalyst must undergo repeated oxidation and reduction cycles. The power of the 3D nanostructure to undergo these cycles without significant degradation is critical for maintaining the performance and stability of the electrocatalyst over time. Furthermore, the redox properties of 3D nanostructures may be modulated by controlling their composition, size, and morphology. For example, by varying the ratio of metal ions in a metal oxide 3D nanostructure, the redox properties can be adjusted to optimize the electrocatalytic performance for a specific reaction. Similarly, by varying the shape and size of a 3D nanostructure, the quantum confinement effects can be tuned to modify the electronic band structure and hence the redox properties of the composite [22].

The heterogeneous catalysts' active surface area is a crucial factor to investigate their catalytic performance, as it influences the active centers and its number which are available for the process. Enhancing the active surface area can be achieved through nanostructuring, which provides more active centers with increment in number of exposed edges, corners, and facets. However, this increase in active sites does not always guarantee a corresponding improvement in the catalyst's performance because the charge diffusion pathway can occasionally be blocked. Hence, to attain superior catalytic performance, it is essential to employ a distinctive hierarchical three-dimensional nanostructure that maximizes the exposure to active sites while upholding a resilient interconnected framework. The 3D hierarchical nanostructure offers several unique benefits compared to other nanostructures, such as maximum exposure to active sites, outstanding charge carrier mobility, and mechanical as well as structural stability. The interconnected design of the three-dimensional

(3D) nanostructure provides concise pathways for mass as well as ions transportation, effectively enhancing the contact region at the interface between the electrode and electrolyte. This attribute guarantees a high current density at a low input voltage, positioning the 3D hierarchical nanostructure as a promising choice for electrocatalytic applications. Kong et al. fabricated three-dimensional hierarchical $NiMo_3S_4$ nanoflowers supported by carbon textiles, demonstrating excellent catalytic performance and stability in basic media [23]. Introducing nickel into Mo-S altered the atomic-level surface structure and generated notable surface defects, which serve as active sites. This enhancement markedly boosts the material's performance and durability. The ample specific surface area of the three-dimensional (3D) hierarchical $NiMo_3S_4$ nanostructure maximizes the contact with active sites, yielding exceptional catalytic activity. Moreover, the three-dimensional hierarchical design amplifies electronic conduction by elevating the density of states at the Fermi level and ensures robust support stability even at elevated anodic potentials. During water electrolysis, the 3D hierarchical nanostructure's advantages are crucial for efficient electrocatalytic activity. In water electrolysis, the anodic and cathodic reactions occur on the electrode's surface, which makes its active surface area as a critical factor that determines the efficiency of the process. The three-dimensional hierarchical nanostructure's maximum exposure to active sites, short diffusion channels for ions and mass transportation, and enhanced electronic conduction all contribute to improving the efficacy of the water-splitting procedure. Furthermore, the interconnected structure of the 3D hierarchical nanostructure enhances the bubbles' desorption rate, which is critical for efficient gas evolution reactions during water electrolysis. Efficient gas evolution reactions require a rapid bubbles generation from the surface of electrode, which prevents the accumulation of bubbles and reduces the reactions over potential. The interconnected structure of the 3D hierarchical nanostructure ensures efficient gas evolution, making it an ideal candidate for water electrolysis applications.

In recent years, 3D hierarchical hybrid electrocatalysts have captivated noteworthy consideration for their potential in water-splitting applications. These materials possess unique features such as high specific surface area, robust support linkage, and effective diffusion channels, which efficiently catalyze both the oxygen evolution reaction (OER) and hydrogen evolution reaction (HER) at low input voltages. One example of a 3D hierarchical hybrid electrocatalyst is the Co_9S_8 nanosheets embedded into vertically aligned Ni_3S_2 nanosheets [24]. The heterointerface configuration encourages reconstruction of interface during adsorption of intermediate and gas desorption stages while maintaining the distinctive attributes of active centers because of robustly linked interconnections. Such robust interfacial interactions sustain the electrocatalytic conversion reaction even when solar input energy is intermittent, rendering them appealing for commercial applications. Theoretical analyses have unveiled that these materials' three-dimensional (3D) structure facilitates interface reconstruction throughout intermediate adsorption and gas desorption phases. This is attributed to the potent interfacial interactions between active sites and the supporting structure, ensuring the preservation of active sites' structural characteristics. Furthermore, the 3D design of these materials establishes efficient diffusion pathways for ions and mass transport, augmenting the contact region at the interfacial contact of electrode and electrolyte.

The Lee group developed a 3D hierarchical oxygenated nickel tungsten selenide electrocatalyst using a hydrothermal technique followed by controlled selenization on a Ni foam substrate [25]. The catalyst displays a nanoporous network architecture rich in oxygenated intermediates and substantial active sites, operating effectively at a low cell voltage to yield higher current densities in water-splitting process. The catalyst with hierarchical 3D infrastructure, complete with uniform mesopores, notably enhances water diffusion and stabilizes intermediates. This, in turn, promotes efficient gas desorption from active sites, creating space for subsequent cyclic reactions. The catalyst's surface structure provides stability to selenium within a metallic lattice, and the selenium-oxygen bond enhances the metallic component in the hybrid material. This results in a reduction in the charge transfer resistance (R_{ct}) at the interface between the electrolyte and electrode. The standalone nature of the hybrid material reduces reliance on a polymeric binder, elevates active site exposure, lowers the likelihood of leaching of metal at elevated current densities, and augments the anodic specific faradaic (ASF) efficiency. Moreover, the three-dimensional hierarchical structure enhances the chemical conversion of reactive elements into their most active phase by promoting extensive contact at the interface of electrode and electrolyte. This dynamic structure bolsters the self-healing feature of active centers, extending it in a three-dimensional manner. Such construction around the precatalyst curbs metal leaching risks under high anodic or cathodic potentials.

In the meantime, a unique three-dimensional hierarchical nanoarchitecture, composed of Co–Ni carbonate hydroxides and copper nitride on a substrate (copper foam), accelerates the dual-phase transition in the process of water splitting. This configuration offers a hydrophilic surface that allows for the penetration of the electrolyte, a substantial electrocatalytic active surface area to facilitate intermediate adsorption, and clear pathways for unobstructed electron conduction at the interface (Fig. 8.2a–j) [26]. The experimental results presented by the authors emphasize the significance of a 3D hierarchical configuration featuring a heterointerface. Such a structure effectively stabilizes nano sites with high entropy, thereby mitigating their accumulation and preserving the structural integrity of the catalyst during electrochemical reactions. The three-dimensional open architecture proves advantageous for unimpeded mass migration, sidestepping the development of nonconductive layers and preventing active center blockage from bubble buildup. Furthermore, this three-dimensional surface architecture accelerates the formation of a passive layer, namely CoOOH and CuO, through a dual-phase transformation. This transformation reduces the rate at which metals are leached and supports the continuity of the electrocatalytic procedure for prolonged durations.

The 3D hierarchical structure provides the following:

- Large active sites.
- Maximum exposure to the electrolyte.
- Efficient channels for mass and ion migration.

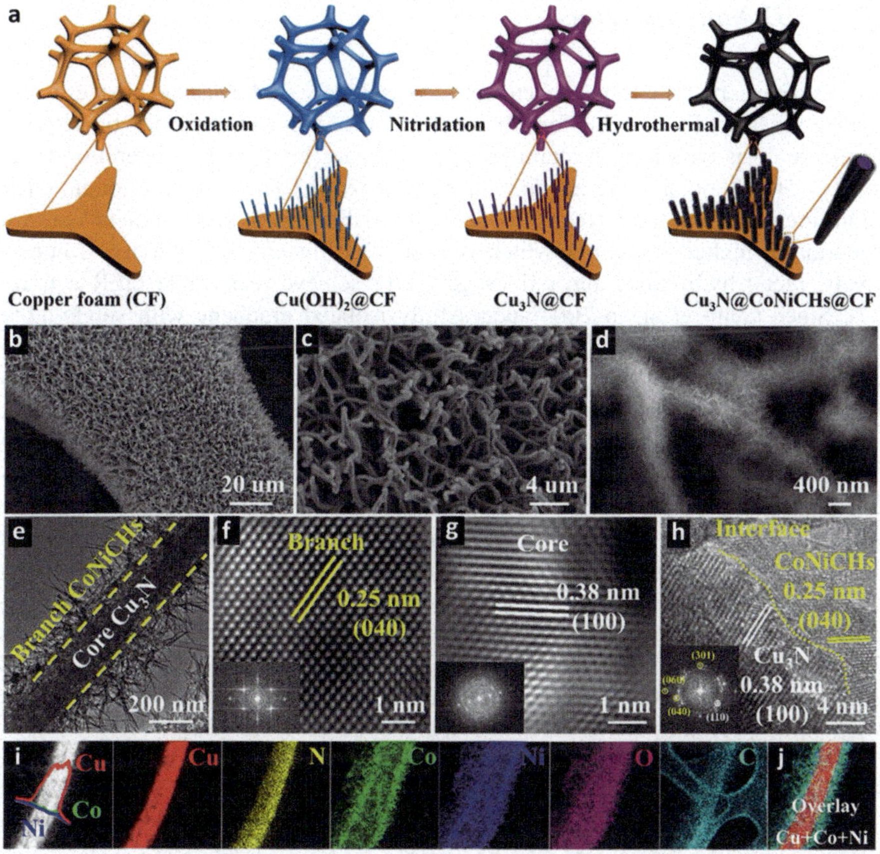

Fig. 8.2 **a** Synthesis scheme for the development of Cu3N@CoNiCHs@CF (**b–d**). SEM images **e** TEM image, **f–h** HRTEM images and **i, j** elemental mapping of Cu_3N@CoNiCHs@CF Reproduced with permission from Ref. [26]. (c) 2021. Elsevier

The architecture improves charge carrier mobility and assures mechanical and structural stability of active sites. In a three-dimensional (3D) hierarchical design, the heterointerface exhibits a substantial specific surface area and robust interlinkage with the underlying support. This design sustains catalytic activity under low input voltages for the HER and OER process. Additionally, the 3D hierarchical arrangement promotes the reconstruction of interfaces during the gas desorption and adsorption of intermediates, which helps to uphold the structural characteristics of active sites through strong coupling interconnections. These robust interfacial interactions sustain the electrocatalytic conversion reaction, even in instances where intermittent input of solar energy plays a role in enhancing their potential commercial viability.

8.2.3 2D Nanomaterials

Initially, rare earth metals have been regarded as widely studied as the most efficient catalyst based on Pt for HER for efficient and outstanding electrocatalytic activity. Electrocatalysts based on Pt attain very low overpotential for H_2 generation reaction as $\leq$50 mV potential to acquire current density of 10 mA cm^{-2}. Though, OER entails greater overpotential than HER; because of the susceptibility of metals toward oxidation at higher potentials, which is unsuitable for efficient activity. Comparatively, metal hydroxides and oxides generally achieve reasonable OER activity [27]. Since Geim et al. in 2004 successfully isolated graphene with single-atom thickness from graphite [28]. 2D composites have attained tremendous attraction as active and efficient electrocatalysts. Apart from the graphene, 2D composites comprise a large family which includes layered double hydroxides (LDHs), black phosphorus (BP), hexagonal boron nitride (h-BN), MXene, transition metal dichalcogenides (TMDCs), (and many others which have displayed outstanding electrochemical features, including superconductors, semiconductors, and insulators [29]. The distinctive 2D layered composites have been investigated widely till now, which consists of transition metal dichalcogenides (TMDs), black phosphorus (BP), hexagonal boron nitride (h-BN), graphitic carbon nitride (g-C3N4), layered double hydroxides (LDHs) layered metal oxides), covalent organic frameworks (COFs), perovskites [30, 31]. On the other side, the flourishing 2D nonlayered composites have gained incessant attracted and attention as well as nitrides, encompassing metal, phosphides, metal oxides, carbides, perovskites, metal chalcogenides, etc. [32].

2D electrocatalyst attains robust covalent bonds and weaker van der Waals forces interactions between layers, allowing such composites' applicability in electrocatalytic energy conversion and for storage purpose. Additionally, 2D composites having variable exposed edge centers may be attained by lateral modulating sizes, which allows catalytic activity modulation. 2D materials have many other advantages, like thickness and lateral sizes, as well as highly larger and active specific surface areas. Such characteristics impart noticeable benefits in electrochemical reactions in which 2D composites with greater specific surface areas may generate greater active centers and confined atomic thicknesses, which may facilitate the enormous transfer of electronic density transfer. Likewise, the interfacial and electronic infrastructure of 2D composites may be modulated via various chemical or physical strategies, which may facilitate active centers with appropriate adsorption and desorption capabilities for numerous intermediates during electrochemical reactions. Furthermore, many porous structures may also be prepared via interlayer spacing enlargement in 2D electrocatalytic, allowing surface pores formation and heterostructures construction, which may fasten gas diffusion and mass transfer, which leads to enhanced catalytic activity. Previously, 2D composites comprised a huge family containing many elements with earth-abundant nature like MoS_2 and graphene, assembling them as favorable materials for electrochemical applications on a larger scale. 2D composites are regarded as an ideal catalyst for efficient electrochemical reactions because of well-defined and clear infrastructures, which are simpler and different from nanoparticle catalysts, for better electrocatalysis mechanisms investigations. Likewise, high costs and time may be saved in computational analysis for experimental investigations because 2D composites are easy to simulate as composites for electrocatalytic processes [33]. Because of all these great merits, 2D composites have been used for electrocatalytic OER and HER activities.

8.2.3.1 2D Heterostructure Composites for HER

Water electrolysis involving hydrogen evolution reaction is well suited to establish an inexpensive, clean, and emission-free renewable path for bulk hydrogen generation as the energy vector. Platinum (Pt) metal has been considered a benchmark HER electrocatalyst, but the limits and high-cost reserves hinder its commercial applications. Thus it's significant to decrease the loading of Pt and enhance its utilization for driving H_2 cost production from water electrolysis. Downsizing nanoparticles of platinum into single atoms or clusters is regarded as an effective route for decreasing the usage of noble metals and enhancing their catalytic performance [34]. Additionally, interfacing nanocrystals (NCs) of platinum with metal-based composites may be the facile approach for exerting Pt efficacy for H_2 generation under the synergistic effects. Instantly, a nanometer-thick 2D platinum layer installation at the nanosheet surface of $Ni(OH)_2$ might construct a high area of the intimate heterointerface $Pt/Ni(OH)_2$ along the 2D developed plane. Moreover, the 2D ultrathin Pt composite might offer the greatest active centers exposure and a noticeably reduced pathway for mass diffusion and charge transfer to attain great utilization of Pt atom with long-term stability. As a proof of concept, Lee et al. synthesized the confined in-solution growth of 2D Pt-extended islands of basal area coverage on the layered double hydroxide (LDH-based) functional template (LDH@2D-Pt) (Fig. 8.3a) [35, 36]. Specifically, Ni–Fe-LDHs composed of colloidal aqueous suspension have been mixed with ascorbic acid and Na_2PtCl_4 solution, maintained under 3.5 pH (acidic) and 70 °C temperature. The synthesized products were chosen as A-LDH/2D-Pt. According to Fig. 8.3b, Pt growth has been confined strictly within the Ni–Fe-LDH 2D plane, which results in the hexagonal monodispersed lamellar sheets-like morphology of Pt with an overall 120 ± 20 nm lateral size. It has been shown that an isolated sheet of A-LDH/2D Pt has an average 9.6 nm thickness, with a thickness of 1.5 nm for 2D Pt which is grown finely on an LDH sheet. Additionally, the composite N-LDH/2D Pt was prepared with the growth of Pt conducted around a neutral pH environment for ensuring LDH component stability. The HER performance of N-LDH/2D Pt, A-LDH/2D Pt, 2D Pt [LDH + 2D Pt] (physical mixture with 1:1 weight percentage), and Pt/C (around 20%) were assessed in a KOH solution of 1.0 M in a standard three-electrode system. On evaluation, N-LDH/2D Pt reached 100 mA cm^2 current density by attaining 61 mV overpotential (η_{100}), significantly much lower compared to [LDH + 2D Pt] ($\eta_{100} = 173$ mV), A-LDH/2D Pt ($\eta_{100} = 111$ mV), 20% Pt/C ($\eta_{100} = 342$ mV), and 2D Pt ($\eta_{100} = 165$ mV) at similar current density (Fig. 8.3c). N-LDH/2D Pt attained an unchanged profile of linear sweep voltammetry (LSV) with only 11 mV overpotential increment for attaining 100 mA cm^2 current density after 5000 cycles of cyclic voltammetry (CV), which indicates outstanding durability for N-LDH/2D Pt (Fig. 8.3d). Also, N-LDH/2D Pt attained stable chronopotentiometry at 10 mA cm^2 and chronoamperometric 0.05 V vs. RHE curves electrolysis of 50h. These kinds of heterostructures of LDH@2D-Pt presented the well-structured 2D–2D tight interface and synergistically integrated active centers on Ni–Fe-LDH composite and ultrathin porous Pt layer, which contributes to outstanding HER activity as well as excellent stability (Fig. 8.3e). Owing to abundant storage and low-cost features, the transition metal-based composites, like disulfides, TM-based oxides, carbides, nitrides, and phosphides, are considered as efficient alternatives compared to expensive rare earth metals composites. Furthermore, the 2D heterostructures based on transition metals presenting higher specific surface area are required structures for electrochemical reaction in liquid-phase by complete exposure of active centers and accessible surfaces [37]. The modulation of the exceptional d-band center of transition metals may be tuned by the components and topologies inside the hybrids, which additionally enhance their electrochemical activity. Hou et al. fabricated a 2D hybrid composite comprising a Mo-doped WC core entrenched by N-doped carbon shells (NCS) by carbonization treatment on a precursor of a 2D MOF template [38]. Because of maximum surface area, the prepared Mo-WC@NCS attained robust and good HER activity in neutral acidic and basic media (Fig. 8.3f–h). The hypothetical calculations validated that the W d-band center in composite Mo-WC@NCS hybrid was shifted down after Mo dopants, which is advantageous for tuning the W centers electronic infrastructure, and thereby H desorption facility for boosting H_2 production (Fig. 8.3i–k).

Recently, MXene-based H_2 evolution composites have attracted enhanced attention, and lots of theoretical and experimental research has been done on them. Jiang et al. proposed an effective HER composite based on the ultrathin O-functionalized Ti_3C_2 MXene [39]. The counterpart F-functionalized MXenes along the basal plane have been unfavorable for H_2 evolution, depressing H_2 adsorption kinetics. The $Ti_3C_2O_x$ composite has been synthesized by $Ti_3C_2T_x$ dispersion in an aqueous solution of KOH and reduction of F-termination with OH groups. The nanosheets of $Ti_3C_2O_x$ have outstanding HER performance with 190 mV overpotential at 10 mA cm^{-2} current density, much lower than $Ti_3C_2T_x$-450 with 266 mV and $Ti_3C_2(OH)_x$ with 217 mV overpotential. Moreover, the 60.7 mV dec^{-1} Tafel slope was evaluated because of O-centers with high activity on $Ti_3C_2O_x$ MXenes basal plane. It laid the basis for enhanced training of MXenes-based HER composites through tuning their terminal surface groups.

Most substantially, MXenes may significantly enhance H_2 evolution activity by supporting binary alloys and metal nanoparticles (NPs). Li et al. prepared Pt_3Ti intermetallic compound (IMC) NPs on $Ti_3C_2T_x$ MXenes, with maximum interaction of Pt with Ti on the $Ti_3C_2T_x$ MXenes through an in-situ process of co-reduction [40]. A quick-freezing approach has been utilized to synthesize $MoS_2/Ti_3C_2T_x$' nano roll-like electrocatalyst. Extraordinary catalytic activity was attained, manifested as a 30 mV overpotential and a low 168 mV overpotential at the 10 mA cm^{-2} current density. The value of the Tafel slope was recorded at 70 mV dec^{-1}. Additionally, there was 25 times more increment in exchange current density compared

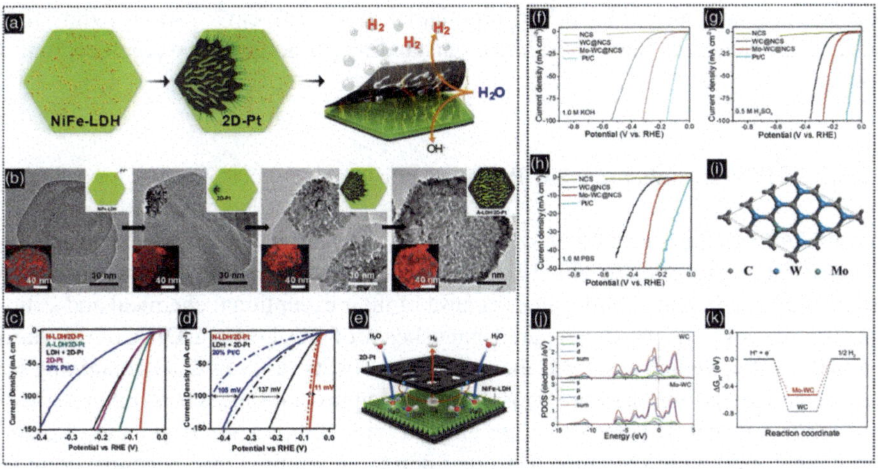

Fig. 8.3 **a** Depicts the pictorial illustration of the synthesis of 2D Pt on LDH for HER. The corresponding images of TEM and HAADF-STEM along with EDS elemental mapping for Pt for various intermediates and the final product are shown in (**b**). The iR-corrected LSV profiles and profiles after the cyclic stability test are presented in (**c**) and (**d**), correspondingly. Figure **e** shows the schematic figure of facilitated HER mechanism at the heterointerface. The polarization curves in alkaline, acid, and neutral media are demonstrated in (**f**), (**g**), and (**h**), respectively. The geometric configuration of 2D Mo-WC (001) with Mo atoms replacing the subsurface W atoms is shown in (**i**). The calculated partial density of states (PDOS) of W atom in WC (001) and Mo-WC (001) are presented in (**j**). Finally, the HER free energy diagrams for WC and Mo-WC are presented in (**k**). Reproduced with permission from Ref. [36]. (c) 2022. Wiley

to MoS_2. The impulsive oxidation may be hindered by mounting carbon-coated nanocrystals of MoS_2 on the Ti_3C_2 MXenes stabilized by carbon. The as-prepared nanocrystals of $MoS_2/Ti_3C_2T_x$@C indicated enhanced HER electrochemical performance and greater durability in acidic media. Almost 135 mV overpotential was determined at the 10 mA cm^{-2} current density, with Tafel slope value of 45 mV dec^{-1}. The in-situ X-ray absorption spectra presented that the Pt atom was changed to IMC with enhanced temperature. The attained Pt/$Ti_3C_2T_x$-550 composite indicated efficient activity, requiring 32.7 mV potential at current density 10 mA cm^{-2}. The specific activity and mass activity of Pt/$Ti_3C_2T_x$-550 was much better to Pt/Vulcan. Wang et al. prepared electrocatalysts FeNi@MXene ($Mo_2TiC_2T_x$)@nickel foam (NF) by Fe^{2+} ions introduction and in-situ combining with surface Ni atoms on NF. FeNi@$Mo_2TiC_2T_x$@NF attained efficient HER activity with overpotential of 165 mV at 10 mA·cm^{-2} because of the synergetic effect of FeNi nanoalloys and $Mo_2TiC_2T_x$ [41]. Du et al. proposed a $Ti_3C_2T_x$ MXenes-based hybrid composites improved by Nb doping and surface Ni/Co alloy [42]. DFT studies have shown that, after doping of Nb, the shifting of Fermi energy level occurred toward the conduction band, which improves conductivity. Additionally, Ni/Co alloy modified the M-H bond on the catalyst surface, thus H adsorption with the lower ΔG_H. The as-prepared $Ni_{0.9}Co_{0.1}$@NTM (Nb-doped $Ti_3C_2T_x$) hybrid composite displayed exceptional HER performance, attaining 43.4 mV overpotential at 10 mA cm^{-2}. Along with HER process, 2D materials have also shown outstanding properties for oxygen evolution reaction.

8.2.3.2 2D Materials for OER Performance

The anodic OER half-reaction in water electrolysis usually undergoes a sluggish four-electron transfer reaction in which 2D heterostructure materials have been considered efficient and promising composites because of their exceptional chemical and structural properties. Zhu et al. prepared nanosheets of 2D Fe-Co LDH/CoO as hard templates. They then transferred them into nanosheets of CoP/ CoO containing Fe via a selective low-temperature phosphorization approach [36]. The synthesized 2D Fe–CoP/CoO catalyst heterostructure aligned with component design and in-plane component modulation attained a strong interfacial coupling effect, contributing to excellent OER activity with 219 mV overpotential at current density of 10 mA cm^{-2}. The CP test of Fe–CoP/CoO attained great stability of 12 h at 10 mA cm^{-2} current density in an electrolytic solution of 1.0 M KOH.

Similarly, the catalyst Fe–CoP/CoO showed very little difference in the polarization curve after 3000 CV cycles compared to the original data, demonstrating greater durability. Regarding greater stability and activity, density functional theory (DFT) studies were accompanied by interfacial mechanism exploration. The comparative outcomes for charge density distribution and density of states (DOS) of Fe–CoP and CoP showed that Fe- introduction was promising for the transfer of charge during the OER process. Also, the O in CoP/CoO interfaces lowers the Gibbs free energy of ΔG_2 from 1.75 to 1.40 eV and considerably improves the rate-determining step from step 2 [ΔG (*O)] to step 3 [ΔG (*OOH)] in comparison to pure CoP. The Fe incorporation optimized the ΔG of the rate-determining step 3 from 1.71 to 1.64 eV, comprising OER easily compared to pure CoP/CoO. The above outcomes determine that 2D heterostructure construction may optimize the compact heterointerfaces for electronic infrastructure modulation, which enhances charge carrier densities, and lowers Gibbs free energies for fastening the OER process.

Song et al. synthesized iridium (Ir) in the form of single atoms and clusters by its incorporation in defect-rich nanosheets of cobalt-based hydroxide (termed as Co-Ir) for OER [43]. The optimized Co-Ir having 9.7 wt% Ir content (designated as CoIr-0.2) displayed extraordinary OER performance with 373 mV overpotential at a current density of10 mA cm^2 in the solution of 1.0 M phosphate buffer. Moreover, the composite CoIr-0.2 also attained a lower 235 mV overpotential for achieving 10 mA cm^2 current density in a 1.0 M KOH electrolyte for OER process. CP tests confirmed the excellent stability of CoIr-0.2 experimented for 10 h at 10 mA cm^2 in alkaline and neutral solutions. The XAFS spectroscopy toward the CoIr-0.2 sample after OER showed α-$Co(OH)_2$ transformation into a phase of β-CoOOH on the surface of the CoIr-0.2 composite along with the formation of high-valence species of Ir with low-coordination structure because of greater oxidation potentials, which were the real active centers for the OER performance.

Moreover, Hou et al. also synthesized Ni-based MOF doped with Ce with O-functionalized electrochemically exfoliated surface of graphite nanosheets (Ce–Ni-BDC/OG) through a hydrothermal approach [44]. The synthesized catalyst Ce–Ni-BDC/OG delivered 265, 311, and 337 mV overpotentials for reaching 10, 50, and 100 mA cm^2 current densities for O2 evolution reaction in basic media. Furthermore, the current density has no such change after the CP test for 10 h at 10 mA cm^2, which suggests great Ce–NiBDC/OG durability for OER. The in-situ electrochemical Raman analysis has also been experimented with, and it showed that there was an in-situ transformation for Ce–NiBDC/OG transformation into the Ce–NiOOH phase, which acted as the real active centers for OER electrocatalysis.

Likewise, Huang et al. proposed an electrocatalytic–chemical exfoliation approach to fabricate 2D-based Co MOFs [45]. Direct growth of 3D Co can be done on nickel foam for attaining 3D-Co@Ni catalyst, which might be changed into 2D-Co-NS@Ni through electrolysis technique for 0.5 h at 10 $mA{\cdot}cm^{-2}$ current density. The ultrathin 2nm thickness, 2D-Co-NS@Ni NS metal–organic layers, indicated great OER performance. Consequently, composites reformed by doping Fe with 2D-Co-MOF (Fe: 2D-Co NS@Ni) showed an ultralow 211 mV overpotential and a 46 $mV{\cdot}dec^{-1}$ Tafel slope at the current density of 10 $mA{\cdot}cm^{-2}$. The O_2 has 99% faradic efficiency. Furthermore, the stability of long-term was sustained at 10 $mA{\cdot}cm^{-2}$ for 96 h. Remarkably; Xu et al. developed a ligand-assisted method to synthesize 2D-based MOFs [46]. The bimetallic electrocatalyst 2D Co–Fe was designed through a solvothermal process using terephthalic acid as the linker and double-layered hydroxide Co–Fe (LDHs) as the template in DMF solvent. The attained product was coated on electrode glassy carbon to test the electrocatalytic activity of the OER in a basic media. It achieved 274 mV lower than overpotential, with 46.7 $mV{\cdot}dec^{-1}$ Tafel slope and 74 h stability at 10 $mA{\cdot}cm^{-2}$ current density. The composite bimetallic 2D Co Fe MOF indicated great OER activity due to the coupling effects of Fe and Fe. Wen et al. fabricated a Ni-doped ZIF-67 NS template on $Ti_3C_2T_x$ MXene surface and was directly transformed to NiCoS/ $Ti_3C_2T_x$ via a pyrolysis treatment in the presence of fine powder of sulfur [47]. The composite NiCoS/$Ti_3C_2T_x$ attained 364 mV overpotential at 10 $mA{\cdot}cm^{-2}$ a current density, superior to IrO_2 with 400 mV overpotential. The 58.2 $mV{\cdot}dec^{-1}$ Tafel slope was lower than IrO_2. Such positive effects from robust interfacial interaction and distinctive infrastructures between $Ti_3C_2T_x$ and NiCoS enhanced the number of active centers and the electrical conductivity, resulting in outstanding OER activity.

Thus, it's obvious that overall, lots of electrocatalysts have been demonstrated as highly active regarding electrocatalysis, and ideal composites designs with long-term durability and higher catalytic activities include many factors. First, 2D composites might possess enhanced specific surface areas along with large, exposed pores in maximizing active centers. Second, the vibrant centers of 2D composites attain appropriate binding strengths to products and reactants in optimizing corresponding electrocatalytic processes, where electronic infrastructures may be modulated to tune binding strengths. Third, 2D composites may conduct electrons toward external circuits and accelerate the kinetics of electrodes. Fourth, 2D electrocatalysts attain porosity regulating gas diffusion and mass transfer in electrolytes. Lastly, 2D electrocatalytic composites having inexpensive and earth-abundant nature compared to rare earth catalysts accomplish great HER and OER electrochemical activities.

In summary, 2D nanomaterials offer the following unique features that make them an ideal candidate for overall water electrolysis.

1. 2D nanomaterials have a large surface area-to-volume ratio, which is beneficial for electrocatalytic applications. The enormous surface area allows for more active centers for the catalytic reaction, leading to increased catalytic efficiency. The high surface area of 2D nanomaterials enhances electrolyte contact, promoting faster electron transfer rates and improving electrocatalytic performance.
2. The electronic properties of 2D nanomaterials can be tuned to achieve optimal electrocatalytic activity for water splitting. The electronic features of these composites may be controlled through changing their composition, thickness, and the defects presence. The electronic band structure of 2D nanomaterials plays a crucial role in electrocatalytic activity, as it governs the surface reactivity and the adsorption of reactants and intermediates. For example, transition metal dichalcogenides (TMDs) such as MoS_2 and WS_2 have tunable electronic properties and have shown promising results as efficient electrocatalysts for water electrolysis.
3. 2D nanomaterials have been reported to exhibit remarkable electrocatalytic activity for water electrolysis compared to their bulk counterparts. The high surface area of 2D nanomaterials provides more active sites for the catalytic reaction, leading to increased catalytic efficiency. Additionally, the unique structure of 2D nanomaterials allows for strong covalent bonds to form between atoms within the plane of the material, resulting in higher catalytic activity. For example, graphene-based materials have shown promising results as electrocatalysts for water electrolysis, exhibiting high activity and stability.
4. 2D nanomaterials also exhibit excellent stability under harsh electrochemical conditions, making them ideal for water electrolysis applications. The stability of these materials is due to their strong covalent bonds between atoms within the plane of the material, which allows for efficient electron transfer and prevents corrosion. Additionally, the high surface area of 2D nanomaterials provides more active sites for the catalytic reaction, leading to increased catalytic efficiency and stability.
5. The facile synthesis of 2D nanomaterials is another advantage for water electrolysis applications. The development of new synthesis methods for 2D nanomaterials has made obtaining large quantities of these materials with controlled morphologies and properties possible. This has led to a better understanding of the structure–activity relationship of 2D nanomaterials and has allowed for the development of more efficient electrocatalysts for water electrolysis.
6. 2D nanomaterials are also advantageous for water electrolysis applications due to their low cost. The production cost of 2D nanomaterials is relatively low compared to other electrocatalysts, such as noble metals. Additionally, the high

electrocatalytic activity and stability of 2D nanomaterials make them a cost-effective alternative to traditional electrocatalysts.

7. The thickness and size of 2D composites can be precisely controlled during synthesis, allowing for the production of nanomaterials with specific applications and characteristics. The control over the size and thickness of 2D nanomaterials is essential for water electrolysis applications as it allows for the optimization of electrocatalytic activity and stability.
8. 2D nanomaterials can also exhibit synergistic effects when combined with other materials, resulting in improved electrocatalytic activity and stability. For example, 2D nanomaterials can be cumulative with other composites, like carbon nanotubes, metal nanoparticles, or conductive polymers, to enhance their electrocatalytic activity for water electrolysis.
9. Some 2D nanomaterials, such as graphene and other composites based on carbon, exhibit excellent flexibility and transparency, making them suitable for flexible and transparent electrodes for water electrolysis applications. The flexibility and transparency of 2D nanomaterials are advantageous for developing flexible and transparent devices that can be integrated into various applications, such as wearable electronics, biomedical devices, and energy storage devices.

8.2.4 Defects Engineering

Defects in electrocatalysts can have a significant impact on their performance. Defects refer to imperfections or irregularities in the lattice structure of the electrocatalyst. These imperfections can occur during the synthesis or fabrication process or can be induced by exposure to harsh operating conditions. The presence of defects can lead to a decrease in the electrocatalyst's catalytic activity and selectivity. Defects can act as sites for unwanted chemical reactions, leading to the formation of undesired byproducts. Defects can also inhibit the activation and adsorption of reactants on the electrocatalyst surface, which leads to a decrease in catalytic performance. In addition to affecting the selectivity and electrochemical performance of the catalyst, defects can also reduce its stability and durability. Defects can act as sites for corrosion and oxidation, leading to the degradation of the electrocatalyst over time. Defects can also lead to the formation of cracks and other structural defects, leading to mechanical failure of the electrocatalyst. To address the issue of defects in electrocatalysts, defects engineering has emerged as a promising strategy. Defects engineering involves modifying the surface structure of the electrocatalyst by introducing defects or impurities in the material's lattice. Introducing these defects alters the material's electronic system and surface properties, which can enhance or inhibit the catalytic activity of the electrocatalyst [48, 49].

Doping is a common method of defect engineering that involves intentionally introducing impurities into the lattice of the material to alter its electronic structure. Doping can be used to tune the material's electronic structure, leading to improved catalytic activity, selectivity, and stability. Doping can be achieved through various methods, including solid-state doping, solution doping, and gas-phase doping. Solid-state doping involves mixing dopants with the precursor material before synthesis, while solution doping involves adding dopants to the precursor solution before synthesis. Gas-phase doping involves introducing dopants into the reaction chamber during synthesis. The selection of dopants is critical to the success of doping as a defective engineering strategy. The dopant must have a similar lattice structure to the precursor material and be compatible with the synthesis process. The dopant should also be chosen based on its ability to alter the material's electronic structure in a way that enhances its catalytic activity, selectivity, and stability [50].

One common dopant used in defect engineering is transition metals. Transition metals are often used to dope noble metals, like gold and platinum, to enhance their stability and electrochemical performance. The addition of transition metals can alter the material's electronic structure, leading to improved catalytic activity and selectivity. Another common dopant in defects engineering is nonmetal elements, such as nitrogen and sulfur. These dopants can be used to modify the material's electronic structure, which not only leads to enhanced catalytic selectivity as well as its activity [51]. Nitrogen, for example, may be utilized to dope composites based on carbon, such as graphene, to enhance their catalytic activity toward oxygen reduction reactions. The concentration of dopants is also essential in doping as a defect engineering strategy. However, the concentration of dopants should be carefully controlled to ensure that the desired electronic structure modification is achieved without compromising the material's structural integrity. Too high a concentration of dopants can lead to the formation of unwanted phases or the degradation of the material. In addition to tuning the material's electronic structure, doping can also be used to introduce active sites for catalysis. For example, introducing nitrogen into carbon-based materials can create functional groups comprising of nitrogen that may behave as active centers for oxygen reduction reactions [52]. One limitation of doping as a defects engineering strategy is that it can be challenging to control the location of the dopants within the lattice of the material. This can lead to the formation of undesired phases or the creation of defects that can inhibit the catalytic performance of electrocatalyst.

Surface modifications are a powerful defect engineering strategy for improving the performance of electrocatalysts. Surface modifications involve introducing functional groups or metal nanoparticles onto the electrocatalyst surface, which can alter the material's electronic structure and surface chemistry. This can lead to improved catalytic activity, selectivity, and stability [49]. One common surface modification technique is the deposition of metal nanoparticles onto the surface of the electrocatalyst. Metal nanoparticles can act as catalytic sites for various reactions, and their high surface area to volume ratio which enhances the number of active centers available for electrocatalytic activity. Metal nanoparticles can be deposited onto the surface of the electrocatalyst through various methods, including chemical reduction, electrochemical deposition, and physical vapor deposition. Another surface modification technique is the introduction of functional groups onto the electrocatalyst surface. Functional groups may vary the material's electronic infrastructure and surface chemistry, which leads to boosted catalytic selectivity as well as activity. For example, introducing carboxyl groups onto the surface of materials based on carbon can increase their electrochemical performance toward oxygen reduction reactions. Surface modifications can also be used to improve the stability of electrocatalysts. Introducing metal oxides or other protective coatings onto the surface of the electrocatalyst can prevent degradation due to harsh reaction conditions, such as acidic or alkaline environments. For example, introducing titanium oxide onto the surface of platinum-based electrocatalysts can enhance their stability toward oxygen reduction reactions [53]. One challenge with surface modifications as a defects engineering strategy is the need to control the location and density of the functional groups or metal nanoparticles on the electrocatalytic surface. The location of the functional groups or metal nanoparticles can affect their catalytic activity, and too high a density can lead to aggregation and loss of surface area. In order to solve this problem, various strategies have been proposed to control the location and density of the functional groups or metal nanoparticles on the surface of the electrocatalyst. For example, self-assembly monolayers can be used to manage the site of functional groups on the surface of the electrocatalyst. This involves the monolayer formation of functional groups on the surface of the electrocatalyst through a self-assembly process, which can control the location and density of the functional groups [54].

Defects annealing is a widely used defects engineering technique for improving the performance of electrocatalysts. Defects annealing involves the thermal treatment of the electrocatalyst at high temperatures, which can modify the material's microstructure and remove or reduce defects that may be present. One of the primary benefits of defects annealing is the removal of surface defects and impurities. Surface defects and contaminants can significantly impact the performance of electrocatalysts, as they can reduce the number of active centers available for catalysis and hinder the diffusion of reactants to the surface of the electrocatalyst [55]. Defects annealing can remove these surface defects and impurities, leading to enhance the number of active centers and an improvement in the overall performance of the electrocatalyst. Another benefit of defects annealing is the reduction of bulk defects. Bulk defects can arise during the synthesis of electrocatalysts and can also have a significant impact on their performance. Bulk defects can reduce the electronic conductivity of the electrocatalyst, which can hinder the transport of electrons through the material and reduce its catalytic activity. Defects annealing can reduce the density of bulk defects, leading to an improvement in the electronic conductivity of the electrocatalyst and an increase in its catalytic activity. Defects annealing can also lead to the formation of new defects that assists in improvement of electrocatalytic activity. Like, the annealing of metal oxide electrocatalysts can lead to the formation of oxygen vacancies, behaving as additional active centers for catalysis. Oxygen vacancies can enhance the reactant molecules adsorption onto the surface of the electrocatalyst and facilitate the reaction kinetics, improving the electrochemical performance of the material. Several factors need to be considered when performing defects annealing, including the temperature and duration of the annealing process. The temperature and duration of the annealing process can significantly affect the microstructure and properties of the electrocatalyst. High temperatures and long annealing times can lead to the sintering of nanoparticles, which can reduce the surface area of the electrocatalyst and decrease its electrochemical activity [52]. Therefore, careful optimization of the annealing temperature and duration is essential to attain the required microstructure and features of the electrocatalyst.

Defect engineering has been shown to induce a high electronic state at the topological region and tune the active site density factor (ASF) [56]. Both theoretical computations and experimental results have indicated that specific types of defects can generate active centers for diverse electrocatalytic processes by optimizing reactant binding energies. Furthermore, defects-induced crystal strain can manipulate electronic states or band gaps of transition metals along the edges or basal planes. This manipulation ultimately heightens the inherent performance and stability of the active sites [57]. Defects can be categorized according to their dimensions, encompassing 0-D point defects (like doping and vacancies), 1-D line defects (such as dislocations), 2-D planar defects (like grain boundaries), and 3-D volume defects. Electrochemical studies can delve deeper into line and surface defects using the point defect model [58, 59].

Moreover, defects have the potential to influence both the activation energy required for the adsorption of reactant molecules and the energy required for the dissociation of the specified product molecules. This influence contributes to the formation and transformation of reaction intermediates during water splitting. Theoretical analyses have also indicated that materials enriched with defects nearby Fermi level may enhance electrical conductivity, electrochemical performance and durability. Point defects in nonmetal compounds or metal materials often fall into three key configurations: defective coordination structures induced by metal atomic doping, nonmetallic vacancies (like S, B, P, and N), and vacancy defects or reconstructions. Atomic defect engineering offers a pragmatic approach to altering the atomic arrangement of host materials, thus enabling control over their electronic structure. As an illustration, earlier investigations have showcased that layered double hydroxides (LDHs) possess brucite-like layers featuring trivalent metal cations, which can be isomorphously substituted with bivalent metal cations, yielding adjustable electronic properties [60]. Layered double hydroxides (LDHs) typically consist of layers resembling brucite, incorporating trivalent metal cations (M^{3+}) that can replace specific previously octahedrally coordinated bivalent metal cations (M^{2+}) that were in the structure. An illustrative case comes from Song et al., who investigated the influence of atomic V replacement on pyrite NiS_2, culminating in transformation toward metal state from a semiconductor state [61]. XAS measurements unveiled that the transfer of electron occurring between the doped V heteroatoms and the Ni centers played a pivotal role in the reconfiguration of electronic structure of NiS_2, resulting in the heightened performance and stability of its active sites [62].

Likewise, the Mu research team engineered a defective RuO_2/TiO_2 heterostructure by substituting a single Ru atom on the surface's uppermost layer. This alteration enhanced sustainability as well as electrochemical activity compared to a counterpart with nondefective nature [63]. DFT calculations unveiled that transfer of electrons was from the bottom TiO_2 layer to the upper RuO_2 layer upon the creation of the heterostructure, resulting in highly charged Ru. This enhancement improved chemical stability of Ru and curtailed leaching of Ru at condition of high anodic potentials.

Inside these defects, the Ru atoms acquired additional electrons with a positive charge because of the Ru vacancy, which encouraged electron redistribution and reduced the activation barrier for the coupling of O–O.

An advanced electrocatalyst of La-doped RuO_2, employ a synergistic approach involving heteroatom doping and oxygen vacancies. The combined influence of oxygen vacancies and La doping plays remarkable role in sustained and remarkable electrocatalytic performance of La-doped RuO_2 in HER and OER processes. DFT calculations established that RuO_2 has 2.46 eV energy barrier, constituting the rate-determining step which involves *OOH to O_2. After introducing La, the binding energy of the intermediate state of reaction was optimized, revealing the third step as the rate-determining step which is $^*O$ to $^*OOH$. In contrast to RuO_2, there is decrement in binding energy of La-doped RuO_2 and for this step it reduces to 1.94 eV, indicating a lower free energy barrier for the reaction to proceed. Surface defect engineering serves as a methodology to amplify the efficiency of water electrolysis process. Defects present in electrocatalysts exert control over their electronic and surface structure, impacting the active component and its environment. This influence accelerates the kinetics of reaction at the interface of electrolyte and electrode, and stabilizes oxygen evolution reaction (OER) intermediates, ultimately heightening OER efficiency. Furthermore, introducing defects increases the unsaturation of electrocatalysts, which leads to the dangling bonds and modification of neighboring metal centers and their valance states. This destabilizes water molecules, thereby reducing adsorption energy and boosting the durability of active centers. A defect known as a vacancy emerges as a lattice defect encompassing substitutional and interstitial atoms, along with vacancies. Driven by fluctuations in atomic energy, there's invariably enough energy for the displacement of bound particles, facilitating their migration and resulting in the disturbance of the initial equilibrium position. Huang and his coworkers have innovated a structure featuring defective NiS_2 doped by P, with S vacancies through a Prussian-blue-analogue-sacrificed approach, along with substitution of phosphorous (Fig. 8.4a–i) [57]. X-ray investigation has unveiled that the presence of sulfur vacancies governs the structure's defects and modifies the catalyst's electronic system. Electrochemical assessments have exposed that the optimal combination of sulfur vacancies' strength and content of P amplifies water oxidation activity, sustaining elevated geometric activity of the catalyst at low input voltages.

In order to interpret the sources of the samples' electrochemical performance, adsorption Gibbs free energies for each step involving OH^*, O^*, and OOH^* have been estimated. A recent study indicated that during the OER process pure NiS_2 has rate-determining step of O^* to OOH^* species conversion. This is illustrated in Fig. 8.4l–p. During OER, the high overpotential primarily results from the robust adsorption of intermediates onto the NiS_2 surface, yielding a substantial energy barrier. The modulation of adsorption of intermediates on Ni sites through S vacancy incorporation and P-doping lessens the requisite energy barrier. Foretold limiting barriers were determined as NiS_2, P-NiS_2, and Vs-P-NiS_2 indicating that Vs-P-NiS_2 exhibited the maximum theoretical OER performance, attributable to defective sites

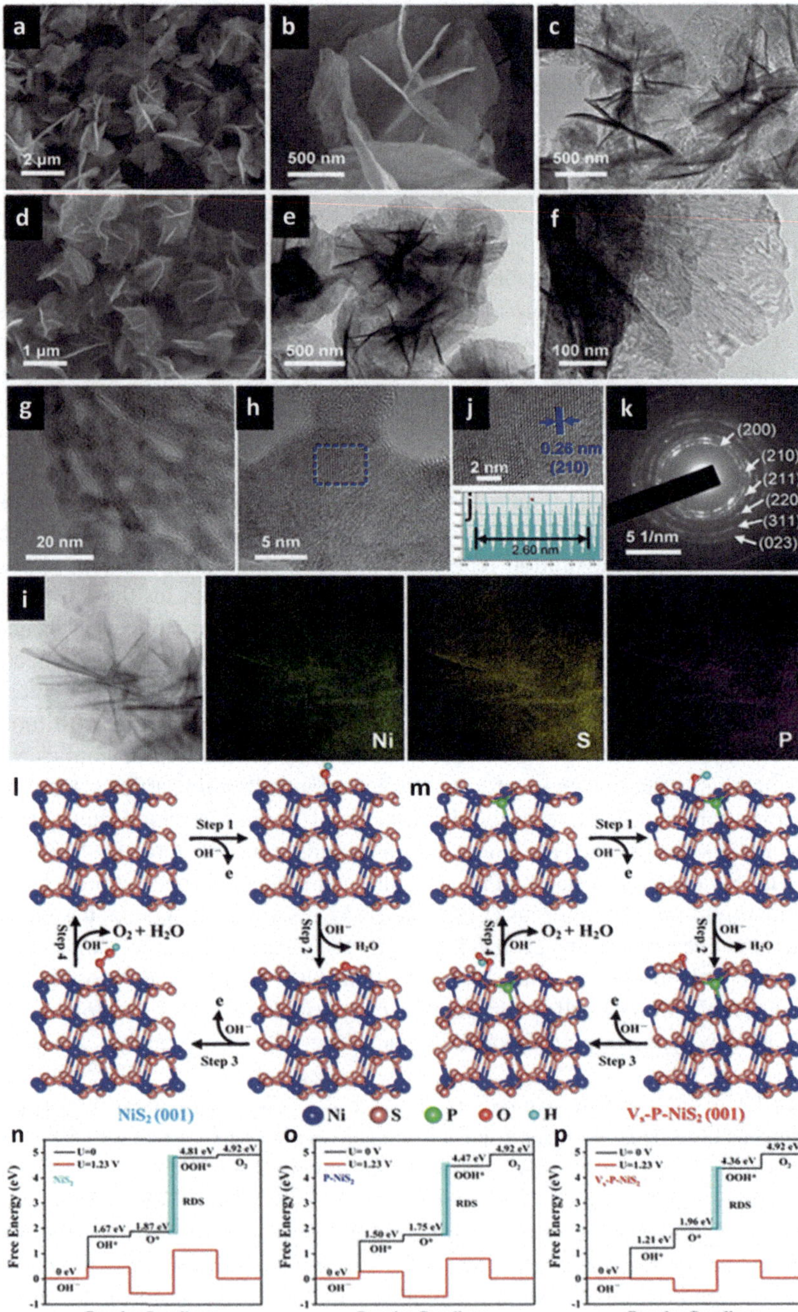
a
2 µm
b
500 nm
c
500 nm
d
1 µm
e
500 nm
f
100 nm
g
20 nm
h
5 nm
j
2 nm
0.26 nm
(210)
j
2.60 nm
k
(200)
(210)
(211)
(220)
(311)
(023)
5 1/nm
i
Ni
S
P
l
Step 1
OH⁻
e
Step 2
OH⁻
H₂O
Step 3
OH⁻
e
Step 4
OH⁻
O₂ + H₂O
m
Step 1
Step 2
Step 3
Step 4
NiS₂ (001)
Ni S P O H
Vₛ-P-NiS₂ (001)
n
U=0
U=1.23 V
NiS₂
0 eV
OH⁻
1.67 eV
OH*
1.87 eV
O*
4.81 eV
OOH*
4.92 eV
O₂
RDS
Free Energy (eV)
Reaction Coordinate
o
U= 0 V
U=1.23 V
P-NiS₂
0 eV
1.50 eV
1.75 eV
4.47 eV
4.92 eV
RDS
Free Energy (eV)
Reaction Coordinate
p
U= 0 V
U=1.23 V
Vₛ-P-NiS₂
0 eV
1.21 eV
1.96 eV
4.36 eV
4.92 eV
RDS
Free Energy (eV)
Reaction Coordinate

◀**Fig. 8.4** The catalyst's structure and morphology were analyzed. **a, b** SEM images of the catalyst. **c** TEM image. **d** SEM image of pure NiS_2. **e, f** TEM images. **g** Magnified TEM images. **h, i** High-resolution TEM. **j** Integrated pixel intensities of P-NiS2-500 (taken from the blue dotted rectangle in panel). **k** SAED patterns and Elemental map of the P-NiS_2-500 catalyst. The process of H_2O activation, OH*, OOH*, O*intermediates formation, and O_2 evolution processes were schematically illustrated. **l** On pure NiS_2 and **m** Vs-P-NiS_2. The calculated Gibbs free-energy presentation diagram for 4-OER steps were shown on **n** pure NiS_2, **o** P-NiS_2, and **p** Vs-P-NiS_2. The light blue part indicates the rate-determining step (RDS). Reproduced with permission from Ref. [57]. Copyright 2021, Elsevier

ensuring robust durability. An exploration of variations in the electronic configuration post-P-doping and insertion of sulfur vacancy was explored using density of states (DOS) analysis. NiS_2 typically possesses a bandgap of approximately 0.38 eV, consistent with semiconductor characteristics. Nevertheless, P-doped NiS_2 bulk records a reduced bandgap of 0.01 eV, showcasing the improved conductivity of the NiS_2 crystal achieved through P doping.

8.2.5 Porous Materials

The term "porous nanostructure" refers to a material with a porous structure at the nanoscale, typically ranging from a few nanometers to several hundred nanometers. The pore size as well as its distribution in porous nanostructures can be controlled through various methods, such as template-assisted synthesis, sol–gel, and hydrothermal methods. The pore size and distribution significantly influence the properties and applications of the resulting material [64]. Porous nanostructures may be divided into two broad groups: ordered and disordered porous nanostructures. Ordered porous nanostructures have a highly ordered and periodic pore arrangement, whereas disordered porous nanostructures have a random and nonperiodic pore arrangement. The pore arrangement significantly influences the material's properties, like mechanical strength, optical properties, and surface area [65].

Porous nanostructures can be fabricated from a different materials, including metals, metal oxides, polymers, and carbon-based materials. Among these materials, metal oxides are commonly used because of their long-term stability, biocompatibility, and diverse properties. Porous metal oxide nanostructures may be developed through many strategies, like template-assisted synthesis, sol–gel, and hydrothermal methods. In the template-assisted synthesis method, a sacrificial template is used to create the pores, which are subsequently filled with the metal oxide precursor. The template is then removed to leave behind a porous metal oxide nanostructure. Sol–gel and hydrothermal methods involve the formation of a gel or a precursor solution that is subsequently transformed into a porous metal oxide nanostructure through hydrolysis and condensation reactions [66]. Micro, meso, and macroporous are three classifications of porous materials that are defined on basis of pore size and structure of the material. Microporous materials are defined as having pores with diameters of less than 2 nm. These materials typically have an enormous surface area and maximum level of adsorption capacity, making them ideal for applications such as gas separation, catalysis, and sensing. Examples of microporous materials include zeolites, metal–organic frameworks (MOFs), and activated carbon. Mesoporous materials have pores with diameters between 2 and 50 nm. These materials have a moderate surface area and are often used as catalysts or as supports for catalytic reactions. They are also used in applications such as drug delivery, separation, and adsorption. Examples of mesoporous materials include mesoporous silica and mesoporous carbon. Macroporous materials have pores with diameters greater than 50 nm. These materials have a low surface area but can be useful in applications such as chromatography, filtration, and tissue engineering. Examples of macroporous materials include foams, sponges, and membranes. It is important to note that the pore size of a material can be tuned through various methods, such as changing the synthesis conditions or by adding pore-forming agents. Additionally, many porous materials can exhibit a combination of micro, meso, and macroporosity, making them highly versatile for a range of applications [67].

Following are the unique features of porous materials:

1. Porous materials attain high surface area because of their porous nature, which provides a large area for interactions with other substances, such as gases or liquids. This property makes them ideal for applications such as gas separation, catalysis, and adsorption.
2. Porous materials can be designed with specific pore sizes, allowing them to selectively separate molecules based on their size or shape. This property is useful in applications such as molecular sieving and gas separation.
3. Porous materials have a high porosity, which is the percentage of the volume that is made up of pores. This property is important for applications such as gas storage, where the amount of gas that can be stored is proportional to the volume of the pores.
4. The surface chemistry of porous materials can be modified to make them more or less reactive with certain molecules or to make them hydrophilic or hydrophobic. This property is useful for applications such as catalysis and adsorption.
5. Porous materials can be designed to be structurally stable, meaning they can maintain their shape and porosity under a variety of conditions. This property is important for applications such as catalysis and gas separation, where the material must be able to withstand high temperatures and pressures.
6. Porous materials can be designed with specific chemical or physical properties that allow them to selectively adsorb or react with certain molecules. This property is important for applications such as catalysis and gas separation, where high selectivity is desired.
7. Porous materials can be designed to release molecules or ions in a controlled manner, making them useful for applications such as drug delivery and sensing.
8. Porous materials can be flexible and deformable, allowing them to change shape in response to external stimuli. This property is useful for applications such as actuation and sensing.

Controlling the porosity and pore size distribution of porous materials is a crucial aspect that determines their properties and potential applications. Several techniques can be employed to achieve desired porosity and pore size distribution in porous materials, including:

Templating

Templating is a widely used approach for pore size regulation and porous materials distribution. It comprises sacrificial template utilization that is later removed to leave behind a porous structure. This technique can be used to create an extensive variety of pore sizes and structures, including ordered, uniform pores or random, irregular pores. The most commonly used templates arc hard templates, soft templates, and colloidal templates. Hard templates are typically used to create ordered pore structures. They involve the use of a preformed template, such as a nanoporous alumina membrane or a silica template, which is coated with the desired material. After the coating process, the template is removed, leaving behind the porous structure. The advantage of hard templating is that it allows for precise control of pore size and pore distribution, as the template itself provides a predefined structure. However, hard templating can be time-consuming and may require specialized equipment [67].

Soft templating involves the use of a surfactant or block copolymer to create a template for the porous structure. The surfactant or block copolymer forms micelles or vesicles that are coated with the desired material. After the coating process, the template is removed, leaving behind the porous structure. The advantage of soft templating is that it allows for the creation of complex pore structures, such as interconnected pores or hierarchical structures. Soft templating is also relatively easy to perform and does not require specialized equipment. Colloidal templating involves the use of colloidal particles as a template for the porous structure. The colloidal particles are coated with the desired material, and the template is removed, leaving behind the porous structure. Colloidal templating allows for the creation of ordered or random pore structures, depending on the arrangement of the colloidal particles. Colloidal templating is relatively easy to perform and does not require specialized equipment. However, it can be challenging to achieve precise control over pore size and distribution [68].

Sol-gel processing

Sol–gel processing is a versatile strategy for producing highly porous materials with precise porosity and pore size distribution. The procedure comprises the conversion of a sol, which is a stable colloidal suspension of nanoparticles, into a gel-like network of interconnected particles. The sol is typically formed by the hydrolysis and condensation of metal alkoxides, such as tetraethyl orthosilicate (TEOS), in the presence of a catalyst, such as hydrochloric acid or ammonia. The condensation reactions and hydrolysis result in the formation of silica nanoparticles that are linked together by siloxane bonds to form a three-dimensional network. The sol–gel method may be modified to control the porosity and pore size distribution of the resulting composites. One approach is to use a templating agent, such as a surfactant or a polymer, to create pores in the sol–gel matrix. The templating agent is typically removed after the sol–gel process by calcination or solvent extraction, leaving behind a highly porous material. The size and shape of the pores may be controlled by changing the kind and concentration of the templating agent, as well as the processing conditions [69].

Another approach is to use a sol–gel precursor that contains preformed nanoparticles or colloids, such as polystyrene or silica particles. The preformed particles act as templates for the formation of pores in the sol–gel matrix. After the sol–gel method, the preformed particles are removed by calcination or solvent extraction, leaving behind a highly porous material with a pore size distribution that is examined through morphology and size of the preformed particles. The sol–gel process can also be modified to produce mesoporous materials with uniform pore sizes in 2–50 nm range. This may be attained by using a surfactant as a templating agent to form micelles in the sol–gel matrix. The shape and size of the micelles may be controlled by varying the type and concentration of the surfactant, as well as the processing conditions. After the sol–gel procedure, the removal of surfactant is by calcination or solvent extraction, leaving behind a highly porous material with uniform mesopores.

Electrospinning

It involves the use of an electric field to create a jet of polymer solution or melt that is then deposited onto a collector. The resulting fibers can have diameters ranging from nanometers to micrometers, depending on the processing conditions. By carefully tuning the electrospinning parameters, such as the polymer concentration, the applied voltage, the distance between the needle and collector, and the solvent composition, the porosity and pore size distribution of the resulting materials can be controlled. The porosity of electrospun materials can be controlled by adjusting the polymer concentration and solvent composition. Increasing the polymer concentration results in a higher degree of entanglement and therefore a lower porosity. On the other hand, increasing the solvent content leads to a decrease in viscosity, which results in larger pore sizes and a higher porosity. The choice of solvent can also have a significant impact on the porosity and pore size distribution. For example, using a solvent that is easily evaporated can result in smaller pore sizes and a higher porosity.

In addition to adjusting the polymer concentration and solvent composition, the electrospinning process itself can also be modified to control the porosity and pore size distribution. By varying the applied voltage, the distance between the needle and collector, and the spinning rate, the diameter and orientation of the fibers can be controlled. This, in turn, affects the pore size and pore size distribution of the final material. Electrospinning may be used in combination with other techniques to create hierarchical porous structures with multiple length scales. For example, electrospinning can be combined with templating to create hierarchical porous structures with both micro- and nanopores. Alternatively, electrospinning can be used to create the larger pores in a hierarchically porous material, while a separate technique is used to create the smaller pores [70].

Freeze drying

Freeze drying, also known as lyophilization, is a popular technique for the preparation of porous materials with magnificent surface area and precise porosity. The process involves freezing a solution or suspension of the material and subsequently removing the frozen solvent under vacuum to obtain a dry, porous structure. The use of this technique can be advantageous as it allows for the preservation of the structural integrity of the material during the drying process, preventing collapse or agglomeration of the particles.

The porosity of the material may be regulated by varying the freezing rate, the concentration of the solution, and choice of solvent. Slower freezing rates lead to larger ice crystals and subsequently larger pores, while faster freezing rates result in smaller pores. Additionally, the concentration of the solution influence the size of the pores, with higher concentrations leading to smaller pores. The solvent choice also plays a significant role to investigate the porosity, as the freezing point of the solvent and its affinity for the solute affect the ice crystal size and pore structure. Freeze drying is commonly used to prepare porous materials such as aerogels, which are lightweight and highly porous materials with excellent thermal insulation properties. These materials find applications in a range of areas like energy storage, catalysis, and drug delivery. Additionally, the technique can also be used to prepare scaffolds for tissue engineering, as the porosity allows for cell infiltration and nutrient transport [71].

Supercritical fluid extraction

Supercritical fluid extraction (SFE) is a promising method to control the porosity and pore size distribution of porous materials. In SFE, a supercritical fluid, such as carbon dioxide, is used to extract a solvent from a polymer or other material, resulting in a porous structure with tunable properties. SFE is a versatile technique that can be used with a number of materials, including polymers, metals, and ceramics. The porosity of the resulting material can be controlled by adjusting the pressure and temperature of the supercritical fluid, as well as the extraction time. Higher pressures and longer extraction times lead to a higher degree of porosity. The pore size distribution can also be controlled by using different solvents, co-solvents, and additives, as well as by varying the processing conditions. One of the main advantages of SFE is that it is a solvent-free process, which eliminates the need for potentially harmful organic solvents. SFE is also a gentle process that can be used for structure preservation and features of the starting composites. Additionally, SFE can be used to create hierarchical structures with multiple levels of porosity, which can be useful for a variety of applications. SFE has been used to produce a wide range of porous materials, like aerogels, membranes, and particles, with applications in fields like catalysis, energy storage, and drug delivery. Overall, SFE is a promising technique for controlling the porosity and pore size distribution of porous materials, with potential for further development and optimization in the future [72].

Gas foaming

Gas foaming is a technique commonly used to fabricate porous materials with a controlled pore size and distribution. This method involves the use of a gas, typically carbon dioxide (CO_2), which is mixed with a polymer solution or melt. When the gas is rapidly released, it creates bubbles that expand and solidify the polymer, leaving behind a porous intact structure. The pore size and its distribution can be controlled by adjusting many parameters, like the gas pressure, temperature, and polymer concentration. Higher gas pressures typically lead to smaller pore sizes, while lower gas pressures lead to larger pore sizes. The use of different polymers or polymer blends can also affect the pore size and distribution. Additionally, the addition of surfactants can help to stabilize the bubbles and improve the pore uniformity. Gas foaming represents a straightforward and cost-efficient method for generating a diverse array of porous materials like foams, membranes, and scaffolds, with applications in various fields including drug delivery, tissue engineering, and filtration [73].

Phase separation

The process involves the separation of two immiscible phases, such as a polymer and a solvent, which are initially mixed together. The mixture is then subjected to a controlled phase separation process, such as cooling, solvent evaporation, or immersion in a nonsolvent, which induces the formation of a two-phase system. The polymer-rich phase forms a continuous matrix, while the solvent-rich phase forms isolated droplets or pores within the matrix. The size and distribution of the pores can be regulated by modulating the concentration of the polymer and solvent, the temperature, and the rate of phase separation. The resulting porous structure can be stabilized by crosslinking or thermal treatment to prevent collapse or deformation of the pores. Phase separation is a versatile technique that can be used to fabricate a number of porous materials, such as foams, membranes, and scaffolds, with tunable features for various practical applications, like tissue engineering, energy storage, and catalysis [74].

Porous materials have unique features for applications in water electrolysis. One of the critical advantages of porous materials is their large surface area, which provides ample space for catalytic reactions. The high surface area also allows for a more significant number of active centers, enhancing the efficiency of the catalyst. Another essential feature of porous materials is their tunable pore size and distribution. This allows for control over the diffusion rate of water molecules and reactant ions, ensuring they are evenly distributed throughout the catalyst. Additionally, the pore structure of porous materials can be tailored to promote gas bubble release, which is essential for the efficiency of gas-evolving reactions. Porous materials also have excellent mass transport properties, allowing for efficient transport of water and ions to the catalytic sites. This is particularly important for the OER process, which can be limited by the mass transfer of products as well as reactants. In addition, porous materials can significantly enhance the OER kinetics by providing a pathway for rapid diffusion. Finally, porous materials can provide high stability and durability in harsh chemical environments. The porous structure allows for easy reactants as well as products access to the active centers while protecting the catalyst from harsh chemical environments. This feature is significant for water electrolysis, as the process typically requires high current densities and extreme reaction conditions.

Ren and colleagues have effectively developed porous S-doped Ni–Fe-OOH on substrate Ni foam under ambient conditions. This material could attain a significant current density of 500 mA cm^{-2} and a minimal 1.83 V cell voltage [75]. The material's porous nature was characterized by highly mesoporous features having pore sizes of range around 20–50 nm, as verified by transmission electron microscopy (TEM) analysis (Fig. 8.5). These minuscule pores within the oxyhydroxide particles maximized the exposure of reactive sites. They facilitated the generation of O_2 bubbles from the catalyst surface, which is particularly advantageous for seawater oxidation applications. With the passage of time, the catalytic activity saw a rise, primarily due to the gradual dissolution of the metal, which exposed a greater number of rough and porous active centers for OH– adsorption. Additionally, these mesopores served as conduits for transportation of electrolyte and retained their surface properties even after prolonged operation in seawater electrolysis, underscoring the material's structural durability. The distinctive attributes of the porous S-doped Ni–Fe-OOH, such as densely distributed active sites, a substantial electrochemically active surface area (ECSA), a wide range of porosity for efficient electrolyte diffusion, rapid oxygen detachment, and its direct production through Ni etching, collectively contributed to improved kinetics, structural robustness, enhanced catalytic efficiency, and selectivity. In a research study led by Ren and his colleagues, they discovered that treating substance B with $NaBH_4$ created a porous structure that facilitated the easy diffusion and transport of electrolyte and reaction intermediates [76]. Consequently, this led to the release of gases, minimized the development of a nonconductive layer on surface of the catalyst, and enhanced the reaction's endurance. Moreover, within the same study, the research team synthesized Ni–Fe LDH on Ni foam. It showcased that the porous configuration of the three-dimensional Ni foam network allowed for substantial Ni–Fe LDH loading, leading to heightened catalytic performance in the context of seawater oxidation. Furthermore, the reaction's efficiency remained consistent with minimal degradation even after a continuous 96-h electrolysis at a high current density. Metal–organic frameworks (MOFs) have emerged as versatile materials suitable for creating porous structures characterized by an enhanced electrochemical surface area (ECSA), which leads to notable activity scaling factors (ASFs) compared to conventional transition metal oxides (TMOs). However, in simple MOFs, the effective diffusion of water molecules can be hindered due to their distinctive metal coordination and limited kinetics for mass transport. To address this challenge, the pyrolysis approach has been utilized to disrupt the coordination within the MOF, consequently unveiling a greater number of porous reactive centers. This enhances the ability for gas desorption as well as ionic diffusion. Nevertheless, it's worth noting that pyrolysis is energy-intensive, constraining the widespread application of MOFs. Chen and colleagues devised a novel approach to overcome these limitations: they developed a trimetallic hollow Fe-Co–Ni-based MOF on substrate of Ni foam utilizing poly-pyrrole and tannic acid, as illustrated in Fig. 8.6a–e [77]. The porous architecture of the MOF enabled composite reconfiguration during the electrolysis process and demonstrated impressive resistance to corrosion in seawater conditions. This surface restructuring led to the deposition of active oxyhydroxide on the surface of catalyst, thereby augmenting the activity and rendering the active sites more resistant to toxicity in electrolytes containing chloride ions [78].

Researchers have determined that the porous configuration of an electrocatalyst plays a pivotal role in determining its specific surface area and mass transfer capability. More precisely, three-dimensional ordered porous structures, with unique

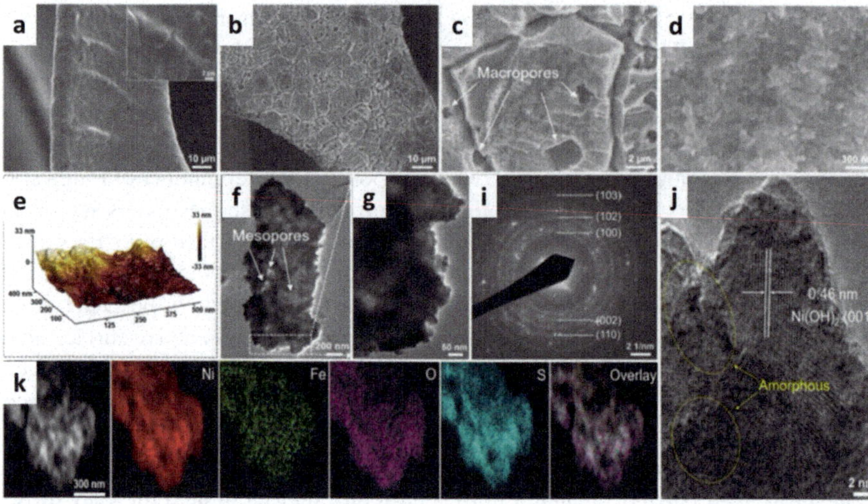

Fig. 8.5 Scanning electron microscopy images (SEM) of **a** nickel foam without modification, and S–(Ni, Fe)-OOH at varying magnifications, **b–d**, with corresponding surface topography (**e**) and TEM images (**f**, **g**) of S–(Ni, Fe)–OOH. Reproduced with permission [75]. Copyright 2020, Royal Society of Chemistry

surface characteristics, optimize the presentation of surface sites. This aids in the efficient interaction with central active centers, eases the penetration of electrolytes to reduce electron transportation and ion diffusion routes, and ensure rapid mass transfer kinetics. The main cause of overpotential is typically attributed to the combined influence of ohmic voltage loss and limitations in mass transfer during the flow of charge. As a result, nanostructures featuring attributes like reduced tortuosity, well-aligned open pathways, and highly porous configurations are in great demand. These characteristics help alleviate the hindrance caused by insoluble precipitates and ions, enhance effective gas desorption, and reduce overpotential. Chen and colleagues have engineered a distinctive catalyst design for seawater electrolysis involving S and P-doped (Ni, Mo, Fe)-OOH/Ni-Mo-P/wood aerogel sandwiches [79]. The sandwich-style arrangement boosts the interaction at the interface between heteroatom-doped multimetal hydroxide and the supporting material, resulting in promising activity, selectivity, and longevity during seawater electrolysis (depicted in Fig. 8.6). The surface's notable roughness and numerous hydroxyl groups facilitate swift electron transfer rates. At the same time, direct and open microchannels enhance electrolyte diffusion and permeation, promoting effective gas desorption.

These explicit and unobstructed microchannels in the catalyst design prevent bubble accumulation and enable the force generated by H_2/O_2 bubble rupture to expel insoluble residues away from the surface of electrode. This phenomenon guarantees the presence of intermediate activity centers. The authors ascribed the improved activity to the template's (wood aerogel) induction effect and proposed a mechanism to elucidate the catalyst's performance. Figure 8.6 showcases the layered sandwich-like structure engenders robust interfacial interaction with Ni–Mo–P alloys.

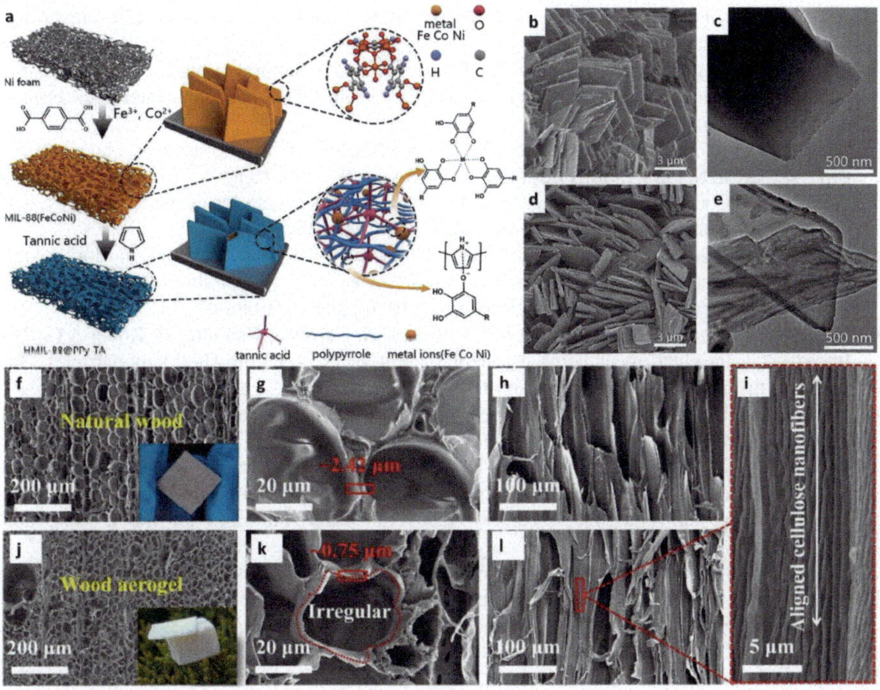

Fig. 8.6 **a** Hollow MIL-88(Fe-Co–Ni) (HMIL-88@PPy-TA) synthesis modified by Polypyrrole (PPy) and Tannic acid (TA) at nickel foam substrate. Scanning electron microscopy images of **b** MIL-88(Fe-Co–Ni), **c**HMIL-88@PPy-TA, and **d** corresponding Transmission electron microscopy images of MIL-88(Fe-Co–Ni) and HMIL-88@PPy-TA. Reproduced with permission Ref. [77]. Copyright 2021, Elsevier. **g–h** SEM images illustrating natural basal wood with mesoporous infrastructure with aligned and open microchannels, while **i–k** Scanning electron microscopy images of wood aerogel which depicts a lamellar arch-layer along with a minor thickness in comparison to natural wood. **l** Scanning electron microscopy images of nanofibers of cellulose having well-oriented structure, with intricate and rough framework. Reproduced with permission Ref. [79] Copyright 2021, Elsevier

References

1. Chatelier C, Garreau Y, Piccolo L, Vlad A, Resta A, Ledieu J, Fournée V, De Weerd M, Picca FE, De Boissieu M et al (2020) From the surface structure to catalytic properties of Al_5Co(21 0): a study combining experimental and theoretical approaches. J Phys Chem C. https://doi.org/10.1021/acs.jpcc.9b09675
2. Tao FF, Ralston WT, Liu H, Somorjai GA (2018) Surface structures of model metal catalysts in reactant gases. J Phys Chem B. https://doi.org/10.1021/acs.jpcb.7b06950
3. Giannakakis G, Flytzani-Stephanopoulos M, Sykes ECH (2019) Single-atom alloys as a reductionist approach to the rational design of heterogeneous catalysts. Acc Chem Res. https://doi.org/10.1021/acs.accounts.8b00490
4. Haq TU, Haik Y, Hussain I, Rehman HU, Al-Ansari TA (2021) Gd-doped Ni-oxychloride nanoclusters: new nanoscale electrocatalysts for high-performance water oxidation through

surface and structural modification. ACS Appl Mater Interfaces 13(1):468–479. https://doi.org/10.1021/acsami.0c17216
5. ul Haq T, Mansour S, Haik Y (2022) Electronic and structural modification of Mn_3O_4 nanosheets for selective and sustained seawater oxidation. ACS Appl Mater Interfaces. https://doi.org/10.1021/acsami.1c24304
6. Somorjai GA, Contreras AM, Montano M, Rioux RM (2006) Clusters, surfaces, and catalysis. Proc Natl Acad Sci USA. https://doi.org/10.1073/pnas.0507691103
7. Somorjai G (2006) Cluster chemistry and dynamics special feature: clusters, surfaces, and catalysis. Proc Natl Acad Sci. https://doi.org/10.1073/pnas.0507691103
8. Kun-udom R, Jantarang S, Du Z, Kitiyanan B, Rirksomboon T, Meeyoo V (2022) Morphology dependent? Elucidating the catalyst structures of $Ni/Ce_{0.75}Zr_{0.25}O_2$ in the steam reforming of acetic acid. Green Chem Eng. https://doi.org/10.1016/j.gce.2021.08.004
9. Niether C, Faure S, Bordet A, Deseure J, Chatenet M, Carrey J, Chaudret B, Rouet A (2018) Improved water electrolysis using magnetic heating of FeC-Ni core-shell nanoparticles. Nat Energy. https://doi.org/10.1038/s41560-018-0132-1
10. Yin X, Yang L, Gao Q (2020) Core-shell nanostructured electrocatalysts for water splitting. Nanoscale. https://doi.org/10.1039/d0nr03719b
11. Wang Y, Li X, Zhang M, Zhou Y, Rao D, Zhong C, Zhang J, Han X, Hu W, Zhang Y et al (2020) Lattice-strain engineering of homogeneous $NiS_{0.5}Se_{0.5}$ core–shell nanostructure as a highly efficient and robust electrocatalyst for overall water splitting. Adv Mater. https://doi.org/10.1002/adma.202000231
12. Nong HN, Oh HS, Reier T, Willinger E, Willinger MG, Petkov V, Teschner D, Strasser P (2015) Oxide-supported IrNiOx core-shell particles as efficient, cost-effective, and stable catalysts for electrochemical water splitting. Angew Chemie Int Ed. https://doi.org/10.1002/anie.201411072
13. Zhu W, Chen W, Yu H, Zeng Y, Ming F, Liang H, Wang Z (2020) NiCo/NiCo–OH and NiFe/NiFe–OH core shell nanostructures for water splitting electrocatalysis at large currents. Appl Catal B Environ. https://doi.org/10.1016/j.apcatb.2020.119326
14. Ge J, Chen RR, Ren X, Liu J, Ong SJH, Xu ZJ (2021) Ferromagnetic-antiferromagnetic coupling core-shell nanoparticles with spin conservation for water oxidation. Adv Mater. https://doi.org/10.1002/adma.202101091
15. Jeong H-Y, Oh J, Yi GS, Park H-Y, Cho SK, Jang JH, Yoo SJ, Park HS (2023) High-performance water electrolyzer with minimum platinum group metal usage: iron nitride-iridium oxide core-shell nanostructures for stable and efficient oxygen evolution reaction. Appl Catal B Environ. https://doi.org/10.1016/j.apcatb.2023.122596
16. Patil SJ, Chodankar NR, Hwang SK, Shinde PA, Seeta Rama Raju G, Shanmugam Ranjith K, Huh YS, Han YK (2022) Co-metal–organic framework derived $CoSe_2$@$MoSe_2$ core–shell structure on carbon cloth as an efficient bifunctional catalyst for overall water splitting. Chem Eng J. https://doi.org/10.1016/j.cej.2021.132379
17. Yang Y, Zhu B, Guo PF, Wang WJ, Wang WT, Wang K, He ZH, Liu ZT (2022) Core-shell trimetallic NiFeV disulfides and amorphous high-valance NiFe hydroxide nanosheets enhancing oxygen evolution reaction. Chem Eng J. https://doi.org/10.1016/j.cej.2021.133047
18. Ligani Fereja S, Li P, Zhang Z, Guo J, Fang Z, Li Z, He S, Chen W (2022) W-doping induced abundant active sites in a 3D NiS_2/MoO_2 heterostructure as an efficient electrocatalyst for urea oxidation and hydrogen evolution reaction. Chem Eng J. https://doi.org/10.1016/j.cej.2021.134274
19. Abid N, Khan AM, Shujait S, Chaudhary K, Ikram M, Imran M, Haider J, Khan M, Khan Q, Maqbool M (2022) Synthesis of nanomaterials using various top-down and bottom-up approaches, influencing factors, advantages, and disadvantages: a review. Adv Coll Interface Sci. https://doi.org/10.1016/j.cis.2021.102597
20. Chandrasekaran S, Khandelwal M, Dayong F, Sui L, Chung JS, Misra RDK, Yin P, Kim EJ, Kim W, Vanchiappan A et al (2022) Developments and perspectives on robust nano- and microstructured binder-free electrodes for bifunctional water electrolysis and beyond. Adv Energy Mater. https://doi.org/10.1002/aenm.202200409

21. Liu Y, Zheng D, Zhao Y, Shen P, Du Y, Xiao W, Du Y, Fu Y, Wu Z, Wang L (2022) Ru-doped 3D porous Ni_3N sphere as efficient bi-functional electrocatalysts toward urea assisted water-splitting. Int J Hydrogen Energy. https://doi.org/10.1016/j.ijhydene.2022.05.268
22. Zhang J, Chen LL, Lu B, Guo Y (2022) 3D hierarchical porous Fe/Ni-P-B as practical bifunctional electrode for alkaline water electrolysis. Chemsuschem. https://doi.org/10.1002/cssc.202200937
23. Kong D, Wang Y, Huang S, Von Lim Y, Wang M, Xu T, Zang J, Li X, Yang HY (2022) Defect-engineered 3D hierarchical $NiMo_3S_4$ nanoflowers as bifunctional electrocatalyst for overall water splitting. J Colloid Interface Sci. https://doi.org/10.1016/j.jcis.2021.10.020
24. Hou Y, Qiu M, Nam G, Kim MG, Zhang T, Liu K, Zhuang X, Cho J, Yuan C, Feng X (2017) Integrated hierarchical cobalt sulfide/nickel selenide hybrid nanosheets as an efficient three-dimensional electrode for electrochemical and photoelectrochemical water splitting. Nano Lett. https://doi.org/10.1021/acs.nanolett.7b01030
25. Singh M, Nguyen TT, Balamurugan J, Kim NH, Lee JH (2022) Rational manipulation of 3D hierarchical oxygenated nickel tungsten selenide nanosheet as the efficient bifunctional electrocatalyst for overall water splitting. Chem Eng J. https://doi.org/10.1016/j.cej.2021.132888
26. Liu SQ, Gao MR, Liu S, Luo JL (2021) Hierarchically assembling cobalt/nickel carbonate hydroxide on copper nitride nanowires for highly efficient water splitting. Appl Catal B Environ. https://doi.org/10.1016/j.apcatb.2021.120148
27. Suntivich J, May KJ, Gasteiger HA, Goodenough JB, Shao-Horn Y (2011) A perovskite oxide optimized for oxygen evolution catalysis from molecular orbital principles. Science (80-) 334(6061):1383–1385. https://doi.org/10.1126/science.1212858
28. Novoselov KS, Geim AK, Morozov SV, Jiang D, Zhang Y, Dubonos SV, Grigorieva IV, Firsov AA (2004) Electric field effect in atomically thin carbon films supplementary. Science (80-.). https://doi.org/10.1126/science.aab1343
29. Gupta A, Sakthivel T, Seal S (2015) Recent development in 2D materials beyond graphene. Prog Mater Sci. https://doi.org/10.1016/j.pmatsci.2015.02.002
30. Ryu B, Wang L, Pu H, Chan MKY, Chen J (2022) Understanding, Discovery, and synthesis of 2D materials enabled by machine learning. Chem Soc Rev. https://doi.org/10.1039/d1cs00503k
31. Fareza AR, Nugroho FAA, Abdi FF, Fauzia V (2022) Nanoscale metal oxides-2D materials heterostructures for photoelectrochemical water splitting—a review. J Mater Chem A. https://doi.org/10.1039/d1ta10203f
32. Yu W, Gong K, Li Y, Ding B, Li L, Xu Y, Wang R, Li L, Zhang G, Lin S (2022) Flexible 2D materials beyond graphene: synthesis, properties, and applications. Small. https://doi.org/10.1002/smll.202105383
33. Kaul AB (2014) Two-dimensional layered materials: structure, properties, and prospects for device applications. J Mater Res. https://doi.org/10.1557/jmr.2014.6
34. ul Haq T, Mansour SA, Munir A, Haik Y (2020) Gold-supported gadolinium doped CoB amorphous sheet: a new benchmark electrocatalyst for water oxidation with high turnover frequency. Adv Funct Mater 1910309:1–11. https://doi.org/10.1002/adfm.201910309
35. Jang SW, Dutta S, Kumar A, Hong YR, Kang H, Lee S, Ryu S, Choi W, Lee IS (2020) Holey Pt nanosheets on NiFe-hydroxide laminates: synergistically enhanced electrocatalytic 2D interface toward hydrogen evolution reaction. ACS Nano. https://doi.org/10.1021/acsnano.0c04628
36. Quan Q, Ho JC (2022) Recent advances in the construction of 2D heterostructures for electrocatalytic water splitting. Adv Energy Sustain Res. https://doi.org/10.1002/aesr.202200059
37. Wang T, Cao X, Jiao L (2021) Ni2P/NiMoP heterostructure as a bifunctional electrocatalyst for energy-saving hydrogen production. eScience. https://doi.org/10.1016/j.esci.2021.09.002
38. Zhang Q, Zhang S, Luo Y, Liu Q, Luo J, Chu PK, Liu X (2022) Preparation of high entropy alloys and application to catalytical water electrolysis. APL Mater. https://doi.org/10.1063/5.0097479

39. Jiang Y, Sun T, Xie X, Jiang W, Li J, Tian B, Su C (2019) Oxygen-functionalized ultrathin $Ti_3C_2T_x$ MXene for enhanced electrocatalytic hydrogen evolution. Chemsuschem. https://doi.org/10.1002/cssc.201803032
40. Li Z, Qi Z, Wang S, Ma T, Zhou L, Wu Z, Luan X, Lin FY, Chen M, Miller JT et al (2019) In situ formed Pt_3Ti nanoparticles on a two-dimensional transition metal carbide (MXene) used as efficient catalysts for hydrogen evolution reactions. Nano Lett. https://doi.org/10.1021/acs.nanolett.9b01381
41. Wang J, He P, Shen Y, Dai L, Li Z, Wu Y, An C (2021) FeNi nanoparticles on $Mo_2TiC_2T_x$ MXene@nickel foam as robust electrocatalysts for overall water splitting. Nano Res. https://doi.org/10.1007/s12274-021-3660-0
42. Du CF, Sun X, Yu H, Liang Q, Dinh KN, Zheng Y, Luo Y, Wang Z, Yan Q (2019) Synergy of Nb doping and surface alloy enhanced on water-alkali electrocatalytic hydrogen generation performance in Ti-based MXene. Adv Sci. https://doi.org/10.1002/advs.201900116
43. Zhang Y, Wu C, Jiang H, Lin Y, Liu H, He Q, Chen S, Duan T, Song L (2018) Atomic iridium incorporated in cobalt hydroxide for efficient oxygen evolution catalysis in neutral electrolyte. Adv Mater. https://doi.org/10.1002/adma.201707522
44. Cheng F, Li Z, Wang L, Yang B, Lu J, Lei L, Ma T, Hou Y (2021) In situ identification of the electrocatalytic water oxidation behavior of a nickel-based metal-organic framework nanoarray. Mater Horizons. https://doi.org/10.1039/d0mh01757d
45. Huang J, Li Y, Huang RK, He CT, Gong L, Hu Q, Wang L, Xu YT, Tian XY, Liu SY et al (2018) Electrochemical exfoliation of pillared-layer metal-organic framework to boost the oxygen evolution reaction. Angew Chemie - Int Ed. https://doi.org/10.1002/anie.201801029
46. Xu J, Zhu X, Jia X (2019) From Low-to high-crystallinity bimetal-organic framework nanosheet with highly exposed boundaries: an efficient and stable electrocatalyst for oxygen evolution reaction. ACS Sustain Chem Eng. https://doi.org/10.1021/acssuschemeng.9b03952
47. Wen Y, Wei Z, Ma C, Xing X, Li Z, Luo D (2019) MXene boosted CoNi-ZIF-67 as highly efficient electrocatalysts for oxygen evolution. Nanomaterials. https://doi.org/10.3390/nano9050775
48. Xu X, Shao Z, Jiang SP (2022) High-entropy materials for water electrolysis. Energ Technol. https://doi.org/10.1002/ente.202200573
49. Zhou B, Gao R, Zou JJ, Yang H (2022) Surface design strategy of catalysts for water electrolysis. Small. https://doi.org/10.1002/smll.202202336
50. Zhang Y, Yun S, Dang J, Dang C, Yang G, Wang Y, Liu Z, Deng Y (2022) Defect engineering via ternary nonmetal doping boosts the catalytic activity of ZIF-derived carbon-based metal-free catalysts for photovoltaics and water splitting. Mater Today Phys. https://doi.org/10.1016/j.mtphys.2022.100785
51. Wu J, Yang X, Zhang J, Guan S, Han J, Wang J, Li K, Zhang G, Guan T (2022) Surface engineering of Ni_2P/CoP nanosheet heterojunctions by the formation of F-doped carbon layers for boosting urea-rich water electrolysis. J Power Sources. https://doi.org/10.1016/j.jpowsour.2022.232065
52. Lin L, Piao S, Choi Y, Lyu L, Hong H, Kim D, Lee J, Zhang W, Piao Y (2022) Nanostructured transition metal nitrides as emerging electrocatalysts for water electrolysis: status and challenges. EnergyChem. https://doi.org/10.1016/j.enchem.2022.100072
53. Wang Y, Jiang Y, Zhao Y, Ge X, Lu Q, Zhang T, Xie D, Li M, Bu Y (2023) Design strategies of perovskite nanofibers electrocatalysts for water splitting: a mini review. Chem Eng J. https://doi.org/10.1016/j.cej.2022.138710
54. Tomboc GM, Venkateshalu S, Ngo QT, Choi S, Pollet BG, Lee H, Lee K (2023) Defect-induced electronic modification and surface reconstruction of catalysts during water oxidation process. Chem Eng J. https://doi.org/10.1016/j.cej.2022.140254
55. Zhao J, Zhang Y, Guo H, Ren J, Zhang H, Wu Y, Song R (2022) Defect-rich $Ni(OH)_2$/NiO regulated by WO_3 as Core-shell nanoarrays achieving energy-saving water-to-hydrogen conversion via urea electrolysis. Chem Eng J. https://doi.org/10.1016/j.cej.2022.134497
56. Singh TI, Rajeshkhanna G, Pan UN, Kshetri T, Lin H, Kim NH, Lee JH (2021) Alkaline water splitting enhancement by MOF-derived Fe–Co–Oxide/Co@NC-MNS heterostructure:

boosting OER and HER through defect engineering and in situ oxidation. Small. https://doi.org/10.1002/smll.202101312
57. Huang S, Jin Z, Ning P, Gao C, Wu Y, Liu X, Xin P, Chen Z, Jiang Y, Hu Z et al (2021) Synergistically modulating electronic structure of NiS_2 hierarchical architectures by phosphorus doping and sulfur-vacancies defect engineering enables efficient electrocatalytic water splitting. Chem Eng J. https://doi.org/10.1016/j.cej.2020.127630
58. Wang L, Peng C, Lin H, Zhao B (2022) Unraveling the role of defects in electrocatalysts for water splitting: recent advances and perspectives. Energy Fuels. https://doi.org/10.1021/acs.energyfuels.2c02017
59. Lei W, Yu Y, Zhang H, Jia Q, Zhang S (2022) Defect engineering of nanostructures: insights into photoelectrochemical water splitting. Mater Today. https://doi.org/10.1016/j.mattod.2021.10.028
60. Li H, Wu S, Hood ZD, Sun J, Hu B, Liang C, Yang S, Xu Y, Jiang B (2020) Atomic defects in ultra-thin mesoporous TiO_2 enhance photocatalytic hydrogen evolution from water splitting. Appl Surf Sci. https://doi.org/10.1016/j.apsusc.2020.145723
61. Liu H, He Q, Jiang H, Lin Y, Zhang Y, Habib M, Chen S, Song L (2017) Electronic structure reconfiguration toward pyrite NiS_2 via engineered heteroatom defect boosting overall water splitting. ACS Nano. https://doi.org/10.1021/acsnano.7b06501
62. Li W, Zhang H, Hong M, Zhang L, Feng X, Shi M, Hu W, Mu S (2022) Defective RuO_2/TiO_2 nano-heterostructure advances hydrogen production by electrochemical water splitting. Chem Eng J. https://doi.org/10.1016/j.cej.2021.134072
63. Wu Y, Yao R, Zhao Q, Li J, Liu G (2022) La-RuO_2 nanocrystals with efficient electrocatalytic activity for overall water splitting in acidic media: synergistic effect of la doping and oxygen vacancy. Chem Eng J. https://doi.org/10.1016/j.cej.2022.135699
64. Liu X, Li Y, Chen Z, Yang H, Cai Y, Wang S, Chen J, Hu B, Huang Q, Shen C et al (2023) Advanced porous nanomaterials as superior adsorbents for environmental pollutants removal from aqueous solutions. Crit Rev Environ Sci Technol. https://doi.org/10.1080/10643389.2023.2168473
65. Tripathy SP, Subudhi S, Ray A, Behera P, Parida K (2022) Metal organic framework-based janus nanomaterials: rational design, strategic fabrication and emerging applications. Dalton Trans. https://doi.org/10.1039/d1dt04380c
66. Sajid M (2022) Nanomaterials: types, properties, recent advances, and toxicity concerns. Curr Opin Environ Sci Health. https://doi.org/10.1016/j.coesh.2021.100319
67. Du C, Li P, Zhuang Z, Fang Z, He S, Feng L, Chen W (2022) Highly porous nanostructures: rational fabrication and promising application in energy electrocatalysis. Coord Chem Rev. https://doi.org/10.1016/j.ccr.2022.214604
68. Kaur A, Bajaj B, Kaushik A, Saini A, Sud D (2022) A review on template assisted synthesis of multi-functional metal oxide nanostructures: status and prospects. Mater Sci Eng B Solid-State Mater Adv Technol. https://doi.org/10.1016/j.mseb.2022.116005
69. Hasegawa G (2022) Free-standing and binder-free porous monolithic electrodes prepared via sol-gel processes. J Sol-Gel Sci Technol. https://doi.org/10.1007/s10971-022-05862-5
70. Liu R, Hou L, Yue G, Li H, Zhang J, Liu J, Miao B, Wang N, Bai J, Cui Z et al (2022) Progress of fabrication and applications of electrospun hierarchically porous nanofibers. Adv Fiber Mater. https://doi.org/10.1007/s42765-022-00132-z
71. Liang H, Mirinejad MS, Asefnejad A, Baharifar H, Li X, Saber-Samandari S, Toghraie D, Khandan A (2022) Fabrication of tragacanthin gum-carboxymethyl chitosan bio-nanocomposite wound dressing with silver-titanium nanoparticles using freeze-drying method. Mater Chem Phys. https://doi.org/10.1016/j.matchemphys.2022.125770
72. Akhgari A, Nosrati F, Rahmanian-Devin P, Hadizadeh F, Sardou HS, Kamali H (2022) Enhancement of valsartan dissolution rate by the increased porosity of pellets using supercritical CO_2: optimization via central composite design. J Pharm Innov. https://doi.org/10.1007/s12247-022-09685-3
73. Ilkhani M, Esfandyari Bayat A, Harati S (2022) Applicability of methane foam stabilized via nanoparticles for enhanced oil recovery from carbonate porous media at various temperatures. J Mol Liq. https://doi.org/10.1016/j.molliq.2022.120576

74. Cao X, Chen W, Zhao P, Yang Y, Yu DG (2022) Electrospun porous nanofibers: pore—forming mechanisms and applications for photocatalytic degradation of organic pollutants in wastewater. Polymers. https://doi.org/10.3390/polym14193990
75. Yu L, Wu L, McElhenny B, Song S, Luo D, Zhang F, Yu Y, Chen S, Ren Z (2020) Ultrafast room-temperature synthesis of porous S-doped Ni/Fe (Oxy)hydroxide electrodes for oxygen evolution catalysis in seawater splitting. Energy Environ Sci. https://doi.org/10.1039/d0ee00921k
76. Wu L, Yu L, Zhu Q, McElhenny B, Zhang F, Wu C, Xing X, Bao J, Chen S, Ren Z (2021) Boron-modified cobalt iron layered double hydroxides for high efficiency seawater oxidation. Nano Energy 83(January):105838. https://doi.org/10.1016/j.nanoen.2021.105838
77. Chen Y, Shen L, Wang C, Feng S, Zhang N, Zhang K, Yang B (2022) Utilizing tannic acid and polypyrrle to induce reconstruction to optimize the activity of MOF-derived electrocatalyst for water oxidation in seawater. Chem Eng J. https://doi.org/10.1016/j.cej.2021.132632
78. Li W, Liu J, Guo P, Li H, Fei B, Guo Y, Pan H, Sun D, Fang F, Wu R (2021) Co/CoP heterojunction on hierarchically ordered porous carbon as a highly efficient electrocatalyst for hydrogen and oxygen evolution. Adv Energy Mater 2102134. https://doi.org/10.1002/aenm.202102134
79. Chen H, Zou Y, Li J, Zhang K, Xia Y, Hui B, Yang D (2021) Wood aerogel-derived sandwich-like layered nanoelectrodes for alkaline overall seawater electrosplitting. Appl B Environ. https://doi.org/10.1016/j.apcatb.2021.120215

Chapter 9
Single-Atom Catalyst for Electrochemical Water Splitting

9.1 Introduction

The most fundamental environmental issues of the twenty-first century are the environment and energy. Because of rapid second industrial revolutions, fossil fuels based on carbon like natural gas, petroleum, and coal have been regarded as dominant energy resources. They have rapid utilization in support of the quickly increasing population and constant economic development. Unfortunately, these are non-renewable resources and deplete rapidly. The enhanced non-renewable energy resource consumption has resulted in environmental damage by increasing pollution to a large extent [1]. In response to the ecological crisis and sustainable and green energy demand, there is a dire need to explore clean, renewable energy resources, where power conversion is a crucial component. Among the several energy conversion processes, water electrolysis for oxygen and hydrogen evolution reactions is a good process with zero carbon emission. This electrochemical procedure is controlled with the help of an efficient and active catalyst. Various efforts have been made to design and fabricate highly selective composite at low production.

Catalysts have a major contribution to the global economy. Over 90% of chemical procedures need catalysts, and there is great demand for developing inexpensive, green, and efficient catalysts to attain a sustainable future for the environment [2]. Berzelius defined a catalyst in 1835, according to which "a catalyst is a substance which helps to accelerate chemical reaction rate without being consumed" [3]. Catalysts have been divided into two broad classes homogeneous (having the same phase as reactants like enzymes and soluble metals) and heterogeneous catalysts (having different phases than reactants like zeolites and supported metals). Around 80% of catalysts being used on an industrial scale are heterogeneous catalysts despite some challenges of homogeneous ones, including thermal instability [4]. Typically heterogeneous catalysts are highly sensitive to surface structural and heterogenic properties.

T. u. Haq and Y. Haik, *Electrochemical Water Splitting*, Materials Horizons: From Nature to Nanomaterials, https://doi.org/10.1007/978-981-99-9860-9_9

First, Zhang et al. introduced heterogeneous single-atom crystal SACs in which metal atoms are majorly active sites [5]. They developed the catalysts of SACs in 2011 by incorporating isolated Pt atoms in iron oxide (Pt/FeO_x) for CO oxidation, which showed extraordinary properties for water electrolysis applications [6]. Certain termination of atom coordination affects the electrocatalytic metal properties, including staking faults, kinks, vacancies, twin boundaries, and step sites which affect the catalytic activity. Furthermore, the size of metal particles is a crucial factor in governing performance and has driven nanoparticle utilization in the form of catalysts. Though, the fabrication of two nanoparticles is almost impossible because of surface defects. To address this concern, the solution is to reduce the metal particle size to the atomic level. A single atom serves as a catalyst SAC's with no surface heterogeneity because the specific activity per metal atom frequently enhances by metal particle size reduction. At the same time, there is an increase in the surface free energy of metals by decreasing particle size, which promotes small clusters agglomeration. This could be overcome by using an appropriate support material that has robust interaction with the metal atoms preventing aggregation and creating homogeneously dispersed metal atoms with efficient catalytic performance, a technique used by industry for a longer time. Single-atom crystals SAC's reduce the cost regarding noble metals and enhance the surface area for activity. The SACs are synthesized on supported materials like carbon nitrides, metal oxides, 2D materials, and graphene. They attain 100% atomic efficiency because of the accessibility of each active catalytic site [3].

9.2 Unique Features of Single-Atom Catalysts (SAC's)

Single-atom catalysts (SACs) have been regarded as the most active frontiers in heterogeneous catalysis because their highest atom-utilization efficiency presents greatly tunable active sites with compact and well-defined electronic structures and coordination geometries comparable to specific enzymes or homogeneous catalysts. The reduction of metal nanoparticles size to nanometer/atomic scale may solve such problems by significantly enhancing the number of active sites on the catalyst surface where all individual metal atoms have ideal accessibility toward reactants [7]. SACs are closer in properties to heterogeneous catalysts, although they are immobilized on supports that may be separated easily and reused. SACs have more advanced features than other heterogeneous composites because the coordination environment of the single atom could be tuned by changing different types of coordinating atoms on selective support. These strategies enhance the features of SACs and help attain optimal active sites to enhance catalytic activity and metal atom efficiency [8]. SACs attain outstanding properties, which include: (i) enhances metal–support interaction, (ii) atomically dispersed metal centers, (iii) confinement of electrons with complete analysis, (iv) presence of coordinatively unsaturated metal catalytic active centers, and (v) quantum size effects. Such unique and excellent characteristics of SACs make them catalytically active and super active and selective for electrochemical reactions [9]. The SAC's features are enlisted in detail as follows:

1. The SACs have defined structural uniformity, which is considered one of the essential properties to connect both homogeneous and heterogeneous catalysis. It serves as the ultimate model to explore the structure–activity relationship [10].
2. The anchoring of SACs on 2D atomic crystals may be efficiently and systematically tuned through external electric fields with specified orientations. Outstanding electrochemical activity has been attained in SAC's under electrostatic modulations. Theoretical studies propose a worldwide "onsite electrostatic polarization" mechanistic approach, in which charge distribution and polarization are made at sites of SACs and modify the kinetics of the rate-determining steps, which leads to enhanced catalytic activity [10].
3. The SACs are synthesized on specified support. They usually have robust metal-support interactions (SMSIs), which play a vital role in the electrocatalytic performance of a catalyst. They attain both structural and synergistic promotions, stabilizing metal centers and maintaining energetic interactions between metal active components and support. The intercalated regions between support and metal atoms represent new active metal entities that catalyze the reaction and its activity. This is because of electronic density transfer between the active metal center and support which directly stimulates the activation energy and promotes the activity [11].
4. SACs have been regarded as a conceptual bridge between heterogeneous and homogeneous catalysts because of their low-coordination environment of active centers and homogeneity advantages.
5. They have quantum size effects features, where confinement of electrons results in different energy levels and assists in widening the HOMO–LUMO gap (Kubo gap).
6. The SACs have heteroatom effects, which help to modulate charge density and stimulate asymmetrical spin [12].
7. The SACs have enhanced unsaturated bonds because of surface effects that sustain the electronic and geometric structures [13].
8. They attain cluster configurations, where specified arrangement and alignment of the number and position of atoms inside a cluster may drastically transform physicochemical characteristics [14].
9. SACs have a linear relationship with metal loading to assess catalytic activity effectively. SACs only contain isolated atoms, so they have constant turnover frequency (TOF) and specific activity per metal atom (site) regardless of metal loading, making them a unique candidate for electrocatalyst selection. This helps large surface area supports utilization (Fig. 9.1) [15].

These outstanding properties of SACs have urged several researchers to design further and modulate their distinctive features. The perception of single-atom catalysts (SACs) was conceived initially to utilize every atom of precious metal, but now it has been broadly extended to various kinds of heterogeneous systems, among which most of them have established the confined selectivity, long-term durability,

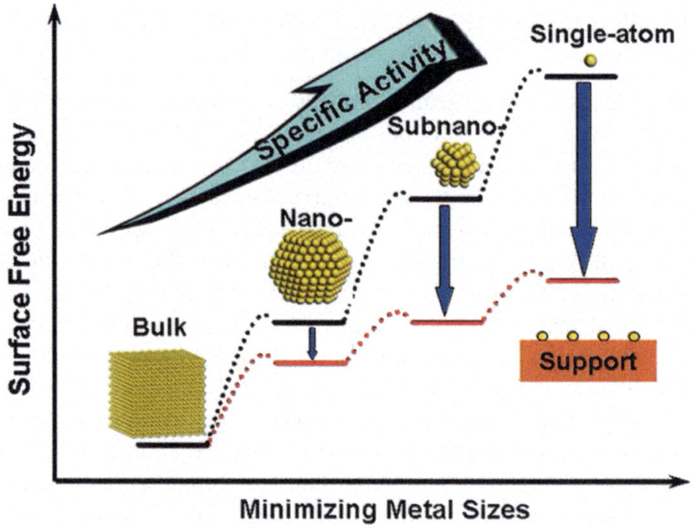

Fig. 9.1 SAC production from bulk catalyst is shown, along with the linear relationship between surface free energy and smaller metal sizes

and greater efficiency for the desired electrocatalytic process. The resultant single-atom catalysts are then characterized through unique catalytic, physical, and chemical features.

9.3 Effects of Support Materials on Single-Atom Catalysts

The activity and efficiency of single-atom catalysts (SACs) not only depend on single-atom nature but also on support and interaction between SAC's and support. In SACs, the solid support plays a vital role in stabilizing metal atoms, increasing catalytic performance, reducing the use of metal, and assisting in avoiding the aggregation of metal atoms. In this regard, various kinds of supported materials have been used, like zeolites, graphene, metal oxide framework (MOFs), transition metal dichalcogenides (TMDs), and metal oxides [16]. Thus, selecting proper support for specified SAC is crucial to achieving catalytic selectivity and desired performance. The metal species can interact chemically or bond with nearby atoms in support, increasing catalytic performance and stabilizing SACs [17]. The synergistic interaction between the solid support and a single atom may also increase charge transfer. Furthermore, using various supports may enhance metal–support active centers or increase the redox properties, which are usually responsible for attaining outstanding catalytic activity [18]. Lately, two-dimensional (2D) materials with remarkable chemical and physical features have achieved substantial attention as a new class of favorable supports for SACs. Among various supports, graphene has been regarded unique supported material because of its sp^2 hybridized structure with one atom thick layer with C==C and C–C carbon bonds arranged in the form of a hexagonal crystal lattice structure. It also has the feature of a large surface area which provides high catalytic loading, maximum conductivity, and long-term stability. Because of these attractive features, graphene as a supported material also has wide applications for water electrolysis applications [19].

2D graphene nanosheets with an enhanced specific surface area have greater defects and anchored active sites for trapping single metal atoms on the surface. The graphene surface as support may be doped by various heteroatoms like phosphorous (P), boron (B), nitrogen (N), and sulfur (S) [20]. These doped heteroatoms, especially S and N, may behave as binding centers for single metal atoms because of their metal coordination tendency [21]. Incorporating SACs in graphene may modulate the electronic and geometric infrastructure of neighboring C atoms, which participate as additional active centers for catalysis applications. Apart from graphene, MoS_2 basal inert surface may be activated by specified confinement of Pt single atoms on a two-dimensional surface where Mo atoms replaced Pt [22]. The support of graphene has been developed by Sun and coworkers by synthesizing single atoms of Pt on graphene by ALD technique, also shown in Fig. 9.2 [9]. Such supported materials with SACs have a special role in water electrolysis. To take complete advantage of different kinds of supports, carbon has also been used in this regard. As Mo SACs anchored on N-doped carbon have been developed by Chen and his coworkers for hydrogen evolution activity (HER) [23]. Likewise, Fei et al. deposited Co metal atoms at N-graphene support (CoeNG) for efficient HER performance [24]. SACs may be developed on 2D-supported materials like layered double hydroxides, g-C_3N_4, MoS_2, etc. Lately, Co single atoms have been doped in MoS_2 (Co–SMoS_2) monolayers [25]. EXAFS results demonstrated Co–S formation with a 2.21 Å coordination peak. The outcomes of the HAADF-STEM image of the Co–SMoS_2 catalyst showed the presence of Co single atoms on the Mo side.

The 3D structures of graphene have also been used as support for single-atom catalysts because of greater porosity and well-distributed pore size to confirm that the large density catalyst loading is due to the large surface area of support. It also readily transports associated electrolytic ions during electrochemical reactions. The 3D structure has higher electrocatalytic stability and great electric conductivity. Various 3D graphene structures have been reported in the literature, like 3D porous graphene developed via CVD through the porous template of nickel (Ni foam or dealloyed Ni), which may fulfill all the requirements. During the process of CVD, the growth of graphene on Ni occurs by C atoms dissolution into ligaments of Ni and, upon rapid thermal cooling, C atoms segregation to Ni surface ligaments for successive graphene layer growth [26]. The nanoporous 3D graphene a homogeneous structure with sterically continuous sheets of graphene and well-interconnected pores. Aside from Ni, other nanoporous templates such as Cu and porous oxides (NiO, CaO, Al_2O_3, SiO_2, MgO, etc.) may be used as CVD templates [27]. Compared to metal substrates, the oxides generally produce porous 3D graphene with minimum crystallinity because of their inferior catalytic performance for graphene growth.

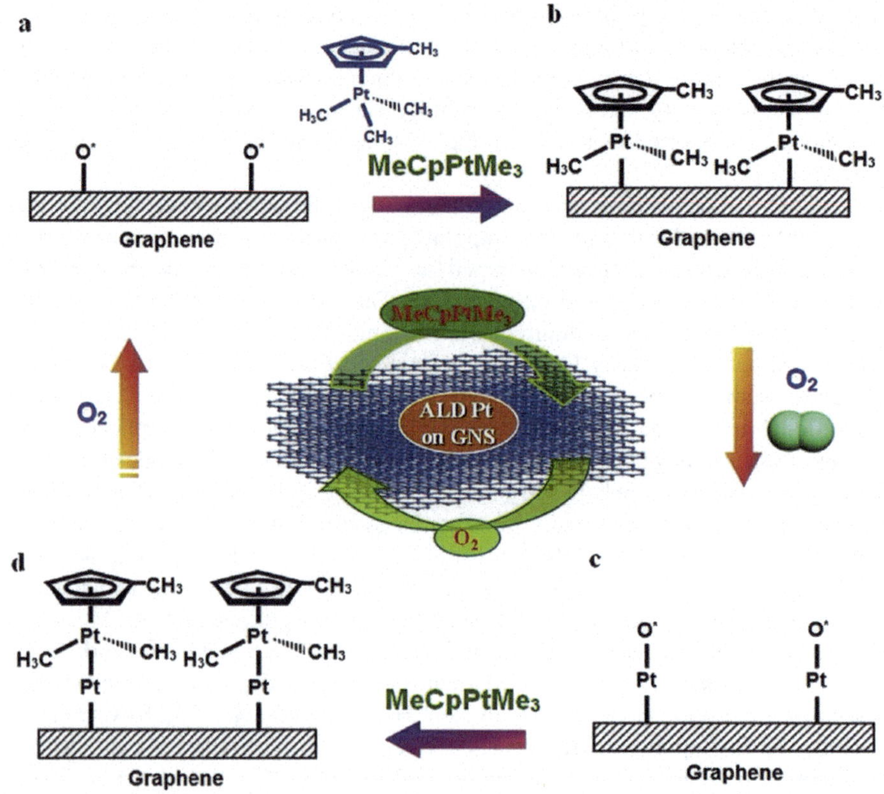

Fig. 9.2 ALD-deposited Pt atoms on graphene nanosheets. Reproduced with permission from Ref. [9]. © 2028 Elsevier

The chemical interaction between support and metal atoms plays an essential role in determining catalytic activity and stability based on the loading content of metal atoms. The metal–support interaction reduced the free energy of single metal atoms with homogeneous and continuous metal atoms dispersions which enhanced the features of SACs.

9.4 Chemical Natures of SACs

Compared to active centers heterogeneity developed from deviation of surface sites and particle size variations in nanocatalysts, the well-distributed and perfectly defined SACs attain well-dispersed distribution with a similar geometric and electronic structure on selective support. Furthermore, the robust metal–support interaction developed from the ionic or covalent bonding between support and metal atoms not only assists in attaining higher electrochemical reactivity of single-atoms crystals but also assures thermodynamic durability of SACs active centers on the surface of the catalyst. SACs may be regarded as a compact bridge between heterogeneous and homogeneous catalysts. The geometric infrastructure of SACs is highly dependent on the coordination environment of support, vigorous variation during the catalytic reaction conditions, and atmospheric exposure. When metal oxide is used as support, the cationic vacancies usually host the SACs because of the facile oxidation of precious metal atoms through uncapped oxygen atoms. Such oxo-ligands can fall apart, partially engendering metal-to-metal contributions. For example, during the analysis of Pt_1 atoms on hematite, the Pt_1 anchored atoms are confirmed to be iron vacancies surface with three dangled atoms of oxygen involved in chemical bonds [6]. Some of the contributions of Pt–Fe may be analyzed by density functional theory and X-ray absorption (XAS) technique under a reducing environment. Recently, Christopher and his coworkers worked on species of (PtO_2)ads at TiO_2 anatase(101) terrace has been identified clearly via DFT study by experimental X-ray absorption near-edge structure (XANES) spectra and aberration-corrected scanning electron microscopy (AXSTEM) as in Fig. 9.3a–d [28]. In the same way, PtO_4 with square planner geometry has been suggested for single atoms of Pt on CeO_2, as in Fig. 9.3e. These fragments of PtO_4 were also evidenced as highly stable for single atoms of Pt on unsaturated penta-Al^{3+} centers with γ-Al_2O_3 surface as the support [29]. Many other SACs, including Ni and Pd, are also assessed to be highly durable through metal bonding centers of O_4 on support in the form of oxide by DFT study [30]. Aside from cationic vacancies on support, surface O may be reduced by O vacancies which act as a binding center for SACs. Geometrically, it's different from metal oxygen architecture on cationic vacancies and acts as doped metal via oxygen sites of support. These structures were regarded as Pd_1 and Pt_1 inside oxygen vacancies of (001) surface of halite-type oxides (MeO, Me=Ni, Mn, Mg, and Fe) [31]. For SACs with other supported materials, precious metals SACs also depend on vacancy types on the surface. Like, with sulfur defects on monolayer MoS_2, Pt atoms may be trapped into anionic vacancies (Fig. 9.3f) and attain doping to nearby vacancies through energy by an electronic beam [32].

On carbon-based materials, the atoms of metals got captured through C vacancies which are generated through heteroatoms dopants S, N, or P as anchored centers [33]. Aside from anionic and cationic defects, Sacs may be adsorbed on the defective or perfect surface with excessive oxygen atoms to form suitable size hollow centers [34]. Apart from second metal as the supported material like single metal alloys (SAAs), the surface morphology of active centers is relative to types and exposed crystallographic planes of host metals (Fig. 9.3g) [35]. These diverse and selective geometric infrastructures of SACS support vigorous behavior in the presence of different circumstances, providing great opportunities to modulate catalytic activity. While because of investigative techniques in time and atomic resolution are essential to attain a local coordination environment of active centers of single atoms. Still, it's a great challenge to accurately detect active species' geometric and electronic structure in SACs, especially in non-ideal environments.

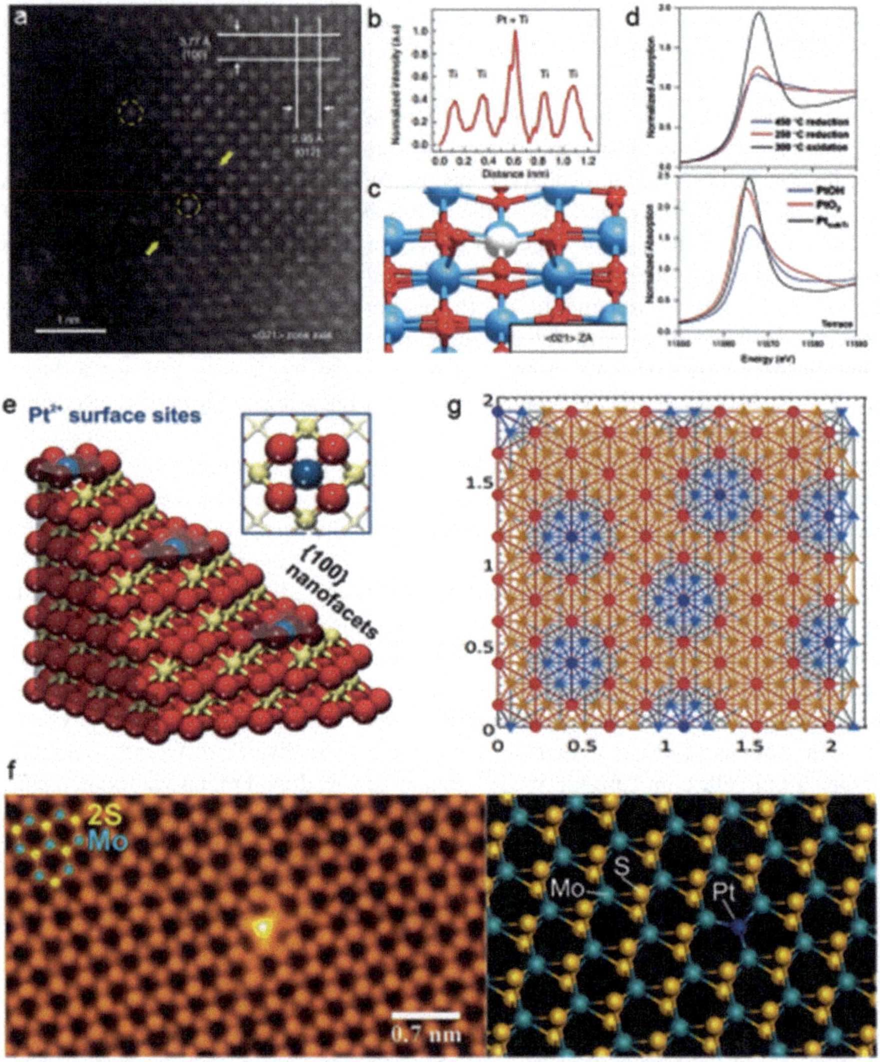

Fig. 9.3 **a** Pt_1/TiO_2 AC-STEM. **b** Normalized intensity profile along yellow arrows. **b, c** Pt_1/TiO_2 DFT model. **d** DFT-predicted structure-based calculations of observed XANES spectra. $(PtOH)_{ads}$, $(PtO_2)_{ads}$, and Pt substituting Ti_6c on the TiO_2 (101) terrace were utilized to produce severe, mild, and oxidation spectra. **e** Pt_1/CeO_2 DFT atomic model. **f** Pt_1/MoS_2 DFT relaxed atomic model and atomic resolution picture. **g** Pt/Cu SAA (111) lattice model schematics. Reproduced with permission from Ref. [35]. © 2019 Wiley

Because of the robust interaction of support and metal, the electronic infrastructure of single-atom active centers can easily get accommodated by the support and its nature through ionic/covalent bonds between them. It leads to the alternation of charge states of single atoms regarding diverse coordination environments on different types of supports. Like, according to the DFT study, single Au atoms may be charged positively (on CeO_2 and TiO_2) or negatively charged (on HfO_2, ThO_2, and ZrO_2) [36]. Even using similar support, various kinds of vacancies like anionic or cationic will result in different charge states of SAC active centers. According to geometries of single atoms on support, the positive charge is possessed by a single atom at cationic vacancy to behave as a cation on the vacancy. However, it will emerge as a negatively charged anion on O vacancy sites. Therefore, in SAC of Pt/FeO_x with the location of Pt1 on Fe-center vacancy, it indicates positive charges at Fe-vacancy centers. Precious metals like Au and Ir have also validated such SACs with a positive charge on cationic vacancies, and Pd is supported on different oxides like TiO_2, CeO_2, and many others [37].

With the reduction in the size of particles, the free energy of metal species usually increases. The precious metals are usually unstable and start agglomerating to form larger nanoparticles, particularly on heating at elevated temperatures. The interaction between support and metal contributes to modulating the electronic structure of SACs to govern and enhance reactivity in catalysis and determines the durability of SACs on respective support. In this way, Tang and coworkers controlled metal–support interaction by developing durable SACs of silver with greater depletion of the 4d electronic state of Ag atoms, leading to greater catalytic performance and facile reducibility [38].

Generally, on downsizing metal nanoparticles to single atoms, both electronic and geometric structures of SACs may be modified by factors like metal atoms' interaction with neighboring atoms and unsaturated coordination environment. Similarly, single atoms show different variations concerning support's chemical and mechanical nature.

9.5 Stability and Durability of Single-Atom Catalysts

With homogeneous SACs dispersion, the coordination chemistry and bonding of support and active sites help to attain identical and uniform local configuration compared to nanoparticles with various activity centers. Such SACs are not termed "singles" because they usually form a bond with a support substrate for achieving maximum stability with high selective catalytic activity and lower surface energy. The electrocatalytic performance is indicated by specified atoms in a definite coordination environment and bonds instead of the single atoms (SA) themselves. Such bonded atoms may boost scaling relations during electrocatalysis involving multistep reactions. They are usually exposed to reactants during catalysis and show unusual specific catalytic activity because of unsaturated coordination [34]. To maximize the stability of single atoms with stabilize these single atoms with the designed coordination chemistry with minimum aggregation of particles, one of the effective strategies is precise control of the loading of metal atoms to a low level. In this regard, the coprecipitation approach and wet impregnation methods facilitate SACs formation by small metal atoms deposition on the surface of the substrate. The clustering and aggregation of metal atoms may be eliminated effectively using such processes. In general, in the first described SAC report, Pt (SA) content in electrocatalyst Pt/FeO_x was around 0.17 wt%, while 2.5 wt% in the case of counterpart nanoclusters of Pt [6]. Moreover, the matrix of support should have a higher specific surface area and great porosity that is adequate for the transfer of individual metal atom centers on its surfaces. Generally, the support with maximum porosity displays mesopores/micropores, usually derived from multiple inorganic/organic precursors like metal–organic frameworks (MOFs), carbon nitrides based on graphite, and polydopamine [39]. Aside from metal content loading control, multiple processing parameters and precursors are used for anchoring single atoms on the support surface. Defect engineering is also regarded as an imperative approach for stabilizing metal atoms. The support matrix's defects, like corners, edges, steps, and vacancies, may modulate the electronic structure and surrounding coordination atmosphere, forming unsaturated coordination sites. Such defects act as "traps" for metal atoms capturing during the conversion/synthesis procedure, creating robust metal–support interaction comparison surfaces with no defects [10]. Like, reducible oxides V_2O_5, TiO_2, and CeO_2 have been used as active supports, and their step edges and surface vacancies make a great difference in stability, catalytic performance, and dispersion of the metal atoms. For example, for single atoms of Au at the surface of CeO_2 (111) (defective support), the sequence of stability was like this; defect-free surface < O vacancies < cation vacancies < steps [40]. For oxides resistant to reduction, like MgO, SACs have promise adsorption on charged O or surface neutral vacancies denoted as Fs^+ and Fs, correspondingly, instead of O_2^- centers (the Fs is left after neutral O atoms removal and formation of Fs^+ with the removal of O^- anion) [41]. For the optimal bond formation of 2D materials like substrate and metal single atoms, they are made defective deliberately either through chemical or mechanical destruction, like newly formed bonding centers are formed either above or within the layer structure. In previously reported studies, specified additives or surfactants were introduced during SACs synthesis for fixation of such single atoms, enhancing metal atoms and supporting substrate interaction. For example, surfactant Pluronic F127 has been used commonly for developing SACs for improving the dispersion of metal and inhibiting undesirable aggregation [42]. During the procedure, the very charged functional group of F127 may have a strong attraction with +vely charged ions of metals which hinders aggregation of metals. The synthesized composite attains a high catalytic performance and magnificent surface area as F127 is regarded as a soft template creating mesopores during the consequent carbonization procedure of the initial precursor. A basic solution is also used as an additive for SACs formation in numerous add-on procedures. Adding alkali makes metal atoms dispersion stabilized in an oxidized form, with OH groups in the surrounding. An appropriate K^+ and Na^+ addition may considerably reduce the size of Pt clusters on support SiO_2, promoting single

Pt–OH site formation, which is highly active for the water–gas shift catalytic reaction (adsorption of CO and activation of H_2O) [43]. A common strategy to stabilize SACs is by constructing robust interaction of support and metal with minimum metal nanoparticle accumulation. The metal–support interactions are categorized into two types based on the nature of the support matrix nature. For supports containing metal carbides, metal oxides, and single metal atoms, the single-atom sites with minimum concentration automatically create a stable metal-to-metal alloy bond with metal atoms in supports material. To a certain report, evaporation and absorption of Pd atoms into single crystal sites of Cu (111) surface, and then trapping at nearby step edge, where exchange of positions may lead to alloying into Cu supported material [44]. For supports without any metal like carbon-based composites, the single atoms may coordinate with atoms of carbon steadily by forming σ bonds and M–π interactions. Also, with the incorporation of some atomic species with lone pair electrons like S, O, and N, the individual single atoms may get stabilized with maximum coordination by covalent bonds. N may generate multiple surface groups in carbonaceous materials doped with N, like pyrrolic N, graphitic N, and pyridinic N, which facilitates maximum bonding sites and greater coordination for single metal atoms [45].

9.6 Noble Metal SACs for OER

Water oxidation is the reaction of prime importance to reduce energy consumption and overpotential in electrocatalytic water electrolysis for hydrogen evolution reaction because of sluggish thermodynamics of oxygen evolution reaction (OER). OER is a multielectron transfer reaction that requires extra energy to overcome the 4 electron transfer mechanism, which results in a large overpotential. It means that input energy is greater than the outcome, which is unfavorable economically. For any specified reaction, OER comprises numerous intermediates formation during each step. Like in the first step, water adsorption occurs, which forms the first intermediate $^*OH$ (* indicates active metal centers), which further oxidizes in the second step and forms $^*O$. During the third step, $^*O$ interacts again with the H_2O molecule and generates $^*OOH$ specie which further forms $*O_2$ at the end. These intermediates also generate electrons and protons that need to be transferred to the cathode for HER reaction. This theory lies behind this sluggish OER kinetics which requires an efficient catalyst for water electrolysis. The catalysts should have high stability and outstanding performance and stability, which are imperative demands to overcome such issues. To improve the intrinsic performance of OER catalysts, active sites' modulated structural and electronic properties have been tailored specifically through interactions between support and single metal atoms of noble metals [46]. Two approaches have been designed to increase the SACs OER activity. One of them is the incorporation of new active centers which boosts the performance of original active centers. After anchoring new active centers on the support surface for OER activity, they may also be regarded as new active sites. An electrocatalyst dispersed single atom Ru–N–C was analyzed for OER activity in acidic media, which displayed 267 mV overpotential at the current density of 10 mA/cm^2 along with a maximum turnover frequency of 13,392 O_2/h at 300 mV overpotential [47]. Owing to the insignificant OER performance of carbon support doped with heteroatoms, like N–C-based-SACs, the major active centers for water oxidation were usually considered individual metal atoms. Also, the electrocatalysts with efficient OER performance have been employed as the supports for anchoring single nobel metal atoms to increase the OER activity of PMSACs. The LDH of CoFe, which is supported on the surface of Ru-SAC, has been developed and displayed a very lower overpotential, around 198 mV, at the current density of 10 mA/cm^2 [48]. The increment in activity has been attributed to the durable synergetic coupling of electrons between LDHs substance and single atomic Ru. The electrode comprising nanosheets of NiO decorated with Ir with great loading of Ir (18 wt%) fabricated, which displayed only 215 mV overpotential at 10 mA/cm^2 current density for OER in alkaline media [49]. The increased activity may be attributed to maximum loading and formation of new bonds of Ir–O stimulated by automated dispersed Ir. Introducing noble metal atoms into supports showed a superior activity, like NiFe-LDH, which may increase the OER performance through electronic modulation of original active centers. The single atoms of Au on NiFe-LDH support needed 280 mV overpotential at 10 mA/cm^2 current density for OER in a basic environment [50]. The Au atoms incorporation may induce the net charge distribution in Au-to-LDH, which is transferred to adjacent Fe, O, and Ni atoms. The tuned electronic catalytic structure might facilitate OH– adsorption, which modifies the adsorption energies of certain intermediates like OOH* and O*, which results in better OER activity. Likewise, the PMSACs OER performance may be increased with the help of two approaches. A composite with single atoms of Ir showed 267 mV overpotential anchored at the surface of nanosheets of Co nanosheets at 10 mA/cm^2 current density in 1 M KOH [51]. The dinuclear sites of Ir–Co were regarded as active centers in which maximum adsorption of OH^- groups occurs on Co sites which further get dissociated into adsorbed O*. It was subsequently transferred further to Ir–O* sites by an intermediate of Co–O–Ir. The PMSAC's stability may be enhanced greatly by tailoring the electronic catalytic structural properties by single atomic interactions between support and metal. The most promising electrode for applications of OER activity is RuO_2 on the commercial level. The worst OER stability is a major concern because of the dissociation of Run+ in a high oxidation state in the electrolyte [52]. An alloy of Pt–Cu with Ru single atoms decorated composite illustrated the enhanced stability of the Pt skin with compressive strain. The electronic donation toward different intermediates is facilitated by Pt–Cu, which reduces overoxidation and Ru single atoms dissolution. Hydroxides may stabilize the Ru atoms with metal-free elements anchored firmly on CoFe-LDH by interactions between support and metal [48]. During the mechanical process of OER, the Ru oxidation state in Ru/CoFe-LDHs catalysts was lesser than +4, which resists the dissociation of Ru with improved stability. Furthermore, the PMSACs are fabricated as bifunctional composites for HER and OER activity. According to recent literature studies, SACs based on iridium have been regarded as the most favorable candidate for efficient bifunctional water electrolysis in alkaline as well as acidic media

[53]. Dispersion of single atoms of Ir on carbon doped with N and particles of Co through Ir-ZIF-67 precursor pyrolysis only attains 1.603 V cell voltage potential to attain a current density of 10 mA/cm^2 in alkaline 1.0 M KOH electrolytic media [54]. The fabrication of structured composites as supports results in incrementing PMSACs water electrolysis activity. The decoration of Ir single atoms on composite $Co_{0.8}Fe_{0.2}Se_2$ nanosheets synthesized on substrate Ni foam by electrocatalytic deposition only acquires 1.39 V cell voltage to achieve 10 mA/cm^2 in a basic medium [55]. To this end, its far clear that nobel metal-based with very low percentages forming single-atom catalysts (SACs) have gained much attraction by reducing the overpotential for OER intermediates to maximize the hydrogen evolution reaction with future perspectives of electrochemical energy storage and its conversion.

9.7 Noble Metal SACs for HER

Hydrogen evolution is the basic reaction for hydrogen production through the electrocatalytic water electrolysis approach. Among the variously reported composites, electrocatalysts based on noble metals unveil outstanding electrocatalytic water electrolysis performance and great stability for HER reaction. Composites based on Pt are considered efficient electrodes for hydrogen evolution in an acidic environment, which has been used widely in industrial applications. Compared to OER, the hydrogen evolution reaction mainly comprises 2 electron transfer reactions, making it kinetically favorable than OER's. It involves protons adsorption by H_2O molecules decomposition to form the first intermediate $^*H$ called the Volmer reaction. While the second step involves a reaction between dissociated H^+ and intermediate $^*H$, called the Herovsky reaction, or two $^*H$ species adsorption, called the Tafel process, to generate a molecule of H_2. To overcome the energy barrier, stable and effective bifunctional catalysts are exceedingly desirable for accelerating water electrolysis by reducing overpotential. The hydrogen evolution, an electron transfer process, may be conducted in a wide range of pH, like acid involving H_2SO_4, the alkaline in KOH, and neutral in the presence of PBS electrolytes. According to the description of the HER mechanism, two pathways are involved: Volmer–Heyrovsky process or Volmer–Tafel procedure. The HER catalytic reaction generally consists of the appropriate proton or H_2O molecules adsorption, rapid protons combination, and faster transportation of electrons accessing the active centers during the procedure. These are the main features for improving HER activity. Regarding PMSACs, most studies have been reported on Pt-based single atoms anchored firmly on multiple conductive supports [56]. For SACs based on Pt involving HER activity, the Pt centers in the atomic form are considered widely as actual active sites. During hydrogen evolution, the mechanical reaction of Pt-based single atoms depends on synthetic strategies and supports species. The electrochemical potential-cycling process prepares the SACs of Pt and generally shows a comparatively lower Tafel slope, following the Volmer–Tafel mechanism. Like, a nanoporous 3D Co0.85Se electrocatalyst loaded with single atoms of Pt, attained by electrochemical technique, displays 55 mV overpotential at a current density of 10 mA/cm^2 and 35 mV/dec Tafel slope [57]. To further enhance single atoms loading, the electrocatalysts with greater specific surface areas are used as supports for anchoring the targeted noble metal atoms. A potential-cycling strategy has developed the nanotubes of the CoP electrode anchored with Pt with 24 mV overpotential at 10 mA/cm^2 and 30 mV/dec Tafel slope [58]. The activation of support during electrode fabrication of Pt-SACs results in enhancing active centers during the Volmer–Tafel mechanism of such Pt-SACs can be inferred that the supports could be activated during the fabrication process, which fastens H adsorption on the surface of the electrode. Further, the decoration of single atoms on a binder-free 3D-supported substrate might increase electronic density transfer, leading to outstanding reaction dynamics. The HER Volmer–Heyrovsky mechanism has also been discovered regarding SACs based on noble metal Pt, in which relatively poor performance is because of support. The decoration of Pt-SACs on surface graphdiyne by a wet chemical technique, exhibited 23.64 A/mg current density at −0.1 V potential and 46.6 mV/dec Tafel slope, which is even greater than state-of-art Pt/C [59]. Also, a unique hydrogen spillover effect mechanism has been used commonly over a single atom Pt/m-WO_{3-x} composite [60]. In Pt/m-WO_{3-x} SACs, there is a reduction in hydrogen diffusion length because of the interfacial effect of tungsten oxide and Pt, which boosts the kinetics of the hydrogen extraction/insertion and improved hydrogen spillover effect. The hydrogen extraction/insertion acceleration may induce the rapid Pt surface re-exposure and allows another H^+ adsorption, therefore, fastening HER mechanism. In comparing bulk counterparts with particles, the SACS of Pt demonstrates the novel electronic structure is indeed derived through strong metal support interactions. The PDOSs of Pt 5d-orbitals exhibit maximum electronic densities along the Fermi level, facilitating hydrogen ions adsorption and enlightening conductivity than bare supports. Likewise, Pt SACs have a specified Gibbs free energy of reaction intermediate which is closer to zero, showing exceptional HER intrinsic performance compared to Pt bulk. The unique electronic structure and high Pt SACs atomic result in the tremendous specific massive activity of Pt SACs. Aside from composites of Pt-based SACs, noble metal-based SACs like Rh and Ru have also been fabricated for electrochemical hydrogen generation. The electrocatalyst based on MoS_2 SAC decorated with Ru has been developed and analyzed for HER in alkaline media [61]. The Ru atoms incorporation in MoS_2 composite caused phase transformation from 2H to 1T along with S vacancies generation, minimizing energy barriers in the Volmer step during HER. It also facilitates optimal adsorption and desorption of H^*, enhancing catalytic activity instantaneously. Also, dispersion of N, S-coordinated Ru SACs on the surface of Mxene presented 76 mV overpotential at the current density of 10 mA/cm^2 and 90 mV/dec Tafel slope [62]. The Rh atoms have been confined atomically into the MoS_2 composite, which displayed outstanding HER activity in 0.5 M acidic H_2SO_4. The Mo atoms are replaced by Rh atoms with dispersion in the lattice with distance. The HER performance of S-sites in-plane, nearby atoms of Rh, might be tuned by varying confined Rh atoms distance, which results in inert plane activation in MoS2 composite, thus, enhancing HER activity. Some SAA composites have also been used for HER activity [63]. A unique RuAu SAA composite has been used as an efficient electrode for HER in alkaline media. The composite RuAu SAA attained 24 mV overpotential at a current density of 10 mA/cm^2 and 37 mV/dec low Tafel slope. These in-depth analyses of HER activities have shown that precious metals with appropriate structural and electronic modulation have a crucial role in catalytic activity.

9.8 Transition Metal-Based SACs for HER

Single-atom catalysts have attractive and unique characteristic features to modulate electrocatalytic performance and selectivity. SACs based on Ru have been utilized as unique composites for hydrogen evolution reaction (HER). Among energy resources, hydrogen has been regarded as a significant energy vector for developing fuel cells. Because of its characteristic features, SAC has been used for H_2 generation via electrochemical reaction applications. Hydrogen evolution is a two-step reaction at the cathode during water electrolysis and a half-reaction at the cathode during the water splitting. To accomplish the process, the frequently available transition metals have also been used along with precious metal-based SACs, which have shown more excellent catalytic activity. Traditionally, the intrinsic HER catalytic activity of composites having precious metals is unacceptable due to variable electronic structural properties compared to precious metals. During the HER process, the different intermediates' adsorption, like H^*, has specified Gibbs free energy ΔGH^* and optimum electrocatalytic performance, which is only achieved when Gibbs free energy is zero. Many efforts have been made to improve the HER process using transition metal-based SACs by modulating the electronic structural and configurational properties. Active centers with specific density and chemical nature play a vital role in the HER process and the modulation of electronic features.

Moreover, HER composite fabrication in harsh alkaline conditions is crucial to address stabilities and the reaction kinetics challenges. Certain parameter variations like composition, morphology, porosity, active sites, and electronic features may assist in tuning surface chemistry to overcome problems. Using the maximum volume-to-surface area ratio of single-atom catalysts by downscaling the size of particles allows enormous atomic dispersion. Thus to modulate electronic structure for HER catalysis in alkaline media, various transition metal-based catalysts like Fe, Co, Ni, V, and Cr have been explored recently. Similarly, transition metal chalcogenides (TMCs) and phosphides (TMPs) have also been studied widely as efficient HER composites because of their unusual activity. Subsequently, SACs supported over TMCs, and TMPs will likely display outstanding HER activity like Luo's and his coworker's developed Pt-SACs on nanotube arrays of CoP (Pt-SACs/CoP) via a facile electrodeposition strategy. The composite Pt SACs/CoP with 1.76 wt% loading of Pt has shown remarkable HER performance compared to commercial Pt/C(20 wt%) [58]. Deng et al. fabricated Pt SACs as shown in Fig. 9.4a–c with homogeneous dispersion on MoS_2 (Pt–MoS_2) presenting considerably boosted HER performance than that of MoS_2(original) Fig. 9.4d [64].

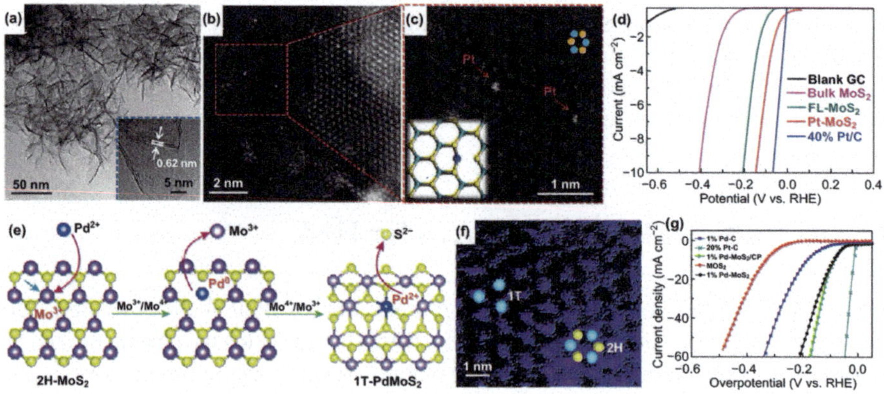

Fig. 9.4 **a** Pictures of Pt–MoS_2 from TEM and HAADF-STEM. **c** The red dashed rectangle in (**b**) was used to enlarge the domain. **d** The polarization curves for bulk MoS_2, FL–MoS_2, Pt–MoS_2, and 40% Pt/C. **e** A diagram of the redox reaction occurring spontaneously between MoS_2 and Pd(II). **f** A STEM image of the 1% Pd–MoS_2 taken in the dark. **g** Polarization curves of 1% Pd–MoS_2/CP, 1% Pd–S_2, pristine MoS_2, 20% Pt/C and 1% Pd–C

Xing and his coworkers also developed various single Pd atoms loading doped with MoS_2 (Fig. 9.4e, f). The Pd–MoS_2 with 1% ratio only has shown boosted HER performance with 89 mV overpotential at 10 mA cm^{-2} current density Fig. 9.4g [65]. Guan et al. have also prepared and designed distinct Ganoderma-like heterostructures of MoS_2/NiS_2 with a uniform dispersion of Pt atoms that indicated remarkable HER activity [66]. Aside from such marvelous activities of SACs based on Pt, industrial commercialization on a larger scale has been hindered till now because of its expensive nature and scarcity. Therefore, considering such factors, it's worthwhile to design electrocatalysts of SACs based on transition metals because of their cheaper nature and frequent availability. Likewise, Fan et al. developed Ni–C-based composite that may be activated to attain Ni SACs at graphitic carbon surface after consequent 4000 cyclic voltammogram cycles, which has shown great durability and activity for hydrogen evolution reaction [67]. In another work by Chen's group, Ni-based SACs embedded in nanoporous graphene also displayed greater catalytic stability and activities in acidic media by water electrolysis reactions. Fe-SACs have been reported as well by many researchers because of their abundance and most inexpensive nature among transition metals.

Furthermore, the identification of active sites during HER reaction has recently been reported at the atomic level. They validated the formation of HO–Co1–N_2 moiety in high-valence form by binding individual Co1–N_4 centers with OH^{-1} groups in the electrolyte. This further unravels the formation of desired adsorption of water molecules by intermediate as H_2O–(HO–Co1–N_2). Such outcomes are critical to commercial water–alkali electrolyzers, which are indescribable and are part of intense research [68]. Fe, W, and Mo-based composites have also shown outstanding HER activity. So, it's quite fascinating to explore W and Mo-based-SACs for HER workout. Indeed, Wang's group revealed that the MOF-derived strategy is used to attain W-SACs with maximum HER activity. XAFS spectroscopic analysis and HAADF-STEM have shown the development of W-SACs, which are anchored along three atoms of carbon and one atom of N ($W_1N_1C_3$). DFT studies have further elaborated that the unique structural features of W1N1C3 moiety have played a vital role in maximizing HER performance [69]. Furthermore, Mo-SACs have also been developed by chitosan by a similar research group. Thus, catalytic Mo-SACs structure has been probed through XAFS spectroscopic analysis and AC-HAADF-STEM, which suggests Mo-SACs formation, anchored with two atoms of C and one N atom ($Mo_1N_1C_2$). The Mo1N1C2 composite has displayed exceptional HER activity concerning bulk Mo_2C and MoN with better stability than commercial Pt/C 20 wt%. DFT studies have further revealed that $Mo_1N_1C_2$ is more advantageous to proceed with HER catalysis than Mo_2C and MoN [70]. Recently, Du's group established V-SACs, supported on the surface of graphene with great HER activity, also validated by DFT studies [71]. It has been recommended that V-SACs anchored on graphene may be tuned substantially, which further modulates the free energy of H adsorption (ΔGH^*) with optimum free energy value as ΔGH^* implies to -0.01 eV, that's improved compared to Pt composite. It's worth noting that in a few years, SACs have been studied with faster growth. All types of transition metal-SACs have been developed as high-performance HER electrocatalysts.

Recently, Zhang et al. demonstrated that electrocatalytic deposition, either anodic or cathodic, may be used as a general approach to develop SACs, as demonstrated by the successful fabrication of SACs which consists of numerous types of metals like Ru Pt, Au, Ag, Rh, Pd, and Ag with different supports NC, MnO_2, $Co_{0.8}Fe_{0.2}Se_2$, and MoS_2 [72]. Significantly, the single atoms loading may be controlled facilely through concentration variations of precursors of metals, the scanning cycle numbers with time control, and the scanning rate. Like, during the facile cathodic electrodeposition of Ir in the presence of 1 M alkaline KOH electrolytes, Ir is deposited with specific mass loading, which keeps on increasing with an increment in Ir concentration. The Ir with 150 μM concentration, Ir-SACs, has still been attained with 3.5% mass loading though clusters of Ir emerged with 4.7% mass loading with further enhancement of Ir 200 μM concentration. This mass loading suggested that SACs formation attains an upper mass loading limit between 3.5% and 4.7%. Remarkably, SACs synthesized by the strategy of cathodic electrodeposition have displayed outstanding potential to catalyze HER in alkaline media along with Ir-SACs on nanosheets of $Co_{0.8}Fe_{0.2}Se_2$ delivering 8 mV overpotential to attain 10 $mAcm^{-2}$ current density in 1 M basic KOH electrolyte. Likewise, Liu et al. developed an electrochemical pulse voltammetry approach to preparing U-single atoms on nanosheets of MoS_2 (U/MoS_2) via radioactive wastewater [73]. The U-single atoms mass loading has been controlled with pulse cycle increment. The obtained composite at 100 pulse cycles has suitable 5.2% U loading with optimum HER performance in 1 M KOH. It has shown a lower 72 mV overpotential at 10 $mAcm^{-2}$.

A comprehensive analysis has assisted in elaborating that single-atom catalysts (SACs), with maximum atom-utilization efficiency, unveil many advantages, predominantly for HER application with enormous activity and long-term stability.

9.9 Transition Metal-Based SACs for OER

Compared to precious metal counterparts, experimental and theoretical investigations have been done on transition metal-based SACs in oxygen evolution reactions. This sparks great research interest due to their typical and efficient oxygen evolution reaction (OER). Significantly, many efforts have been devoted to developing energetic composites M–N–C with characteristic SAC properties and investigating the OER activity. Along with verifying SACs and appropriate catalytic active moieties, some other imperative SAC aspect and their research lie in metal species effects on generating the active sites and comprehensive mechanistic reaction. Though, because of variations in the physicochemical features of supports depending on metals and the conformed active centers elucidate the role of various metals in difficult ways. To address these challenges, with specified considerations, the profuse pyridine-N in C_3N_4 delivers maximum lone pairs of electrons for metal ions accommodation. Qiao et al. assembled a molecule-level electrocatalyst Co–C_3N_4/CNT highlighting ring of CoN_3C_2 ring via direct transition metal graphting on g-C_3N_4 surface, which aims to propose highly active OER electrocatalyst [74]. Besides the efficient electrochemical performance of Co–C_3N_4 composite for OER, metal species impacts on oxygen evolution activity have been explored to explore the OER volcano plot. Based on widespread investigations of metal ions and C_3N_4 interactions, the Song group has synthesized a dual metal doped electrocatalyst NiFe-C_3N_4/CNT. The superior OER activity was observed than that of single metal counterparts due to the synergistic intercalated effect between metal ions and C_3N_4 [75]. Lately, Duan et al. described a general method to numerous single non-noble metal atoms Ni, Co, and Fe on frameworks of N-doped holey graphene (M-NHGFs) [76]. The similar moiety of MN_4C_4, regardless of metal species, has been confirmed unambiguously by systematic STEM and XAS analysis. On the other hand, similar nitrogen with a 5 atom % ratio and metal with 0.05 atom % ratio contents with similar morphological features makes the assessment trustworthy. As shown in Fig. 9.5a, the OER catalytic performance has been estimated through DFT studies [77]. They investigated a single-site mechanistic approach applied to electrocatalyst Fe(Co)-NHGFs in which intermediates strongly bind at the M site concerning the C site. A dual-site mechanistic approach has been involved in the Ni-NHGF with OER catalysis, in which OH^* and O^* favor staying at the C site. However, favorable adsorption of OOH^* occurs on the M atom (Fig. 9.5b). The energy diagrams have clearly shown that the rate-determining step of Ni–NHGF via OER catalysis with the mechanism on dual sites has a lesser barrier of 0.42 eV. These have activity trends in order of Fe < Co < Ni (Fig. 9.5c). The consequent electrocatalytic experimentations have shown that Ni–NHGF attained excellent OER activity at a 10 mA cm^{-2} current density with lower tafel slope of 331 mV (Fig. 9.5d).The calculated turnover frequency (TOF) of Ni–NHGF was 0.72 s^{-1}. This higher value concerns previously reported OER catalysts; Ni–NHGF is regarded as a highly active OER electrocatalyst. The trend of activity was inconsistent with the DFT analysis. Along with metal species investigations, some studies have been focused on selective supports, which aim to attain insights into the specific mechanism at the atomic level. The creation of high-density defects on graphene accommodated more single metal atoms [78]. In the same way, Yao et al. also used the defective graphene as subsequent support for anchoring Ni single atom, with 1.24 wt % ratio [78] loading, which is greater concerning pristine graphene. The synthesized SACs may act as an efficient bifunctional composite for OER. It has also been demonstrated that single atom Ni@defect with various electronic configurations are dominant active centers for OER. Wang and his group successfully grafted molecular Co^{2+} on various heteroatoms like O, N, S- doped graphene (Fig. 9.5e) [79]. The systematic characterizations and its analysis using electron paramagnetic resonance and XAS have depicted that heteroatom having functional groups in SNG in combination with one acac–group may immobilize Co^{2+} efficiently. The heteroatom doping effect has been investigated, which showed the highest activity for SNG-Co^{2+} with the following trend; SNG-OG-Co^{2+} < NG-Co^{2+} < SGCo^{2+} < Co^{2+} (Fig. 9.5f). The active center comprising C–SO coupling with Co^{2+} has contributed greater toward OER activity, as shown in electrochemical and spectral characterizations. Sequential Co^{2+} oxidation with a single Co-center mechanistic procedure has been proposed to proceed OER process (Fig. 9.5g), which enhanced deep insight into the mechanism study of SACs. Also, FeNx species dispersed atomically and embedded on the N, S-codoped with carbon has been demonstrated. These FeNx moieties codoped with N, S have far better activity than on N-doped carbon, which paves an innovative route for optimizing and tuning supports that betters OER activity [80]. Compared to SACs supported by carbon, Wang's group suggested a pathway for fabricating unsaturated coordinatively metal centers in conventional ZIF-67 by facile dielectric barrier discharge plasma treatment [81]. Most of the MOF structure's low conductivity and completely coordinated electronic configuration are the foremost concerns that limit electrochemical performance. The increment in OER performance and reduced charge transfer resistance has been observed after treatment to pristine ZIF-67. The complete DFT studies have shown that major active centers comprise unsaturated ZIF-67 moiety having three ligands on Co centers. Song et al. also described the Ir-SACs and Ir clusters dispersion onto nanosheets of $Co(OH)_2$ nanosheets by one-step facile $NaBH_4$ reduction [82]. The subsequent species of Ir incorporated into defect-rich hydroxide nanosheets attained excellent OER performance with faster kinetics for reaction. The electrocatalyst with specified optimizations exhibited 373 mV overpotential at 10 mA cm^{-2} in the presence of 1.0 M PBS, much slower than commercial IrO_2 electrocatalyst. Significantly, the comprehensive characterization like XAFS and XRD analysis has verified greater OER activity because exclusive β-CoOOH in high-valence as well as Ir species having lower coordination structure when oxidation potential is applied.

Because of incredible characteristics and features, SACs can accomplish high activity and durability because of their specified catalytic nature and ability to utilize maximum atom efficiency to practically use low-cost and efficient electrodes for catalysis. SACs have revealed insightful potential in electrocatalytic water electrolysis. Though, OER and HER activities of most reported SACs-based electrodes have shown lower potential than state-of-art commercial precious metal composites (Table 9.1).

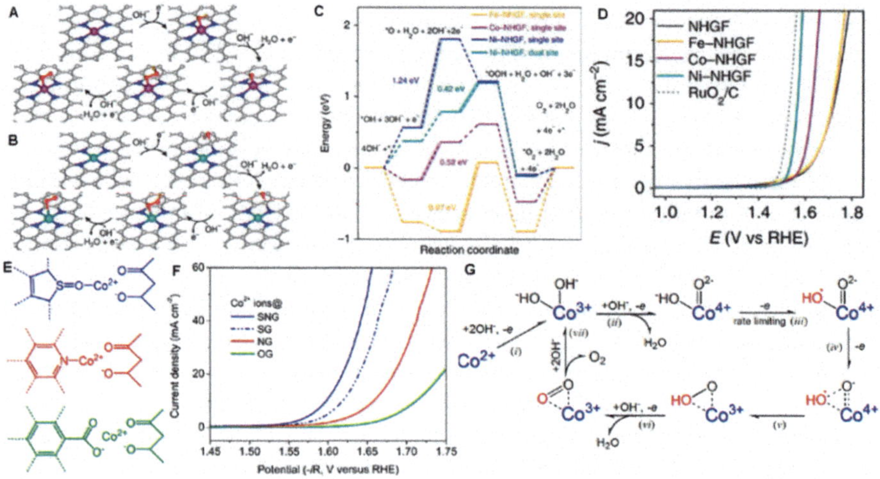

Fig. 9.5 **a** The OER reaction through single-site and **b** dual-site reaction. **c** A diagram of the free energy for OER over FeNHGF, CoNHGF, and NiNHGF, which have different ways of reacting. **d** Different catalysts' OER polarization curves in 1 M KOH. **e** The plan for how Co^{2+} ions stick to S-, N-, and O-doped by graphene. **f** OER CV plots of SNG-Co^{2+}, SG-Co^{2+}, NG-Co^{2+}, and OG-Co^{2+} in 1 M alkaline KOH electrolyte. **g** Co^{2+} Oxidation in steps using a single Co site. Reproduced with permission from Ref. [77]. (c) 2018. American Chemical Society

Table 9.1 Comparison between nanoparticles and single-atom catalyst

Characteristic	Single-atom catalysts	Nanoparticles
Size	Consist of individual atoms	Consist of clusters of atoms with diameters typically between 1 and 100 nm
Active sites	All atoms contribute to catalytic activity	Only surface atoms contribute to catalytic activity
Reactivity	High reactivity due to high coordination number and strong metal–support interactions	Lower reactivity due to reduced coordination number and weaker metal–support interactions
Stability	High stability due to uniformity and strong metal–support interactions	Lower stability due to the presence of defects and surface reconstruction
Selectivity	High selectivity due to precise control over active sites	Lower selectivity due to random distribution of active sites
Toxicity	Lower toxicity due to the absence of bulk metal	Higher toxicity due to the presence of bulk metal
Durability	High durability due to uniformity and resistance to sintering	Lower durability due to the tendency to sinter and agglomerate
Synthesis	More challenging synthesis process	Relatively easier synthesis process

(continued)

Table 9.1 (continued)

Characteristic	Single-atom catalysts	Nanoparticles
Tunability	Easily tunable by changing support and coordination environment	Less tunable by changing support and size
Mass Transfer	Low mass transfer limitations due to small size and high dispersion	High mass transfer limitations due to larger size and lower dispersion
Efficiency	Higher catalytic efficiency due to maximum utilization of active sites	Lower catalytic efficiency due to incomplete utilization of active sites
Chemical activity	Higher chemical activity due to uniformity and high surface area to volume ratio	Lower chemical activity due to lower surface area to volume ratio
Electronic structure	More tunable electronic structure	Less tunable electronic structure
Precious metals	Enables use of lower cost and more abundant metals	Requires precious metals for high catalytic activity
Thermal stability	Higher thermal stability due to absence of defects and low thermal mass	Lower thermal stability due to the presence of defects and high thermal mass
Size distribution	Monodisperse, with precisely controlled size and distribution	Polydisperse, with varied sizes and distributions
Catalyst loading	Lower catalyst loading due to high efficiency and activity	Higher catalyst loading required for comparable activity
Poisoning	Resistant to poisoning due to unique atomic structure	Susceptible to poisoning due to larger size and higher surface area
Solvent effect	Sensitive to solvent effect due to exposed coordination sites	Less sensitive to solvent effect due to fewer exposed sites
Ligand effect	Ligand-free, enabling precise control over surface properties	Presence of ligands can affect surface properties
Mass production	Difficult to mass-produce due to complex synthesis process	Relatively easier to mass-produce
Heterogeneity	Homogeneous, with uniform and precisely controlled active sites	Heterogeneous, with varied and randomly distributed active sites
Support material	Less dependent on support material due to strong metal–support interactions	Dependent on support material for stability and activity
Environmentally friendly	More environmentally friendly due to lower metal usage and toxicity	Less environmentally friendly due to higher metal usage and toxicity
Scalability	Less scalable due to complex synthesis process and small size	More scalable due to simpler synthesis process and larger size
Redox chemistry	Facilitates redox chemistry due to unique atomic structure	Less conducive to redox chemistry due to larger size and limited coordination number

References

1. Chu S, Majumdar A (2012) Opportunities and challenges for a sustainable energy future. Nature. https://doi.org/10.1038/nature11475
2. Li J, Stephanopoulos MF, Xia Y (2020) Introduction: heterogeneous single-atom catalysis. Chem Rev. https://doi.org/10.1021/acs.chemrev.0c01097
3. Yang X, Liu Y, Ta HQ, Rezvani E, Zhang Y, Zeng M, Fu L, Bachmatiuk A, Luo J, Liu L et al (2021) Single-atom catalytic growth of crystals using graphene as a case study. npj 2D Mater Appl. https://doi.org/10.1038/s41699-021-00267-4
4. Sun R, Liao Y, Bai ST, Zheng M, Zhou C, Zhang T, Sels BF (2021) Heterogeneous catalysts for CO_2 hydrogenation to formic acid/formate: from nanoscale to single atom. Energy Environ Sci. https://doi.org/10.1039/d0ee03575k
5. Kaiser SK, Chen Z, Faust Akl D, Mitchell S, Pérez-Ramírez J (2020) Single-atom catalysts across the periodic table. Chem Rev. https://doi.org/10.1021/acs.chemrev.0c00576
6. Qiao B, Wang A, Yang X, Allard LF, Jiang Z, Cui Y, Liu J, Li J, Zhang T (2011) Single-atom catalysis of CO oxidation using Pt1/FeO_x. Nat Chem. https://doi.org/10.1038/nchem.1095
7. Back S, Kulkarni AR, Siahrostami S (2018) Single metal atoms anchored in two-dimensional materials: bifunctional catalysts for fuel cell applications. ChemCatChem. https://doi.org/10.1002/cctc.201800447
8. Zhang B, Fan T, Xie N, Nie G, Zhang H (2019) Versatile applications of metal single-atom @ 2D material nanoplatforms. Adv Sci. https://doi.org/10.1002/advs.201901787
9. Alarawi A, Ramalingam V, He JH (2019) Recent advances in emerging single atom confined two-dimensional materials for water splitting applications. Mater Today Energy. https://doi.org/10.1016/j.mtener.2018.10.014
10. Chen Y, Ji S, Chen C, Peng Q, Wang D, Li Y (2018) Single-atom catalysts: synthetic strategies and electrochemical applications. Joule. https://doi.org/10.1016/j.joule.2018.06.019
11. Liu JJ (2011) Advanced electron microscopy of metal-support interactions in supported metal catalysts. ChemCatChem. https://doi.org/10.1002/cctc.201100090
12. Zhang H, Liu G, Shi L, Ye J (2018) Single-atom catalysts: emerging multifunctional materials in heterogeneous catalysis. Adv Energy Mater. https://doi.org/10.1002/aenm.201701343
13. Van Hardeveld R, Hartog F (1969) The statistics of surface atoms and surface sites on metal crystals. Surf Sci. https://doi.org/10.1016/0039-6028(69)90148-4
14. Gates BC (1995) Supported metal clusters: synthesis, structure, and catalysis. Chem Rev. https://doi.org/10.1021/cr00035a003
15. Yang XF, Wang A, Qiao B, Li J, Liu J, Zhang T (2013) Single-atom catalysts: a new frontier in heterogeneous catalysis. Acc Chem Res. https://doi.org/10.1021/ar300361m
16. Xue Y, Huang B, Yi Y, Guo Y, Zuo Z, Li Y, Jia Z, Liu H, Li Y (2018) Anchoring zero valence single atoms of nickel and iron on graphdiyne for hydrogen evolution. Nat Commun. https://doi.org/10.1038/s41467-018-03896-4
17. Zhang C, Sha J, Fei H, Liu M, Yazdi S, Zhang J, Zhong Q, Zou X, Zhao N, Yu H et al (2017) Single-atomic ruthenium catalytic sites on nitrogen-doped graphene for oxygen reduction reaction in acidic medium. ACS Nano. https://doi.org/10.1021/acsnano.7b02148
18. Jackson C, Smith GT, Inwood DW, Leach AS, Whalley PS, Callisti M, Polcar T, Russell AE, Levecque P, Kramer D (2017) Electronic metal-support interaction enhanced oxygen reduction activity and stability of boron carbide supported platinum. Nat Commun. https://doi.org/10.1038/ncomms15802
19. Li J, Zhao Z, Ma Y, Qu Y (2017) Graphene and their hybrid electrocatalysts for water splitting. ChemCatChem. https://doi.org/10.1002/cctc.201700175
20. Choi CH, Park SH, Woo SI (2012) Binary and ternary doping of nitrogen, boron, and phosphorus into carbon for enhancing electrochemical oxygen reduction activity. ACS Nano. https://doi.org/10.1021/nn3021234
21. Wang X, Sun G, Routh P, Kim DH, Huang W, Chen P (2014) Heteroatom-doped graphene materials: syntheses, properties and applications. Chem Soc Rev. https://doi.org/10.1039/c4cs00141a

22. Deng J, Li H, Xiao J, Tu Y, Deng D, Yang H, Tian H, Li J, Ren P, Bao X (2015) Triggering the electrocatalytic hydrogen evolution activity of the inert two-dimensional MoS_2 surface via single-atom metal doping. Energy Environ Sci. https://doi.org/10.1039/c5ee00751h
23. Chen W, Pei J, He CT, Wan J, Ren H, Zhu Y, Wang Y, Dong J, Tian S, Cheong WC et al (2017) Rational design of single molybdenum atoms anchored on N-Doped carbon for effective hydrogen evolution reaction. Angew Chemie Int Ed. https://doi.org/10.1002/anie.201710599
24. Fei H, Dong J, Arellano-Jiménez MJ, Ye G, Dong Kim N, Samuel ELG, Peng Z, Zhu Z, Qin F, Bao J et al (2015) Atomic cobalt on nitrogen-doped graphene for hydrogen generation. Nat Commun. https://doi.org/10.1038/ncomms9668
25. Liu G, Robertson AW, Li MMJ, Kuo WCH, Darby MT, Muhieddine MH, Lin YC, Suenaga K, Stamatakis M, Warner JH et al (2017) MoS_2 monolayer catalyst doped with isolated Co atoms for the hydrodeoxygenation reaction. Nat Chem. https://doi.org/10.1038/NCHEM.2740
26. Yu Q, Lian J, Siriponglert S, Li H, Chen YP, Pei SS (2008) Graphene segregated on Ni surfaces and transferred to insulators. Appl Phys Lett. https://doi.org/10.1063/1.2982585
27. Zhang R, Wen S, Wang N, Qin K, Liu E, Shi C, Zhao N (2018) N-Doped Graphene modified 3D porous Cu current collector toward microscale homogeneous Li deposition for Li metal anodes. Adv Energy Mater. https://doi.org/10.1002/aenm.201800914
28. DeRita L, Resasco J, Dai S, Boubnov A, Thang HV, Hoffman AS, Ro I, Graham GW, Bare SR, Pacchioni G et al (2019) Structural evolution of atomically dispersed Pt catalysts dictates reactivity. Nat Mater. https://doi.org/10.1038/s41563-019-0349-9
29. Kwak JH, Hu J, Mei D, Yi CW, Kim DH, Peden CHF, Allard LF, Szanyi J (2009) Coordinatively unsaturated Al^{3+} centers as binding sites for active catalyst phases of platinum on γ-Al_2O_3. Science (80-.). https://doi.org/10.1126/science.1176745
30. Neitzel A, Figueroba A, Lykhach Y, Skála T, Vorokhta M, Tsud N, Mehl S, Ševčíková K, Prince KC, Neyman KM et al (2016) Atomically dispersed Pd, Ni, and Pt species in ceria-based catalysts: principal differences in stability and reactivity. J Phys Chem C. https://doi.org/10.1021/acs.jpcc.6b02264
31. Kropp T, Mavrikakis M (2019) Transition metal atoms embedded in graphene: how nitrogen doping increases CO oxidation activity. ACS Catal. https://doi.org/10.1021/acscatal.9b01944
32. Li H, Wang S, Sawada H, Han GGD, Samuels T, Allen CS, Kirkland AI, Grossman JC, Warner JH (2017) Atomic structure and dynamics of single platinum atom interactions with monolayer MoS_2. ACS Nano. https://doi.org/10.1021/acsnano.7b00796
33. Hejazi S, Mohajernia S, Osuagwu B, Zoppellaro G, Andryskova P, Tomanec O, Kment S, Zbořil R, Schmuki P (2020) On the controlled loading of single platinum atoms as a Co-catalyst on TiO_2 anatase for optimized photocatalytic H_2 generation. Adv Mater. https://doi.org/10.1002/adma.201908505
34. Liu JC, Tang Y, Wang YG, Zhang T, Li J (2018) Theoretical understanding of the stability of single-atom catalysts. Natl Sci Rev. https://doi.org/10.1093/nsr/nwy094
35. Li X, Yang X, Huang Y, Zhang T, Liu B (2019) Supported noble-metal single atoms for heterogeneous catalysis. Adv Mater. https://doi.org/10.1002/adma.201902031
36. Tang Y, Zhao S, Long B, Liu JC, Li J (2016) On the nature of support effects of metal dioxides MO_2 (M = Ti, Zr, Hf, Ce, Th) in single-atom gold catalysts: importance of quantum primogenic effect. J Phys Chem C. https://doi.org/10.1021/acs.jpcc.6b05338
37. Lin J, Wang A, Qiao B, Liu X, Yang X, Wang X, Liang J, Li J, Liu J, Zhang T (2013) Remarkable performance of Ir1/FeO_x single-atom catalyst in water gas shift reaction. J Am Chem Soc. https://doi.org/10.1021/ja408574m
38. Yang J, Li W, Wang D, Li Y (2020) Electronic metal-support interaction of single-atom catalysts and applications in electrocatalysis. Adv Mater. https://doi.org/10.1002/adma.202003300
39. Yang L, Cheng D, Xu H, Zeng X, Wan X, Shui J, Xiang Z, Cao D (2018) Unveiling the high-activity origin of single-atom iron catalysts for oxygen reduction reaction. Proc Natl Acad Sci USA. https://doi.org/10.1073/pnas.1800771115
40. Liu JC, Wang YG, Li J (2017) Toward rational design of oxide-supported single-atom catalysts: atomic dispersion of gold on ceria. J Am Chem Soc. https://doi.org/10.1021/jacs.7b01602

41. Neyman KM, Inntam C, Matveev AV, Nasluzov VA, Rösch N (2005) Single D-metal atoms on F_s and F_s^+ defects of MgO(001): a theoretical study across the periodic table. J Am Chem Soc. https://doi.org/10.1021/ja052437i
42. Yang L, Shi L, Wang D, Lv Y, Cao D (2018) Single-Atom cobalt electrocatalysts for foldable solid-state Zn-air Battery. Nano Energy. https://doi.org/10.1016/j.nanoen.2018.06.023
43. Zhai Y, Pierre D, Si R, Deng W, Ferrin P, Nilekar AU, Peng G, Herron JA, Bell DC, Saltsburg H et al (2010) Alkali-stabilized Pt-OH_x species catalyze low-temperature water-gas shift reactions. Science (80-.). https://doi.org/10.1126/science.1192449.
44. Kyriakou G, Boucher MB, Jewell AD, Lewis EA, Lawton TJ, Baber AE, Tierney HL, Flytzani-Stephanopoulos M, Sykes ECH (2012) Isolated metal atom geometries as a strategy for selective heterogeneous hydrogenations. Science (80-.). https://doi.org/10.1126/science.1215864
45. Du Z, Chen X, Hu W, Chuang C, Xie S, Hu A, Yan W, Kong X, Wu X, Ji H et al (2019) Cobalt in nitrogen-doped graphene as single-atom catalyst for high-sulfur content lithium-sulfur batteries. J Am Chem Soc. https://doi.org/10.1021/jacs.8b12973
46. Luo X, Wei X, Zhong H, Wang H, Wu Y, Wang Q, Gu W, Gu M, Beckman SP, Zhu C (2020) Single-atom Ir-anchored 3D amorphous NiFe nanowire@nanosheets for boosted oxygen evolution reaction. ACS Appl Mater Interfaces. https://doi.org/10.1021/acsami.9b17476
47. Cao L, Luo Q, Chen J, Wang L, Lin Y, Wang H, Liu X, Shen X, Zhang W, Liu W et al (2019) Dynamic oxygen adsorption on single-atomic ruthenium catalyst with high performance for acidic oxygen evolution reaction. Nat Commun 10(1). https://doi.org/10.1038/s41467-019-12886-z
48. Li P, Wang M, Duan X, Zheng L, Cheng X, Zhang Y, Kuang Y, Li Y, Ma Q, Feng Z et al (2019) Boosting oxygen evolution of single-atomic ruthenium through electronic coupling with cobalt-iron layered double hydroxides. Nat Commun 10(1):1–11. https://doi.org/10.1038/s41467-019-09666-0
49. Wang Q, Huang X, Zhao ZL, Wang M, Xiang B, Li J, Feng Z, Xu H, Gu M (2020) Ultrahigh-loading of Ir single atoms on NiO matrix to dramatically enhance oxygen evolution reaction. J Am Chem Soc. https://doi.org/10.1021/jacs.9b12642
50. Zhang J, Liu J, Xi L, Yu Y, Chen N, Sun S, Wang W, Lange KM, Zhang B (2018) Single-atom Au/NiFe layered double hydroxide electrocatalyst: probing the origin of activity for oxygen evolution reaction. J Am Chem Soc 140(11):3876–3879. https://doi.org/10.1021/jacs.8b00752
51. Babu DD, Huang Y, Anandhababu G, Wang X, Si R, Wu M, Li Q, Wang Y, Yao J (2019) Atomic Iridium@cobalt nanosheets for dinuclear tandem water oxidation. J Mater Chem A. https://doi.org/10.1039/c9ta01088b
52. Suen NT, Hung SF, Quan Q, Zhang N, Xu YJ, Chen HM (2017) Electrocatalysis for the oxygen evolution reaction: recent development and future perspectives. Chem Soc Rev 46(2):337–365. https://doi.org/10.1039/c6cs00328a
53. McCrory CCL, Jung S, Ferrer IM, Chatman SM, Peters JC, Jaramillo TF (2015) Benchmarking hydrogen evolving reaction and oxygen evolving reaction electrocatalysts for solar water splitting devices. J Am Chem Soc. https://doi.org/10.1021/ja510442p
54. Lai WH, Zhang LF, Hua WB, Indris S, Yan ZC, Hu Z, Zhang B, Liu Y, Wang L, Liu M et al (2019) General π-electron-assisted strategy for Ir, Pt, Ru, Pd, Fe, Ni single-atom electrocatalysts with bifunctional active sites for highly efficient water splitting. Angew Chemie Int Ed. https://doi.org/10.1002/anie.201904614
55. Zhou M, Dick JE, Bard AJ (2017) Electrodeposition of isolated platinum atoms and clusters on bismuth—characterization and electrocatalysis. J Am Chem Soc. https://doi.org/10.1021/jacs.7b10646
56. Li Y, Sun Y, Qin Y, Zhang W, Wang L, Luo M, Yang H, Guo S (2020) Recent advances on water-splitting electrocatalysis mediated by noble-metal-based nanostructured materials. Adv Energy Mater 10(11):1–20. https://doi.org/10.1002/aenm.201903120
57. Jiang K, Liu B, Luo M, Ning S, Peng M, Zhao Y, Lu YR, Chan TS, de Groot FMF, Tan Y (2019) Single platinum atoms embedded in nanoporous cobalt selenide as electrocatalyst for accelerating hydrogen evolution reaction. Nat Commun. https://doi.org/10.1038/s41467-019-09765-y

58. Zhang L, Han L, Liu H, Liu X, Luo J (2017) Potential-cycling synthesis of single platinum atoms for efficient hydrogen evolution in neutral media. Angew Chemie Int Ed. https://doi.org/10.1002/anie.201706921
59. Yin XP, Wang HJ, Tang SF, Lu XL, Shu M, Si R, Lu TB (2018) Engineering the Coordination environment of single-atom platinum anchored on graphdiyne for optimizing electrocatalytic hydrogen evolution. Angew Chemie Int Ed. https://doi.org/10.1002/anie.201804817
60. Park J, Lee S, Kim HE, Cho A, Kim S, Ye Y, Han JW, Lee H, Jang JH, Lee J (2019) Investigation of the support effect in atomically dispersed Pt on WO_{3-x} for utilization of Pt in the hydrogen evolution reaction. Angew Chemie Int Ed. https://doi.org/10.1002/anie.201908122
61. Zhang J, Xu X, Yang L, Cheng D, Cao D (2019) Single-atom Ru doping induced phase transition of MoS_2 and S vacancy for hydrogen evolution reaction. Small Methods. https://doi.org/10.1002/smtd.201900653
62. Ramalingam V, Varadhan P, Fu HC, Kim H, Zhang D, Chen S, Song L, Ma D, Wang Y, Alshareef HN et al (2019) Heteroatom-mediated interactions between ruthenium single atoms and an MXene support for efficient hydrogen evolution. Adv Mater. https://doi.org/10.1002/adma.201903841
63. Chen CH, Wu D, Li Z, Zhang R, Kuai CG, Zhao XR, Dong CK, Qiao SZ, Liu H, Du XW (2019) Ruthenium-Based single-atom alloy with high electrocatalytic activity for hydrogen evolution. Adv Energy Mater. https://doi.org/10.1002/aenm.201803913
64. Pu Z, Amiinu IS, Cheng R, Wang P, Zhang C, Mu S, Zhao W, Su F, Zhang G, Liao S et al (2020) Single-Atom catalysts for electrochemical hydrogen evolution reaction: recent advances and future perspectives. Nano-Micro Lett. https://doi.org/10.1007/s40820-019-0349-y
65. Luo Z, Ouyang Y, Zhang H, Xiao M, Ge J, Jiang Z, Wang J, Tang D, Cao X, Liu C et al (2018) Chemically activating MoS_2 via spontaneous atomic palladium interfacial doping towards efficient hydrogen evolution. Nat Commun. https://doi.org/10.1038/s41467-018-04501-4
66. Guan Y, Feng Y, Wan J, Yang X, Fang L, Gu X, Liu R, Huang Z, Li J, Luo J et al (2018) Ganoderma-like MoS_2/NiS_2 with single platinum atoms doping as an efficient and stable hydrogen evolution reaction catalyst. Small. https://doi.org/10.1002/smll.201800697
67. Fan L, Liu PF, Yan X, Gu L, Yang ZZ, Yang HG, Qiu S, Yao X (2016) Atomically isolated nickel species anchored on graphitized carbon for efficient hydrogen evolution electrocatalysis. Nat Commun. https://doi.org/10.1038/ncomms10667
68. Cao L, Luo Q, Liu W, Lin Y, Liu X, Cao Y, Zhang W, Wu Y, Yang J, Yao T et al (2019) Identification of single-atom active sites in carbon-based cobalt catalysts during electrocatalytic hydrogen evolution. Nat Catal 2(2):134–141. https://doi.org/10.1038/s41929-018-0203-5
69. Chen W, Pei J, He CT, Wan J, Ren H, Wang Y, Dong J, Wu K, Cheong WC, Mao J et al (2018) Single tungsten atoms supported on MOF-derived N-doped carbon for robust electrochemical hydrogen evolution. Adv Mater. https://doi.org/10.1002/adma.201800396
70. Dong G, Fang M, Wang H, Yip S, Cheung HY, Wang F, Wong CY, Chu ST, Ho JC (2015) Insight into the electrochemical activation of carbon-based cathodes for hydrogen evolution reaction. J Mater Chem A. https://doi.org/10.1039/c5ta02551f
71. He T, Zhang C, Du A (2019) Single-atom supported on graphene grain boundary as an efficient electrocatalyst for hydrogen evolution reaction. Chem Eng Sci. https://doi.org/10.1016/j.ces.2018.03.028
72. Zhang Z, Feng C, Liu C, Zuo M, Qin L, Yan X, Xing Y, Li H, Si R, Zhou S et al (2020) Electrochemical deposition as a universal route for fabricating single-atom catalysts. Nat Commun. https://doi.org/10.1038/s41467-020-14917-6
73. Liu M, Chen H, Tang X, Liu H, Tu B, Guo W, Zheng Y, Liu Y, Tang Y, He R et al (2022) Synthesis of uranium single atom from radioactive wastewater for enhanced water dissociation and hydrogen evolution. Small. https://doi.org/10.1002/smll.202107444
74. Zheng Y, Jiao Y, Zhu Y, Cai Q, Vasileff A, Li LH, Han Y, Chen Y, Qiao SZ (2017) Molecule-level g-C_3N_4 coordinated transition metals as a new class of electrocatalysts for oxygen electrode reactions. J Am Chem Soc 139(9):3336–3339. https://doi.org/10.1021/jacs.6b13100
75. Liu D, Ding S, Wu C, Gan W, Wang C, Cao D, Rehman ZU, Sang Y, Chen S, Zheng X et al (2018) Synergistic effect of an atomically dual-metal doped catalyst for highly efficient oxygen evolution. J Mater Chem A. https://doi.org/10.1039/c8ta00550h

76. Fei H, Dong J, Feng Y, Allen CS, Wan C, Volosskiy B, Li M, Zhao Z, Wang Y, Sun H et al (2018) General synthesis and definitive structural identification of MN_4C_4 single-atom catalysts with tunable electrocatalytic activities. Nat Catal. https://doi.org/10.1038/s41929-017-0008-y
77. Zhu C, Shi Q, Feng S, Du D, Lin Y (2018) Single-atom catalysts for electrochemical water splitting. ACS Energy Lett 3(7):1713–1721. https://doi.org/10.1021/acsenergylett.8b00640
78. Zhang L, Jia Y, Gao G, Yan X, Chen N, Chen J, Soo MT, Wood B, Yang D, Du A et al (2018) Graphene defects trap atomic Ni species for hydrogen and oxygen evolution reactions. Chem. https://doi.org/10.1016/j.chempr.2017.12.005
79. Wang J, Ge X, Liu Z, Thia L, Yan Y, Xiao W, Wang X (2017) Heterogeneous electrocatalyst with molecular cobalt ions serving as the center of active sites. J Am Chem Soc. https://doi.org/10.1021/jacs.6b10307
80. Chen P, Zhou T, Xing L, Xu K, Tong Y, Xie H, Zhang L, Yan W, Chu W, Wu C et al (2017) Atomically dispersed iron-nitrogen species as electrocatalysts for bifunctional oxygen evolution and reduction reactions. Angew Chemie Int Ed. https://doi.org/10.1002/anie.201610119
81. Tao L, Lin CY, Dou S, Feng S, Chen D, Liu D, Huo J, Xia Z, Wang S (2017) Creating coordinatively unsaturated metal sites in metal-organic-frameworks as efficient electrocatalysts for the oxygen evolution reaction: insights into the active centers. Nano Energy. https://doi.org/10.1016/j.nanoen.2017.09.055
82. Zhang Y, Wu C, Jiang H, Lin Y, Liu H, He Q, Chen S, Duan T, Song L (2018) Atomic Iridium incorporated in cobalt hydroxide for efficient oxygen evolution catalysis in neutral electrolyte. Adv Mater. https://doi.org/10.1002/adma.201707522

Chapter 10
Emerging Electrocatalytic Strategies for Hydrogen Production from Water

10.1 Introduction

In the foreseeable future, the energy demand will continue to increase along with population growth and economic expansion. Natural gas, coal, and oil are just a few examples of the fossil fuels that comprise most current energy sources. However, the exhaustion of nonrenewable fossil fuels and the negative environmental effects of fossil fuel utilization has sparked vigorous research into novel technologies for storing and converting clean and sustainable energy resources, like wind and solar. Electrocatalytic hydrogen generation from water electrolysis has been regarded as an appealing approach to the intermittent nature of renewable energy. Traditional water electrolysis is often conducted under acidic electrolytic media with a proton exchange membrane (PEM) or in basic conditions using a diaphragm to reduce charge transport losses during electrochemical processes (Fig. 10.1a) [1].The two half-reactions for splitting water may be expressed differently based on reaction conditions (Fig. 10.1b). Electrochemical water splitting requires a 1.23 V thermodynamic potential under ideal conditions, equivalent to a 237.1 kJ/mol energy input of Gibbs free energy. Unfortunately, to achieve large current density, the sluggish kinetics of the H2 evolution reaction (HER) and the O_2 evolution reaction (OER) require high overpotentials, leading to relatively low energy conversion efficiencies. Hence, in practical electrolyzers, the input potential of water splitting is substantially higher than 1.23 V. Moreover, electrocatalysts are frequently required, like state-of-the-art composites for acidic electrolytes (such as Pt for HER and RuO_2/ IrO_2 for OER) and transition metal-based composites for alkaline media (such as Ni electrodes for HER and stainless steel composites for OER). The inability to produce large amounts of H_2 through water electrolysis is due to the prohibitive price, limited supply, inferior stability of noble metal-based catalysts, and their mediocre performance [2]. Indeed, significant attempts have been made to develop transition metal catalysts that perform better for HER and OER. Chalcogenides, borides, metal alloys,

T. u. Haq and Y. Haik, *Electrochemical Water Splitting*, Materials Horizons: From Nature to Nanomaterials, https://doi.org/10.1007/978-981-99-9860-9_10

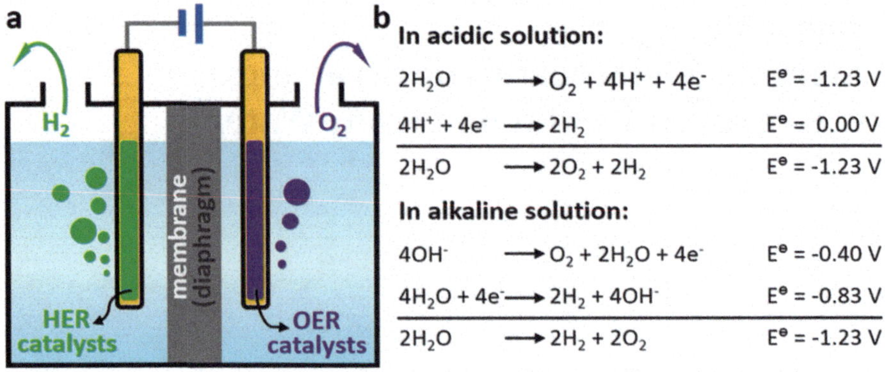

Fig. 10.1 **a** Conventional water electrolysis scheme and **b** water electrolysis in acidic and basic medium. Reproduced with permission from Ref. [1]. (c) 2018. American Chemical Society

carbides, phosphides, and nitrides have been developed as inexpensive and competitive electrocatalysts for HER in acidic media [3]. Though lots of transition metal-based OER composites of (oxy)hydroxides and metal oxides have been explored with outstanding performance in harsh alkaline conditions. Nonetheless, several challenges still exist: (1) Since HER and OER electrocatalysts are typically created under different conditions, the common methodologies for developing nonprecious catalysts frequently result in their incompatibility or problematic integration [4]. (2) Due to the close coupling between OER and HER in conventional water electrolysis, an explosive mixture of gases H_2/O_2 could occur [5]. In addition to posing safety concerns, the gas crossover lowers energy efficiency because O_2 would be converted back to water at the cathode. Meanwhile, the presence of electrocatalysts and gas mixtures may produce reactive oxygen species, which can weaken the membrane and cause premature failure of the device [6]. (3) Subsequently, OER having a four-electron mechanism requires greater overpotential concerning H2 evolution reaction to attain a similar current density. Ironically, O_2, the OER product, is not that much valuable [7]. (4) Finding affordable H_2 storage and delivery systems for on-demand consumption is still difficult.

10.2 Conventional Water Electrolysis

Water electrolysis using conventional proton exchange membrane (PEM) has been known for almost sixty years and was explored by General Electric, in which the electrolyte is a solid sulfonated polystyrene membrane. This approach is also entitled polymer electrolyte membrane electrolysis. The proton exchange membrane (PEM) is an electrolyte and a gas separator. Deionized water is only introduced into the cell during electrolysis, with no electrochemical additives. The setup of conventional water electrolysis is shown in Fig. 10.1. The acidic and alkaline general mechanism reactions are also offered for water electrolysis. PEM electrolysis has many benefits, including a wide operating temperature range (20–80 °C) high current density, improved energy efficiency, low gas permeability, and ease of handling and maintenance. Although Pt, Ru, and Ir are the most commonly used composites for PEM electrolysis, the high price of PEM electrolysis makes them a comparatively more excellent investment setup than alkaline electrolysis. In this regard, scientists have worked incredibly hard to lower the cost and increase the intrinsic catalytic performance of PEM catalysts. During electrolysis via PEM, a dilemma for HER is developing stable, efficient, high-activity electrocatalysts for the cathode. Presently, Pt and Pd-based electrocatalysts have been regarded as ideal composites for PEM electrolyzers at the cathode for excellent electrocatalytic performance for hydrogen evolution reactions in acidic media [8]. Unfortunately, such electrocatalysts' high price and limited reserves prevent their widespread use. Due to this, significant effort has been put into investigating alternative strategies to minimize the mass loading of Pt and the utilization of Pt-free composites. Lim and coworkers developed a Pt@N-containing carbon core–shell composite on carbon nanofibers (Pt@CS/CNF), which presented outstanding durability and electrochemical performance [9]. They investigated that pyridinic and graphitic N on the shell of carbon might not only deliver more excellent electrochemical active centers for H_2 generation but also hamper the dynamic Pt core from agglomeration and dissolution, with improved catalytic stability. Several transition metals have recently proven to be excellent cathodic electrodes for HER. Nørskov and coworkers explored MoS_2 for HER activity and developed an MEA based on a Nafon PEM exhibiting 175 mV overpotential at 10 mA cm^{-2} lower current density. [10]

Because of OER's crucial function in water electrolysis, the energy sector has made significant efforts to utilize excellent OER catalysts. The acidic and intensely oxidative conditions make it difficult to create highly-active and reliable composites for OER in PEM electrolyzers. To date, Ru and Ir-based composites are the best OER catalysts in the presence of acidic media. However, Ru shows great OER performance but has unsatisfactory stability concerning Ir-based electrocatalysts. Due to its low overpotentials and exceptional stability, Ir metal has attracted enormous interest as an essential composite for OER in acidic conditions. The shape, crystal structure, particle size, and substrate materials are primarily responsible for efficient catalytic activity. Several Ir-metal electrocatalysts have been investigated to enhance OER performance and lower the amount of noble metal. Transition metal-based catalysts have also been explored along with precious metal composites [11]. Although these results have significantly contributed to hydrogen production, they still face several challenges.

10.2.1 Challenges of Conventional Water Electrolysis

Polymer electrolyte membrane water electrolysis (PEM-WE) has many benefits, including greater energy densities, facile nature, impulsive response, compact design, ultrapure H_2 generation, which may be pressurized electrochemically, and a small footprint. The concept of PEM-WE was first demonstrated and investigated in the 1960s. Since then, much effort has been put into improving the various PEM WE components, and this approach is getting close to the commercial markets [12]. The acidity of PEM WE makes it difficult to utilize on a large scale; hence noble metals like Ir, Pt, or Ru must be used as electrocatalysts. In addition, acidic conditions are desirable since reactant protons concentration is greater. Another barrier is the high price of the polymeric membrane. Given that the power is generated using renewable sources, the capital investment cost (CAPEX) for a PEM WE system is $1500 per kWh (kW electricity input), and the price per kilogram of hydrogen is $7.1 [13]. Comparatively, the cost of H_2 by steam reforming of methane is only $1.40, and the expected break-even year for renewable PEM hydrogen generation based on survey is around 2033. The same report emphasizes that PEM's direct cost is its electricity use [14]. This is closely related to the electrode's overpotential or ability to generate particular current densities, which is necessary for proficient water splitting. Platinum group metals (PGMs) are regarded as the most effective and stable materials for electrocatalysts in PEM electrolyzers; nonetheless, the issue is how much it will cost. For instance, the entire demand for platinum in 2017 was 2.2 105 kg, whereas the annual global Pt production was 1.7 105 kg. The annual Pt production barely satisfied the whole demand if recycled Pt is considered. As a result, the total demand for PGMs will significantly increase through the widespread installation of Pt-Ir-based PEM electrolyzers. For example, a Terra Watt (TW) hydrogen production system needs 10 and 0.5 years of annual global Ir and Pt production [15]. The fact that Ir is often created as a small by-product of Pt must also be considered [16]. In other words, Pt's production rate also affects Ir's annual output. The increment in demand for Ir will also raise costs because of reliance on Pt mining. In this regard, lots of transition metal-based composites have been explored for the HER in basic, neutral, and acidic media, which comprises metal phosphides metal sulfides, chalcogenides, metal alloys (https://www.nature.com/articles/nmat1752), as well as carbon-based composites substituted by different heteroatoms [17, 18]. Most of them exhibit enhanced efficiencies and outstanding endurance in the presence of an acidic environment whereas many of them are unstable requiring greater onset overpotentials [19]. On the OER side, it is considerably more difficult, where the complex four-electron mechanism that generates protons and oxygen requires larger overpotentials. This reaction is considered the bottleneck in overall water electrolysis. Only precious metal oxides like RuO_2 and IrO_2 are proficient composites for the OER in the presence of acidic electrolytic media, but the RuO_2 rapidly deactivates because of instability [20]. Hence, the main challenge to PEM-based water electrolysis is the need for affordable alternatives to IrO_2.

According to the Nernst equation, HER is favorable in acidic environments, while OER proceeds more quickly in an alkaline medium. As a result, most research and development efforts for OER and HER electrocatalysts are conducted in basic and acidic electrolytes, respectively. But, OER and HER composites are chosen to perform in a similar electrolytic media to accomplish the water-splitting process. The different development circumstances for HER and OER catalysts are expected to lead to challenging integration and subpar activity for overall water electrolysis [20]. In the end, under an optimistic situation, PEM water splitting is predicted to achieve cost parity with fossil fuel-produced H2 around 2030. Presently, the polymer exchange membrane, the precious metal composites, and the greater water electrolysis overpotentials are responsible for the high cost of hydrogen from PEM-WE. A bifunctional electrocatalyst with improved stability and greater electrocatalytic performance in a single electrolytic medium, either acidic or alkaline, is highly desired.

10.3 Approaches to Overcome Conventional Approach

Considering such considerations, there is a much need to fabricate earth-abundant bifunctional catalysts for OER and HER with better activity in the same electrolytic media to attain significant overall water splitting because of the benefits of simplifying the electrolyzer systems and reduced overall cost. The overall efficiency of water electrolysis would be better in primary media than acidic media because of the four-electron process of OER; its overpotential is often significantly higher than that of HER to deliver the same current density. Moreover, the majority of nonprecious OER composites cannot endure highly acidic electrolytes [1].

The different types of strategies have been designed with various advantages to overcome all the aspects of conventional water electrolysis, which utilizes a proton exchange membrane, complicates the whole setup, and increases the average cost. Some other approaches are explained below.

10.3.1 Nonconventional/Overall Water Electrolysis

During water splitting, water reduction at the cathodic side for H_2 generation and oxidation of water at the anode for O_2 generation are kinetically sluggish, leading to lower energy efficiencies. Hence, there is a demand for electrocatalysts to fasten the kinetics of the reaction. Electrocatalysis of water is the major challenge for both PEM and alkaline media electrolysis. Despite significant advancements in recent years, catalyst performance still needs to improve regarding catalytic performance and stability. Water splitting, it's a two-half-cell electrochemical reaction. The two electrochemical reactions include oxygen evolution reaction (OER) at the anode and cathodic hydrogen evolution reaction. Oxidation occurs at the anode and generates O_2. On the evolution of each O_2 molecule, a four-electron transfer reaction occurs. OER is a complicated process. Subsequently, because of sluggish reaction kinetics, there is a high potential to encounter the activation barrier, which is primarily responsible for the inadequacy of electrolytic cells. As a result, there is a lot of room for improving efficiency in the case of the OER by using an appropriate catalyst. A proficient OER electrocatalyst should work in a broader range of pH, with maximum current densities and minimum overpotentials, over a prolonged period with maximum stability and must be synthesized from transition metal-based materials by an inexpensive and straightforward technique. In alkaline conditions, the OER is usually written as:

$$4OH^- \rightarrow 2O_2 + 2H_2O + 4e^- \quad E_{0,\,rev} = 0.401\ V\ (25^\circ\ C,\ pH\ = 14)$$

Similar to OER, HER is quicker in acidic material due to the maximum number of protons. The setup is simpler by using the same electrolyte for the OER and HER reactions, which also has the highest electrocatalytic activity. The pictorial presentation of overall water electrolysis by the bifunctional catalyst is shown in Fig. 10.1a. In the case of HER in alkaline media, the reaction is given as follows:

$$2H_2O + 2e^- \rightarrow H_2(g) + 2OH(aq)$$

Many electrocatalysts have been reported up to this point to carry out both of these reactions utilizing the same electrolyte. The electrodeposition approach has been used to synthesize nickel and cobalt sulfide-based composites [21]. Similarly, potentiodynamic electrodeposition has been conducted to fabricate Co–P film through successive linear scanning in −0.3 to −1.0 V window range using Ag/AgCl as reference electrode at 5 mV s^{-1} scan rate and 500 rpm rotation speed by stirring in the stock solution which comprises of CH_3COONa(100 mM), CoSO4(50 mM) and 500 mM NaH_2PO_2 (500 mM) in water [22]. Scanning electron microscopy (SEM) morphology and the cross-sectional image of electrocatalyst Co–P exposed almost complete Co–P coverage on a substrate of copper with 1–3 μm thickness. The polarization curve via linear sweep voltammetry (LSV) has been experimented to explore the electrochemical OER, HER, and overall water electrolysis activity of electrode in 1.0 M alkaline KOH. Co–P has exhibited -50 mV onset overpotential versus the standard reversible hydrogen electrode (RHE), which is more negative to Pt/C, which has 0 mV potential versus RHE. Co-P has greater current density than Pt/C beyond -167 mV versus RHE, emphasizing its incredible HER performance. The composite Co–P only needs 94 mV overpotential to deliver 10 mA cm^{-2} current density and 42 mV dec^{-1} Tafel slope with faster kinetics, demonstrating a remarkable HER activity under basic media. Even after 24 h of electrolysis (post-HER Co–P), the Co-P morphology confirms the retention of its catalytic structure, indicating long-term HER stability for H_2 generation. Consequently, its OER activity has been evaluated in a similar alkaline electrolyte of 1.0 M KOH solution. The catalyst Co–P attains 345 mV overpotential to deliver 10 mA cm^{-2} current density, much lower than IrO_2 (benchmark) and most earth-abundant OER composites. Compared to the homogeneous morphology of the new and post-HER catalyst, the morphology of post-OER catalyst Co–P showed aggregation of particles. Many other characterizations exposed both P and Co oxidation during oxygen evolution [22]. Thus, the in-situ produced cobalt (oxy)hydroxides and oxides on the bulk are considered the real OER active centers. Finally, the overall electrochemical activity of Co–P as both cathode and anode has also been experimented with using alkaline media in 1.0 M KOH. The catalyst Co–P validated 1.744 cell voltage to deliver a higher 100 mA cm^{-2} current density, much lower than IrO_2 and Pt/C couple. Analogous bifunctional performance for overall water splitting can be obtained from electrodeposited Ni–P film as well. A composite with ideal reaction interfaces is essential for both triphase OER and HER (liquid, solid, gas) to speed up electrocatalytic kinetics and reduce the necessary overpotential. In this situation, it is anticipated that catalysts with a large specific surface area, porous infrastructure, and enhanced accessibility to active centers will improve electrocatalytic performance [23]. Therefore, porous three-dimensional (3D) superstructures of Ni_2P have been synthesized on Ni foam substrate (Ni_2P/Ni/NF) as an earth-abundant bifunctional electrode for overall water electrolysis [23]. The electrocatalyst Ni_2P/Ni/NF has been synthesized by template-free electrodeposition of metallic porous microsphere arrays of nickel on substrate Ni foam (Ni/NF) followed by phosphorization (Fig. 10.2a). Its urchin-like structure and hierarchical macroporosity are confirmed by SEM images. (Fig. 10.2b). The catalyst Ni_2P/Ni/NF has a cell voltage of 1.49 V at 10 mA cm^{-2}, which is very low compared to 1.64 V of composite Co–P and 1.67 V of Ni–P. The catalyst Ni2P/Ni/NF performance is attributed to its distinctive superstructure. Enhanced accessibility of active centers is facilitated by the 3D hierarchically macroporous structure with interconnected porous structures, which also buffers electrolytes to increase gas discussion and transportation of ions [24]. NiSx has also been explored as the bifunctional and efficient electrode for overall water electrolysis. Ni/NF sulfurization at high temperatures can easily generate three-dimensional hierarchically porous superstructures of NiS_x (hp-NiS_x, Fig. 10.2c). The composite hp-NiS_x attains mesoporous and macropore features, compared to Ni_2P/Ni/NF, along with singular macropores. The electrocatalyst hp-NiSx only acquires 1.47 V voltage to attain 10 mA cm^{-2} current density for overall water splitting. Taking advantage from controlled pore texture, enormous specific surface area, and modular inorganic and organic components, the metal–organic frameworks (MOFs) are regarded as favorable candidates to attain competent electrocomposites [25]. The MOF-derived route has been reported to synthesize mesoporous nanopolyhedrons of Co–P/NC composing of nanoparticles of CoP_x embedded in matrices of N-doped carbon as efficient bifunctional catalysts for OER as well as HER process [25]. The catalyst Co–P/NC has been synthesized by direct carbonization of Co-based zeolitic imidazolate frameworks (ZIF-67) and then proceeded by phosphorization (Fig. 10.2d). The morphological structure of Co–P/NC through SEM indicates inheritance of its polyhedron structure of ZIF-67 (Fig. 10.2e). Elemental mapping outcomes endorse the presence of certain elements C, P, Co and N in catalyst Co–P/NC. This mapping also indicates the localization of P with Co, which is inconsistent within the CoPx structure. The Co–P/NC has a greater surface area of around 183 m^2g^{-1} compared to hp-NiSx, which is 11 m^2g^{-1}. [26] CoP/NC composite attains 165 $mAcm^{-2}$ current density at 2.0 V for overall water electrolysis and, as expected, displays outstanding OER and HER performance. The bifunctional composites already reported in the literature at various current densities and using the Pt/C and RuO_2 pair as a reference, Fig. 10.2f compares the cell voltages of overall water splitting. This bifunctional electrode approach for overall water electrolysis may extend to other noble and transition metal-based derivatives and nanocarbons with no metal.

Lots of electrocatalysts have been utilized as bifunctional composites. For example, nanosheets of double-layered hydroxide of NiFe catalyst with sufficient oxygen vacancies have been confirmed. These effectively facilitate the adsorption of intermediates during the catalytic reaction to attain 250 mV OER overpotentials and 240 mV HER overpotentials at 10 mA cm^{-2} [27]. The 3D nanosheets of Fe-doped $Ni(OH)_2$ may modulate the lattice $Ni(OH)_2$ electronic structure and stimulate HER and OER intrinsic catalytic activity to achieve a lower 271 mV overpotential for

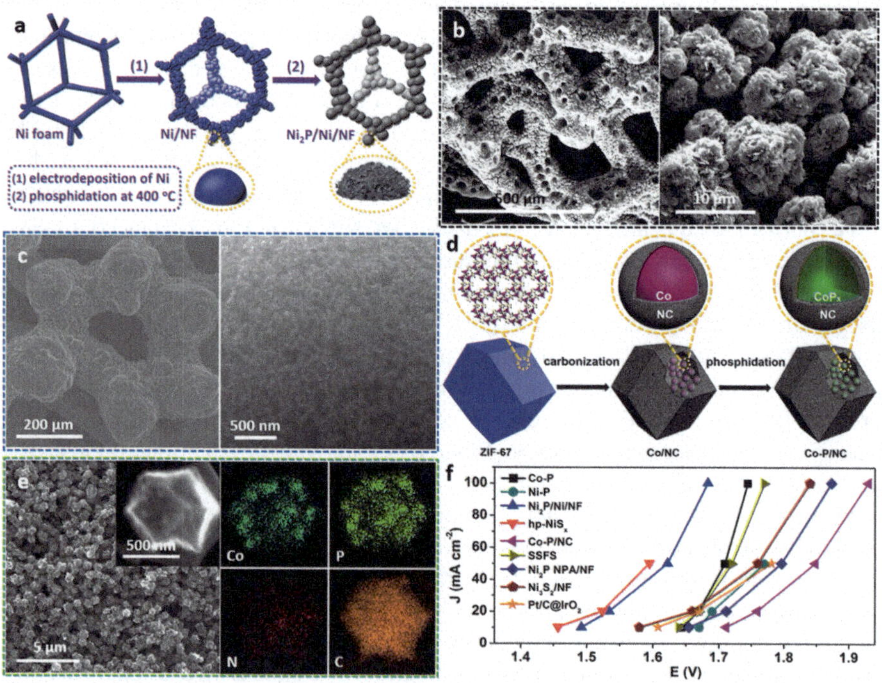

Fig. 10.2 **a** Scheme for preparing Ni2P/Ni/NF. SEM images of **b** Ni2P/Ni/NF and **c** hp-NiS_x show the surface morphology and structure of the two electrocatalysts. **d** Scheme for preparing Co–P/NC. SEM and element mapping images of Co–P/NC are shown in (**e**), which provide detailed information about the morphology and elemental composition of the electrocatalyst. **f** A comparison of cell voltages required to achieve benchmark current densities for various bifunctional electrocatalysts of overall water splitting is presented. This comparison allows for the evaluation of the electrocatalytic performance of Ni_2P/Ni/NF, hp-$_{NiSx}$, Co–P/NC, and other electrocatalysts. Reproduced with permission from Ref. [1]. (c) 2018. American Chemical Society

the OER and 160 mV HER overpotential at 10 mA cm^{-2} [28]. Despite these noteworthy developments, no information is still available regarding the simultaneous OER and HER activities. Meeting the criteria for being a highly active catalyst is still difficult. Fabricating highly efficient and active electrocatalysts for water electrolysis requires maximum intrinsic catalytic sites and an enormous number of active centers with highly active surface area. Extraordinarily active centers for the OER and HER are only located in the redox-active near-surface region of ~1/3 electrocatalyst atomic layer which clarifies that composite beneath the 3rd atomic layer is "inactive" or "dead" [29]. The "inactive area" of such composites has no contributions to HER and OEr reaction, which results in a minimum utilization ratio of catalyst, based on such facts that 2D ultrathin material may depict enormous active centers and excellently escape "dead volume" during HER and OER conditions, fabricating distinctive porous structures constituting 2D ultrathin active composites on conductive 3D substrates with maximum active surface area as well as a greater active phase which may reduce HER and OEr overpotential, alleviating the overall water-splitting challenges.

10.3.2 Hybrid Water Electrolysis

Following the abovementioned approach of overall water splitting, OER will proceed with an oxidation reaction to complete the electrocatalysis cycle, demanding large overpotential, as product O_2 has not high price value. The original purpose of the OER oxidation reaction is to produce and transport electrons toward the cathode for H_2 production, which is a real and desired water electrolysis product. During the whole response, oxygen is just a side product; nonetheless, it's responsible for the maximum energy loss during catalysis.

Hybrid water electrolysis has been considered a unique approach to overcome such challenges in which the OER process is substituted with an alternative oxidation reaction that is receiving growing research interest (Fig. 10.3) [30]. The hybrid water electrolysis system resembles the electrolytic cell used in the chloralkali industry, where H_2 generation occurs at the cathode. The anodic side has a Cl_2 evolution reaction [31]. Many reactions have been reported, such as aldehyde oxidation [32], alcohol oxidation [33], amine oxidation, hydrazine oxidation [34], and urea oxidation, etc., which serves as potentially adequate reactions to work as anodic oxidative reaction replacing OER. Such reactions can lower the overall energy intake for H_2 [35].

In response, hybrid water electrolysis has been proposed, in which OER is replaced by thermodynamically more promising organic oxidation processes and attractive organic substrates produced from biomass [23]. This hybrid water electrolysis approach has at least four benefits: (1) Such a hybrid electrolyzer can give maximum current density with minimum voltage input because of favorable thermodynamics of some organic oxidation reactions, enhancing the efficiency of energy conversion reactions (2) Value-added organic products will be produced at the anode, increasing

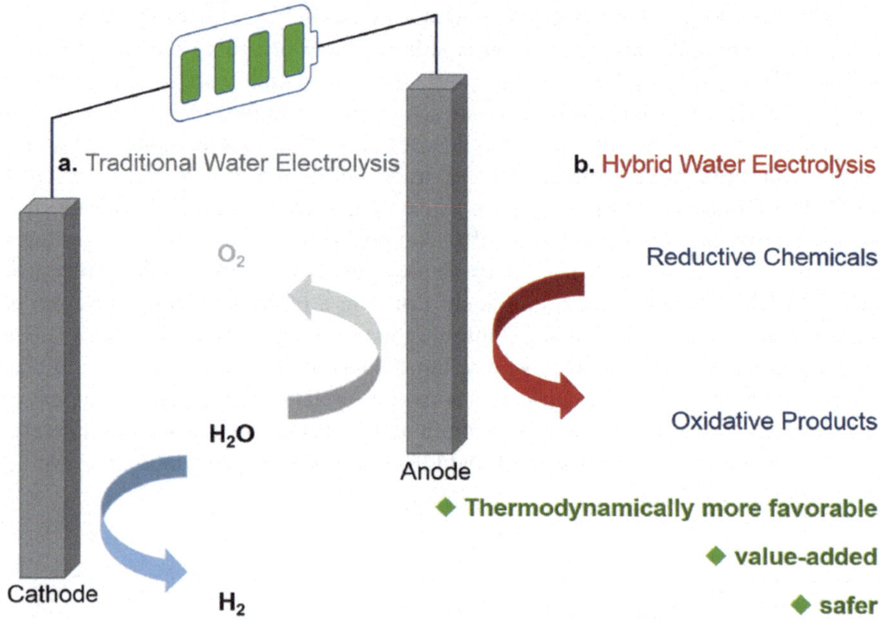

Fig. 10.3 Schematic illustration of **a** traditional water electrolysis and **b** hybrid water electrolysis. Reproduced with permission from Ref. [24]. (c) 2020. Elsevier

the return on invested energy. (3) There would be no O_2/H_2 mixing or reactive oxygen species problems because there won't be any O_2 production. (4) When organic substrates are present, the high selectivity of H_2 generation at the cathode makes single-compartment membrane-free electrolysis possible, significantly lowering the electrolyzer's cost.

A perfect organic substrate should meet the following specifications to attain these benefits:

1. Maximum solubility in water at room temperature
2. Formation of the nongaseous value-added product(s) by oxidation
3. Lower oxidative potential at the anode compared to OER onset potential
4. Negligible competition with the cathodic HER process.

Among most of the intermediates derived from biomass, 5-hydroxymethylfurfural (HMF) has been considered a favorable candidate because of its transformation to numerous valuable products through oxidation, for example, 2,5-furan dicarboxylic acid (FDCA) [36]. Indeed, FDCA has been promoted as terephthalic acid replacement for polyurethanes, polyamides, and polyesters, production [36]. A reported earth-abundant bifunctional efficient electrocatalyst easily catalyzes the oxidation of HMF to FDCA in alkaline 1.0 M KOH for overall water electrolysis. Elemental mapping analysis and SEM images illustrate the elemental composition and the morphology of the model electrode, three-dimensional nanoparticles arrays of Ni2P coated on substrate nickel foam (Ni2P NPA/NF). LSV polarization curves illustrate 10 mM HMF presence; the anodic current takes off at a potential significantly less positive than OER, nearly saving voltage of 200 mV input to attain the 100 mA cm^{-2} current density. The product generation and HMF conversion to time validating 100% FDCA yield and its unity of Faradaic efficiency. The aim of instantaneously H2 generation and FDCA requires maximum electrochemical performance and Ni_2P NPA/NF selectivity toward a generation of H_2 even in HMF presence. The LSV polarization curves of Ni_2P NPA/NF at the cathode exhibit insignificant differences compared to 10 mM HMF addition. The long-term stability and Ni_2P a chronopotentiometry experiment of 12 h validates NPA/NF preference for generation of H2 versus reduction of HMF proceeded at −10 mA cm^{-2} with HMF of 10 mM, displaying Faradaic efficiency of 100% of H_2 generation. The primary outcomes were attained in a three-electrode setup. On employing a two-electrode system Ni_2P NPA/NF as the composite for anode and cathode, instantaneous H_2 and FDCA generation may be realized. HMF oxidative LSV polarity curve integrated along with HER takes off at smaller voltage input than pure water electrolysis. Almost a voltage of 200 mV may be saved universally to attain current densities around 10, 20, and 50 mA cm^{-2}. Electrolysis for a long period with controlled potential experimented at 1.5 V generates H_2 closer to the theoretically predicted yield, revealing of a unity Faradaic efficiency for hydrogen evolution reaction. On continuously running three repetitive cycles of electrolysis using a similar electrocatalyst Ni_2P NPA/NF couple displays no electrocatalytic degradation, retaining enhanced performance, robustness, and selectivity for H_2 generation and oxidation of HMF to FDCA.

Encouraged by the as-mentioned outcomes and delight with the modular system of hybrid water electrolysis, electrocatalytic upgrading of a diverse array of intermediates derived from biomass compounds can be explored (e.g., furfuryl alcohol, benzyl alcohol, ethanol, furfural) and the earth-abundant catalysts utilization like phosphides, sulfides, transition metals, sulfides, etc. [23]. Indeed, the requirement of large anodic potential for OER leaves us with extensive potential window to conduct required organic oxidation processes without O_2 formation in aqueous media. Figure 10.4a–e represents five organic oxidative processes along with corresponding LSV polarization curves on three-dimensional Ni_3S_2/NF (Fig. 10.4f). These reactions occur at less positive potential than OER, inconsistent with more favorable ones thermodynamics in the presence of a basic environment. One remarkable observation is that catalytic currents take off at similar 1.36 V versus RHE potentials, irrespective of the intrinsic thermodynamic variance of individual organic processes. To escape the potential nonmetal element's influence in the composition of the electrocatalyst, we deliberately fabricated porous nickel foam (hp-Ni) as the working electrode (Fig. 10.4g). Consequently, benzyl alcohol has been considered organic substrate with variant electronic substituents (Fig. 10.4h). According to expectations, $-NO_2$ moiety having electron-withdrawing capability should shift oxidative potential towards the more positive side, while CH_3 group having electron-donating ability would enable oxidation towards the more negative potential side. Yet, even though benzyl alcohol has various substituents at the phenyl ring (para

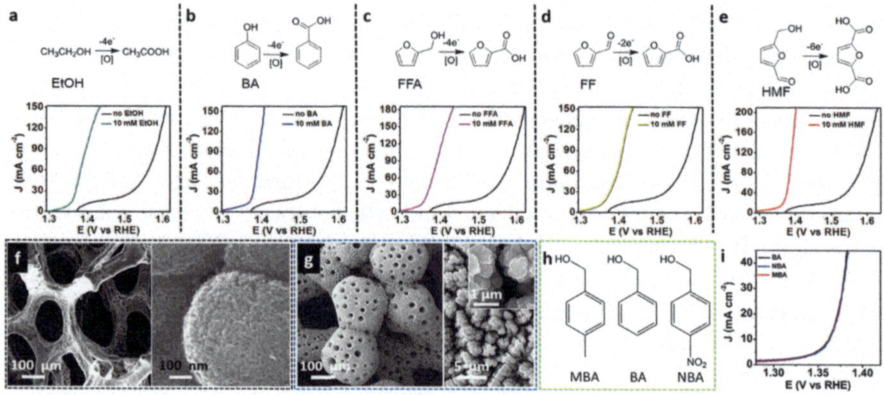

Fig. 10.4 **a–e** This study investigates the oxidation of specific organic compounds, namely benzyl alcohol (BA), furfuryl alcohol (FFA), furfural (FF), and 5-hydroxymethylfurfural (HMF), to produce valuable products. The corresponding LSV curves on Ni_3S_2/NF in 1.0 M KOH were also analyzed. **f** SEM images of Ni_3S_2/NF and **g** hp-Ni were taken to examine their morphology. **h** Additionally, the chemical structures of MBA, BA, and NBA were presented. **i** Finally, LSV curves on hp-Ni in 1.0 M KOH with 10 mM MBA, BA, and NBA were also studied. Reproduced with permission from Ref. [1]. (c) 2018. American Chemical Society

position), these three substrates have very comparable LSV polarization curves catalyst hp-Ni oxidation. (Fig. 10.4i). These preliminary findings lead us to speculate that, compared to general homogeneous oxidation reactions catalyzed by molecular catalysts, alcohol oxidation reactions on solid-state electrocatalysts like hp-Ni are determined primarily by the potential to attain the desired electrocatalyst oxidation state instead of the intrinsic thermodynamics of each organic oxidation.

10.3.3 Decoupled Water Electrolysis

While OER and HER cooccur during general water splitting, the rate of H_2 generation is still strongly constrained by the rate of O_2 production. Though, occasionally, generating H_2 at a higher rate is preferable while using a low-voltage input. When electrolysis is performed utilizing renewable energy sources, where the power inputs are frequently unpredictable or low, the problem of separating the electrolysis product gases becomes more complicated. The low current densities in these circumstances are related to low gas generation rates. These gas production rates may start approaching the gas crossover rates for particular membranes, thereby raising safety concerns. The 10 mA cm^{-2} current density is regarded as a beneficial ideal for solar-driven electrolyzers, which is the estimated current density predictable for a water electrolysis device working at 10% solar-to-fuels efficiency under "1 Sun" illumination (AM 1.5, 100 mW cm^{-2}) [37]. Since hydrogen's lower explosion limit in O_2 is only 4 mol% H_2 in O_2, the crossover of H_2 into the anodic chamber might be an actual possibility and extremely dangerous [38]. Moreover, even if proficient and secure gas separation could be accomplished, any solar-to-hydrogen device with connected half-reactions of water electrolysis will suffer since the sluggish oxygen evolution reaction would still constrain the pace of the comparatively simple HER. In this context, it is challenging and still a problem to harness lower pressures of hydrogen gas safely and effectively from massive solar-to-hydrogen arrays. To this end, redox mediators have been developed recently to "decouple" these processes. A mediator with the right redox potential can connect the OER with the mediator's reduction instead of the direct H_2 generation. The HER can also be carried out without OER assistance by connecting H_2 generation to the mediator reoxidation instead of oxidation of water (Fig. 10.5) [39]. Although each half-reaction happens independently, the kinetics of the HER are much higher than those of traditional water splitting. The flexibility to perform OER and HER at different times ("temporal separation") and in different locations ("spatial separation") greatly improves the efficiency and safety of hydrogen harvesting. It also drastically lessens the requirement for any gas purification processes. An effective mediator must have the features of being stable in both the oxidized and reduced forms and possessing a reversible redox pair with a potential that lies between the onset potentials of the OER and HER. According to the explanation mentioned above, decoupled electrolysis can be described as any process in which the ultimate cathodic and anodic electrolysis products are produced under the conditions listed below:

1. At rates that are not inherently related to each other.
2. At different times to each other.
3. In totally separate electrochemical compartments from one another.

Using redox mediators to produce decoupled water electrolysis allows for the spatial and temporal separation of OER and HER, preventing the mixing of O_2/H_2 and forming reactive O_2 moieties. Theoretically, when HER occurs at the cathode, the redox mediator is oxidized as a counter-reaction rather than OER. After then, the oxidized redox mediator will be reduced to its original condition at the same electrode, which will couple the oxygen evolution at another working electrode. In general, the redox mediator is a reversible electron acceptor/donor. Cronin's group described decoupled water splitting in acidic electrolytic media, utilizing several proton buffers

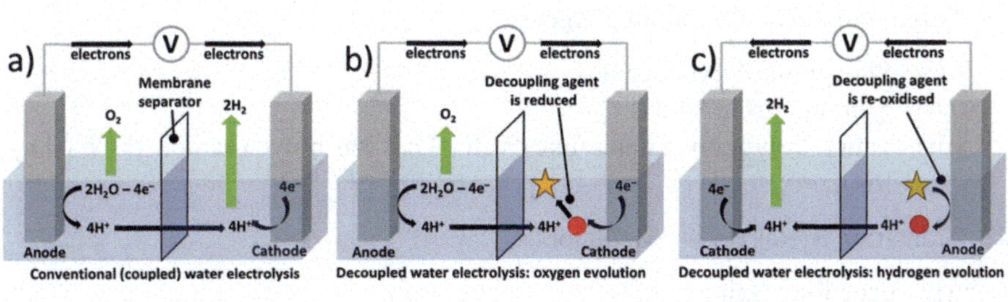

Fig. 10.5 Systematic representation of conventional and decoupled water electrolysis. Reproduced with permission from Ref. [39]. (c) 2020. Wiley

coupled with electrons like derivatives of quinone [40], silicotungstic acid [6], and phosphomolybdic acid [41]. Though, an acidic electrolyte is highly corrosive as well as it prohibits OER nonprecious electrocatalysts utilization. Their group also stated recently that derivative of ferrocene, (ferrocenyl methyl)trimethylammonium chloride (FcNCl), has slight solubility in H_2O, which may behave as an electron reservoir independent of proton to accomplish decoupled water splitting in the 0.5 M Na_2SO_4 neutral electrolytic media. Because of the benign environment, transition metal composites such as Ni_2P, metallic Ni, and Co–P might be used to drive OER and the HER process. Step 1 is to connects the HER electrode and applies a negative voltage bias to start hydrogen production. At the same time, the counterchamber's FcNCl is oxidized since it has a less favorable oxidation potential compared to OER [41]. Ni_2P/Ni/NF displays a 2.2 V onset voltage when FcNCl is not present. When 50 mM FcNCl is added to the counterchamber, the onset voltage drops to 1.2 V, showing that FcNCl oxidation is much simpler than OER's and that approximately 1 V. After this may reduce the voltage input, step 2 involves connecting the OER electrode (a Ni foam) and applying a positive voltage bias to drive O_2 evolution on the Ni foam. In contrast, the oxidized FcNCl + is reduced back to FcNCl in the counterchamber.

$FcNCl^+$ has a more positive reduction potential than HER; hence its presence in the counterchamber lowers the voltage needed for OER by roughly 1.8 V compared to when FcNCl is absent. Gas chromatography (GC) analysis showed that both stages achieved perfect Faradaic efficiency, with exactly one gas product being generated in each step. FcNCl shows remarkable durability and low crossover based on the results of 20 electrolysis cycles (50 h) when used as an electron reservoir for long-term decoupled water electrolysis. On integration with photovoltaic cells, sunlight irradiation is also driven through decoupled water electrolysis with no external voltage. Along with FcNCl, Na4[$Fe(CN)_6$] has also been explored that can work as a reservoir for electrons in both basic and neutral electrolytes. An ideal molecular redox mediator is both inexpensive and extremely soluble in water and possesses a quick and reversible redox pair intermediate between the activation potentials of OER and HER. The redox capacity of $Ni(OH)_2$ for decoupled water electrolysis is dictated in large part by the electrocatalyst mass loading and area, but it can also function as a solid-state redox mediator thanks to molecular complexes [5].

The various kinds of decoupling agents have been used which are listed below;

(1) Polyoxometalate Decoupling Agents

  (i) Phosphomolybdic Acid
  (ii) Silicotungstic Acid
  (iii) Other Polyoxometalates like Wells–Dawson polyoxoanion [$P_2W_{18}O_{62}$] [6–]

(2) Transition Metal Redox Mediators

  (i) Vanadium Salts
  (ii) Iron complexes

(3) Soluble Organic Redox Mediators
(4) Solid State Redox Mediators

(i) $NiOOH/Ni(OH)_2$ Electrodes
(ii) Other Solid Metal Oxides like FeO_x, MnO_2, etc.
(iii) Solar-Thermal Decoupled Electrolysis
(iv) Solid Organic Decoupling Agents.

The lab has extensively used the decoupling electrolysis procedure to produce chemicals. The possibility of using reduced decoupling agents as reducing equivalents for chemical processes apart from hydrogen generation is boosted by the agents' capacity to hold electrons for extended periods. There are two major eye-catching features of this method. First, if synthesized using protons and electrons derived from water, these reduced decoupling agents would be considered sustainable reducing equivalents. Second, electron-coupled proton buffers, decoupling agents that can also store protons, can be utilized as hydrogenation agents in place of gaseous hydrogen. This may assist in replacing dangerous chemical reagents with sustainable water and electricity.

Polyoxometalate silicotungstic acid has been used as a sustainable hydrogenation agent for various organic transformations. MacDonald et al. in 2018. utilized a two-electron reduced form of this compound ($H_6SiW_{12}O_{40}$), produced electrochemically with concomitant water oxidation as electrons and protons source for nitroarene compounds reduction to their aniline analogs with tremendous yields [42]. The silicotungstic acid might be reused, recycled, and rereduced with no significant drop in electrocatalytic activity. Interestingly, it has been investigated that $H_6SiW_{12}O_{40}$ was effective for such hydrogenation reactions with no cocatalyst requirement, facilitating distinctive benefits compared to traditional chemical methods to nitroarene reduction, for example, H_2 gas utilization with a catalyst Pt/C or sacrificial soluble hydrogenation agents usage with numerous cocatalysts. Wu et al. fabricated a similar compound, $H_6SiW_{12}O_{40}$, by decoupling electrolysis with fast water oxidation for acetophenone and phenylacetylene hydrogenation (Fig. 11b) [43]. In this instance, it was discovered that Pd/C or Pt/C electrocatalyst was necessary for reduction. Both the hydrogenation reactions (to increase the solubility of the substrate) and the electrochemical production of $H_6SiW_{12}O_{40}$ used mixtures of water/ethanol as the media. Maximum acetophenone and phenylacetylene conversions were 80% and 100%, respectively, and selectivity over the degree of reduction in the products was attained by changing the reaction conditions. These two findings collectively imply that decreased soluble decoupling agents hold great promise as safe and convenient hydrogenation catalysts for chemical synthesis. Decoupled electrolysis has also been used in processes that include oxidizing organic substrates. Li et al. used the decoupling agent $Na4[Fe(CN)_6]$ to decouple the HER from 5-hydroxymethylfurfural oxidation to 2,5-furan dicarboxylic acid [44]. Authors suggested that the chemical conversion of 5-Hydroxymethylfurfural to 2,5-furan dicarboxylic acid showed a higher-value anodic counterpart for H_2 generation than the water oxidation because it is a readily accessible by-product of the dehydration of some sugars and therefore establishes feedstock chemical. As a result, the authors put together a two-compartment cell with a cation exchange membrane separating the compartments. The decoupling agent was simultaneously oxidized at the carbon counterelectrode, while the HER was first conducted at a CoP on the carbon electrode. During the second step, rereduction of this mediator occurred as oxidation of 5-hydroxymethylfurfural occurred at a Ni anode. A 100% conversion of 5-hydroxymethyl furfural was recorded with mild electrocatalytic conditions, producing 2,5-furandicarboxylic acid with 83% yield. In cases where sustainably produced organic feedstocks are available, this strategy may offer a practical and environment-friendly way to combine the electrolytic production of hydrogen with the creation of valuable compounds.

Finally, we will discuss two applications of decoupled electrolysis for processes other than water splitting; both are related to the chloralkali process. The electrolysis of a sodium chloride solution produces chlorine gas (Cl_2), hydrogen gas (H_2O), and sodium hydroxide (NaOH). The relevance of the chloralkali process to the chemical industry is highlighted by the fact that these chemicals are employed in over half of all industrial activities. Traditionally, H_2, Cl_2, and NaOH have been synthesized in mercury cells according to the Castner–Kellner procedure which is indeed an electrochemical decoupling approach, whereby H_2 and Cl_2 generation decoupling occurs using an electrode of mercury amalgam as the decoupling agent. Concerns about mercury's toxicity have recently caused mercury cells to lose popularity. To separate the H_2 and Cl_2, most modern chloralkali cells employ membranes of fluorine with an ion exchange nature and asbestos diaphragms. On the other hand, Hou et al. claimed that the expensive cost of these separators and their relatively short lifespan during operation enhanced the requirement for modifications. Thus, the authors developed a procedure that did not require a membrane and involved decoupling the stages involved in synthesizing H_2/NaOH and Cl_2. This was accomplished by using the reversible Na-ion intercalation/deintercalation reaction of an electrode based on $Na_{0.44}MnO_2$ [45].

One of the primary benefits of decoupled electrochemistry is the ability to maintain the separation of the electrolysis products, and it's perhaps in this area that the majority of future research will concentrate on it because of the apparent benefits of solar-to-hydrogen devices. Similarly, decoupling the rates of the significant half-reactions made possible by this mechanism might be advantageous for solar-driven electrocatalytic processes of all types. Decoupling agents can be trickle-charged using slow, solar intensity-limited methods, and they can later be employed in faster reactions that are not constrained by the immediate solar flux. Again, the advantages of solar-driven water electrolysis for collecting H_2 are clear and have been emphasized by comprehensive research reviewed in this chapter. In the long term, the capacity to use electrochemical decoupling to produce sustainable reduction equivalents (such as H_2 equivalents from H_2O) will significantly affect chemical synthesis. These strategies might expand on the organic hydrogenation reactions covered above. Still, they might also involve innovative ideas and device designs for crucial supporting mechanisms like reducing N_2 and CO_2. Given the tremendous innovation and rapid advancements that have already been seen in the field, decoupled electrolysis appears to have a promising future.

10.3.4 Tandem Water Electrolysis

Affordable storage and transportation of H_2 is a problem for its widespread implementation. Catalysts such as enzyme utilization for producing value-added chemical fuels like NH_3 or CH_4 during water splitting are a potential solution to this problem. Figure 10.6a schematically describes the perception of tandem water electrolysis, where a two-compartment airtight electrochemical cell is charged with Methanosarcina barkeri (M.barkeri) and the carbon dioxide inside the cathodic chamber [46]. Inorganic catalysts are used to split water using sustainable electricity, solar energy, or both to generate H_2, which M. barkeri consumes in CO_2 presence to produce CH_4. The GC analysis of the gas products after 11.5 h of galvanostatic electrolysis and 0.5 h of sampling showed that CH_4 generation grew linearly and cumulatively under all the applied currents (1.0–7.5 mA), with an average of more than 74% Faradaic efficiency also shown in (Fig. 10.6b). The ratio of CH_4 synthesis to input current indicates that H2 generation via electrocatalysis is the bottleneck step of the entire process. As M. barkeri conducts at a nearly thermodynamic potential, the overpotential is because of HER in a neutral aqueous electrolyte of 1.0 M phosphate buffer. It has been shown that a hybrid bioinorganic system for tandem water electrolysis can last for 7 days with great stability while forming CH4 at a constant rate and with a 75% Faradaic efficiency(Fig. 10.6c). A carbon isotope tagging experiment verified that CO_2 was the true carbon source for CH_4. For example, high-resolution mass spectrometry (Fig. 10.6d) showed that only 13CH4 was observed when $13CO_2$ was used for electrocatalysis, while $12CH_4$ was detected when $12CO_2$ was employed. This hybrid system offers great versatility regarding electrocatalysts for in-situ H2 synthesis. Many other inorganic electrocatalysts, including NiMo and NiS, both widely available on Earth and compatible with living organisms, can be used. This tandem water electrolysis approach can also be used for direct solar-to-chemical conversion once combined with photoelectrodes [46]. Lately, Nocera et al. reported the same hybrid water electrolysis—biosynthetic setup using inorganic HER electrode Co–P and Ralstonia eutropha as the biocatalyst [47]. Compared to photosynthesis, it has higher overall energy conversion efficiency. The same group also reported an alternative tandem water electrolysis system, where H_2-oxidizing autotrophic bacterium Xanthobacter autotrophicus was utilized, and H_2O and N_2 produced NH_3. [48] The optimal reaction media for tandem water electrolysis to be compatible with biological systems is neutral water. Yet, most naturally occurring HER electrocatalysts work poorly in neutral media. For electrocatalytic evolution of H_2 in a neutral aqueous buffer, Sun group recently revealed a broad surface nitrogen modification technique to significantly increase the activity of first-row earth-abundant transition metals like Cu, Ni, Co, Fe, NiCo, and NiFe alloys [49]. Specifically, the as-prepared electrocatalyst of N–Ni competes with commercially available state-of-the-art Pt/C at pH 7 for H_2 evolution. Extensive physical characterizations show that metallic nickel persists in the N-Ni bulk composite. Still, nitrogen is present exclusively in the form of "surface nitrides" on the catalyst's surface. Nitrogen near the surface is likely coupled to three nickel atoms via the

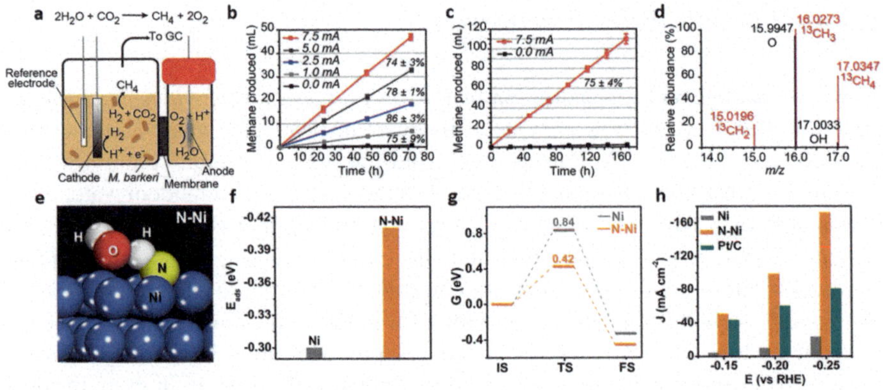

Fig. 10.6 **a** Experimental setup for tandem water electrolysis (**a**). **b**, **c** The CH_4 production and Faradaic efficiency are plotted against varying applied current. **d** Gas products are analyzed using high-resolution mass spectrometer. **e** The optimized structure of water adsorption on N–Ni(111) is shown, along with adsorption energy of water on Ni(111) and N–Ni(111) (**f**). Free energy barrier profiles of water dissociation on Ni(111) and N–Ni(111) are compared (**g**). Finally, the current densities at different overpotentials for N–Ni, Ni framework, and Pt/C for HER under neutral conditions are presented (**h**). Reproduced with permission from Ref. [1]. (c) 2018. American Chemical Society

hollow fcc site, as calculated by density functional theory (DFT) (Fig. 10.6e). This is generally accepted that adsorption and dissociation of water are important processes for HER in both neutral and basic environments. The ideal HER electrocatalyst would have a high affinity for binding water to speed up the first step in the electron transfer process and a high adequate repellency to make water dissociation easier in the second step. Figure 10.6f compares the water adsorption energy (Eads) on Ni (−0.30 eV) versus N–Ni (−0.41 eV). It far clear that N presence in N–Ni facilitates the adsorption of water where the H-bond is invoked naturally. Also, as shown in Fig. 10.6g, theoretical calculations predict that the free energy barrier for water dissociation on N-Ni is just 0.42 eV, significantly less than that on pure nickel, which is 0.84 eV. Thus, practical and theoretical evidence supports surface nitrogen alteration as a viable strategy for increasing HER in neutral environments. Because of its high HER activity at low pH (as shown in Fig. 10.6h), N-Ni is regarded as a great HER electrocatalyst to integrate with many biological systems for tandem water electrolysis and beyond.

It could be concluded from the above-explained aspects that the best chance of realizing conversion efficiencies that could ultimately provide a sustainable and alternative means of supplying chemical fuel to society in place of fossilized carbon sources lies in the process of tandem water electrolysis, which offers the best opportunity for such realizations.

References

1. You B, Sun Y (2018) Innovative strategies for electrocatalytic water splitting. Acc Chem Res 51(7):1571–1580. https://doi.org/10.1021/acs.accounts.8b00002
2. You B, Sun Y (2016) Chalcogenide and phosphide solid-state electrocatalysts for hydrogen generation. ChemPlusChem. https://doi.org/10.1002/cplu.201600029
3. Zhu YP, Guo C, Zheng Y, Qiao SZ (2017) Surface and interface engineering of noble-metal-free electrocatalysts for efficient energy conversion processes. Acc Chem Res. https://doi.org/10.1021/acs.accounts.6b00635
4. Liu X, You B, Sun Y (2017) Facile surface modification of ubiquitous stainless steel led to competent electrocatalysts for overall water splitting. ACS Sustain Chem Eng. https://doi.org/10.1021/acssuschemeng.7b00182
5. Landman A, Dotan H, Shter GE, Wullenkord M, Houaijia A, Maljusch A, Grader GS, Rothschild A (2017) Photoelectrochemical water splitting in separate oxygen and hydrogen cells. Nat Mater. https://doi.org/10.1038/nmat4876
6. Rausch B, Symes MD, Chisholm G, Cronin L (2014) Decoupled catalytic hydrogen evolution from a molecular metal oxide redox mediator in water splitting. Science (80-.). https://doi.org/10.1126/science.1257443
7. Wang W, Kong F, Zhang Z, Yang L, Wang M (2021) Sulfidation of nickel foam with enhanced electrocatalytic oxidation of 5-hydroxymethylfurfural to 2,5-furandicarboxylic acid. Dalt Trans. https://doi.org/10.1039/d1dt02025k
8. Ramakrishna SUB, Srinivasulu Reddy D, Shiva Kumar S, Himabindu V (2016) Nitrogen doped CNTs supported palladium electrocatalyst for hydrogen evolution reaction in PEM water electrolyser. Int J Hydrogen Energy. https://doi.org/10.1016/j.ijhydene.2016.08.195
9. Yoo S, Kim Y, Yoon Y, Karuppannan M, Kwon OJ, Lim T (2021) Encapsulation of Pt nanocatalyst with N-containing carbon layer for improving catalytic activity and stability in the hydrogen evolution reaction. Int J Hydrogen Energy. https://doi.org/10.1016/j.ijhydene.2021.03.225
10. Hinnemann B, Moses PG, Bonde J, Jørgensen KP, Nielsen JH, Horch S, Chorkendorff I, Nørskov JK (2005) Biomimetic hydrogen evolution: MoS_2 nanoparticles as catalyst for hydrogen evolution. J Am Chem Soc. https://doi.org/10.1021/ja0504690
11. Shan J, Ling T, Davey K, Zheng Y, Qiao SZ (2019) Transition-metal-doped RuIr bifunctional nanocrystals for overall water splitting in acidic environments. Adv Mater. https://doi.org/10.1002/adma.201900510
12. Sapountzi FM, Gracia JM, Weststrate CJ, Kee J, Fredriksson HOA, Niemantsverdriet JH (2017) Electrocatalysts for the generation of hydrogen, oxygen and synthesis gas. Progr Energy Combust Sci. https://doi.org/10.1016/j.pecs.2016.09.001
13. James B, Colella W, Moton J, Saur G, Ramsden T (2013) PEM electrolysis H2A production case study documentation. In: PEM electrolysis H2A prod. case study doc, No December, pp 1–27
14. Detz RJ, Reek JNH, Van Der Zwaan BCC (2018) The future of solar fuels: when could they become competitive? Energy Environ Sci. https://doi.org/10.1039/c8ee00111a
15. Paoli EA, Masini F, Frydendal R, Deiana D, Schlaup C, Malizia M, Hansen TW, Horch S, Stephens IEL, Chorkendorff I (2015) Oxygen evolution on well-characterized mass-selected Ru and RuO_2 nanoparticles. Chem Sci. https://doi.org/10.1039/c4sc02685c
16. Fabbri E, Habereder A, Waltar K, Kötz R, Schmidt TJ (2014) Developments and perspectives of oxide-based catalysts for the oxygen evolution reaction. Catal Sci Technol. https://doi.org/10.1039/c4cy00669k
17. Anantharaj S, Ede SR, Sakthikumar K, Karthick K, Mishra S, Kundu S (2016) Recent trends and perspectives in electrochemical water splitting with an emphasis on sulfide, selenide, and phosphide catalysts of Fe Co, and Ni: a review. ACS Catal. https://doi.org/10.1021/acscatal.6b02479
18. Zhou W, Zhou J, Zhou Y, Lu J, Zhou K, Yang L, Tang Z, Li L, Chen S (2015) N-doped carbon-wrapped cobalt nanoparticles on n-doped graphene nanosheets for high-efficiency hydrogen production. Chem Mater. https://doi.org/10.1021/acs.chemmater.5b00331

19. Lu Z, Zhu W, Yu X, Zhang H, Li Y, Sun X, Wang X, Wang H, Wang J, Luo J et al (2014) Ultrahigh hydrogen evolution performance of under-water "superaerophobic" MoS_2 nanostructured electrodes. Adv Mater. https://doi.org/10.1002/adma.201304759
20. Cao L, Luo Q, Chen J, Wang L, Lin Y, Wang H, Liu X, Shen X, Zhang W, Liu W et al (2019) Dynamic oxygen adsorption on single-atomic ruthenium catalyst with high performance for acidic oxygen evolution reaction. Nat Commun 10(1). https://doi.org/10.1038/s41467-019-12886-z
21. Jiang N, Bogoev L, Popova M, Gul S, Yano J, Sun Y (2014) Electrodeposited nickel-sulfide films as competent hydrogen evolution catalysts in neutral water. J Mater Chem A. https://doi.org/10.1039/c4ta04339a
22. Jiang N, You B, Sheng M, Sun Y (2015) Electrodeposited cobalt-phosphorous-derived films as competent bifunctional catalysts for overall water splitting. Angew Chemie Int Ed. https://doi.org/10.1002/anie.201501616
23. You B, Jiang N, Sheng M, Bhushan MW, Sun Y (2016) Hierarchically porous urchin-like Ni_2P superstructures supported on nickel foam as efficient bifunctional electrocatalysts for overall water splitting. ACS Catal. https://doi.org/10.1021/acscatal.5b02193
24. Hanazawa M, Sumiya R, Horikawa Y, Irie M (1992) Thermally irreversible photochromic systems. reversible photocyclization of 1,2-Bis (2-Methylbenzo[b]Thiophen-3-Yl)Perfluorocyclocoalkene derivatives. J Chem Soc Chem Commun. https://doi.org/10.1039/C39920000206
25. You B, Jiang N, Sheng M, Gul S, Yano J, Sun Y (2015) High-performance overall water splitting electrocatalysts derived from cobalt-based metal-organic frameworks. Chem Mater. https://doi.org/10.1021/acs.chemmater.5b02877
26. You B, Sun Y (2016) Hierarchically porous nickel sulfide multifunctional superstructures. Adv Energy Mater. https://doi.org/10.1002/aenm.201502333
27. She ZW, Kibsgaard J, Dickens CF, Chorkendorff I, Nørskov JK, Jaramillo TF (2017) Combining theory and experiment in electrocatalysis: insights into materials design. Science. https://doi.org/10.1126/science.aad4998
28. Ren JT, Yuan GG, Weng CC, Chen L, Yuan ZY (2018) Uniquely integrated Fe-doped $Ni(OH)_2$ nanosheets for highly efficient oxygen and hydrogen evolution reactions. Nanoscale. https://doi.org/10.1039/c8nr01655k
29. Roy C, Sebok B, Scott SB, Fiordaliso EM, Sørensen JE, Bodin A, Trimarco DB, Damsgaard CD, Vesborg PCK, Hansen O et al (2018) Impact of nanoparticle size and lattice oxygen on water oxidation on $NiFeO_xH_y$. Nat Catal. https://doi.org/10.1038/s41929-018-0162-x
30. Du L, Sun Y, You B (2021) Hybrid water electrolysis: replacing oxygen evolution reaction for energy-efficient hydrogen production and beyond. Mater Reports Energy. https://doi.org/10.1016/j.matre.2020.12.001
31. Karlsson RKB, Cornell A (2016) Selectivity between oxygen and chlorine evolution in the chlor-alkali and chlorate processes. Chem Rev 116(5):2982–3028. https://doi.org/10.1021/acs.chemrev.5b00389
32. You B, Liu X, Jiang N, Sun Y (2016) A general strategy for decoupled hydrogen production from water splitting by integrating oxidative biomass valorization. J Am Chem Soc. https://doi.org/10.1021/jacs.6b07127
33. Chen GF, Luo Y, Ding LX, Wang H (2018) Low-voltage electrolytic hydrogen production derived from efficient water and ethanol oxidation on fluorine-modified FeOOH anode. ACS Catal. https://doi.org/10.1021/acscatal.7b03319
34. Liu X, He J, Zhao S, Liu Y, Zhao Z, Luo J, Hu G, Sun X, Ding Y (2018) Self-powered H_2 production with bifunctional hydrazine as sole consumable. Nat Commun. https://doi.org/10.1038/s41467-018-06815-9
35. Babar P, Lokhande A, Karade V, Lee IJ, Lee D, Pawar S, Kim JH (2019) Trifunctional layered electrodeposited nickel iron hydroxide electrocatalyst with enhanced performance towards the oxidation of water, urea and hydrazine. J Colloid Interface Sci. https://doi.org/10.1016/j.jcis.2019.09.012

36. Cha HG, Choi KS (2015) Combined biomass valorization and hydrogen production in a photoelectrochemical cell. Nat Chem. https://doi.org/10.1038/nchem.2194
37. McCrory CCL, Jung S, Ferrer IM, Chatman SM, Peters JC, Jaramillo TF (2015) Benchmarking hydrogen evolving reaction and oxygen evolving reaction electrocatalysts for solar water splitting devices. J Am Chem Soc. https://doi.org/10.1021/ja510442p
38. Tijani AS, Rahim AHA (2016) Numerical modeling the effect of operating variables on faraday efficiency in PEM electrolyzer. Procedia Technol. https://doi.org/10.1016/j.protcy.2016.08.054
39. McHugh PJ, Stergiou AD, Symes MD (2020) Decoupled electrochemical water splitting: from fundamentals to applications. Adv Energy Mater. https://doi.org/10.1002/aenm.202002453
40. Rausch B, Symes MD, Cronin L (2013) A bio-inspired, small molecule electron-coupled-proton buffer for decoupling the half-reactions of electrolytic water splitting. J Am Chem Soc. https://doi.org/10.1021/ja4071893
41. Symes MD, Cronin L (2013) Decoupling hydrogen and oxygen evolution during electrolytic water splitting using an electron-coupled-proton buffer. Nat Chem. https://doi.org/10.1038/nchem.1621
42. Macdonald L, Rausch B, Symes MD, Cronin L (2018) Selective hydrogenation of nitroarenes using an electrogenerated polyoxometalate redox mediator. Chem Commun. https://doi.org/10.1039/c7cc09036f
43. Chen L, Xu Q (2019) Metal-organic framework composites for catalysis. Matter. https://doi.org/10.1016/j.matt.2019.05.018
44. Wallace AG, Symes MD (2018) Decoupling strategies in electrochemical water splitting and beyond. Joule. https://doi.org/10.1016/j.joule.2018.06.011
45. Hou M, Chen L, Guo Z, Dong X, Wang Y, Xia Y (2018) A clean and membrane-free chlor-alkali process with decoupled Cl_2 and H_2/NaOH production. Nat Commun. https://doi.org/10.1038/s41467-018-02877-x
46. Nichols EM, Gallagher JJ, Liu C, Su Y, Resasco J, Yu Y, Sun Y, Yang P, Chang MCY, Chang CJ (2015) Hybrid bioinorganic approach to solar-tochemical conversion. Proc Natl Acad Sci U S A. https://doi.org/10.1073/pnas.1508075112
47. Liu C, Colón BC, Ziesack M, Silver PA, Nocera DG (2016) Water splitting-biosynthetic system with CO_2 reduction efficiencies exceeding photosynthesis. Science (80-). https://doi.org/10.1126/science.aaf5039
48. Liu C, Sakimoto KK, Colón BC, Silver PA, Nocera DG (2017) Ambient nitrogen reduction cycle using a hybrid inorganic-biological system. Proc Natl Acad Sci U S A. https://doi.org/10.1073/pnas.1706371114
49. You B, Liu X, Hu G, Gul S, Yano J, Jiang DE, Sun Y (2017) Universal surface engineering of transition metals for superior electrocatalytic hydrogen evolution in neutral water. J Am Chem Soc. https://doi.org/10.1021/jacs.7b06434

Printed by Printforce, United Kingdom